Chevrolet Corvette Automotive Repair Manual

by Mike Stubblefield and John H Haynes

Member of the Guild of Motoring Writers

Models covered:
Chevrolet Corvette
1984 through 1996
Does not include ZR-1 model information

(24041-6T4)

Haynes Group Limited
Haynes North America, Inc.
www.haynes.com

Acknowledgements

Wiring diagrams originated exclusively for Haynes North America, Inc. by George Edward Brodd. Certain illustrations originated by Valley Forge Technical Information Services. Technical writers who contributed to this project include Tim Imhoff and Jay Storer.

A book in the Haynes Automotive Repair Manual Series

ISBN-10: 1-56392-226-6

ISBN-13: 978-1-56392-226-8

Library of Congress Catalog Card Number 96-77420

While every attempt is made to ensure that the information in this manual is correct, no liability can be accepted by the authors or publishers for loss, damage or injury caused by any errors in, or omissions from, the information given.

Contents

About this manual

Its purpose

The purpose of this manual is to help you get the best value from your vehicle. It can do so in several ways. It can help you decide what work must be done, even if you choose to have it done by a dealer service department or a repair shop; it provides information and procedures for routine maintenance and servicing; and it offers diagnostic and repair procedures to follow when trouble occurs.

We hope you use the manual to tackle the work yourself. For many simpler jobs, doing it yourself may be quicker than arranging an appointment to get the vehicle into a shop and making the trips to leave it and pick it up. More importantly, a lot of money can be saved by avoiding the expense the shop must pass on to you to cover its labor and overhead costs. An added benefit is the sense of satisfaction and accomplishment that you feel after doing the job yourself.

Using the manual

The manual is divided into Chapters. Each Chapter is divided into numbered Sections, which are headed in bold type between horizontal lines. Each Section consists of consecutively numbered paragraphs.

At the beginning of each numbered Section you will be referred to any illustrations which apply to the procedures in that Section. The reference numbers used in illustration captions pinpoint the pertinent Section and the Step within that Section. That is, illustration 3.2 means the illustration refers to Section 3 and Step (or paragraph) 2 within that Section.

Procedures, once described in the text, are not normally repeated. When it's necessary to refer to another Chapter, the reference will be given as Chapter and Section number. Cross references given without use of the word "Chapter" apply to Sections and/or paragraphs in the same Chapter. For example, "see Section 8" means in the same Chapter.

References to the left or right side of the vehicle assume you are sitting in the driver's seat, facing forward.

Even though we have prepared this manual with extreme care, neither the publisher nor the author can accept responsibility for any errors in, or omissions from, the information given.

NOTE

A **Note** provides information necessary to properly complete a procedure or information which will make the procedure easier to understand.

CAUTION

A **Caution** provides a special procedure or special steps which must be taken while completing the procedure where the Caution is found. Not heeding a Caution can result in damage to the assembly being worked on.

WARNING

A **Warning** provides a special procedure or special steps which must be taken while completing the procedure where the Warning is found. Not heeding a Warning can result in personal injury.

Introduction to the Chevrolet Corvette

The Chevrolet Corvette is the luxury sports car from General Motors first introduced in 1953. Over the years the car has developed from a six-cylinder with automatic transmission "personal" car to the high-performance sports car against which all other US sports cars are compared.

The Corvette comes with a high-performance version of the Chevrolet 350 cubic inch V8 engine as standard equipment. The engine is fueled by a "Crossfire" throttle body injection system (1984 models) or Tuned Port Injection (1985 and later), both controlled by a sophisticated computer system.

Three transmissions are available - a four speed manual transmission with a separate overdrive unit, a six-speed manual transmission and a four speed automatic transmission.

The suspension is fully independent front and rear, makes extensive use of aluminum components, and features a "plastic" single leaf rear spring.

Vehicle identification numbers

Modifications are a continuing and unpublicized process in automotive manufacturing. Because spare parts manuals and lists are compiled on a numerical basis, the individual vehicle numbers are essential to correctly identify the component required. Vehicle identification number (VIN)

This very important identification number is located on a plate attached to the top left corner of the dashboard of the vehicle **(see illustration)**. The VIN also appears on the Vehicle Certificate of Title and Registration. It contains valuable information such as the vehicle's manufacturing location and the date of its completion.

Engine identification number

The engine identification number is located on a machined pad which is actually an extension of the block head mating surface, just in front of the right head **(see illustration)**. An optional engine identification number location is on the left rear block flange just behind the oil filter mount.

Automatic transmission identification number

The transmission identification number, which indicates the year of the transmission and when and where it was built, is located on the pan mounting flange at the right rear of the transmission **(see illustration)**.

The transmission VIN (serial) number is stamped on the right side of the torque converter housing, with an optional location on the left side of the tailshaft mounting flange.

Alternator numbers

The alternator ID number is on top of the drive end frame.

Starter numbers

The starter ID number is stamped on the outer case towards the rear.

Battery numbers

The battery ID number is on the middle of the cell cover at the top of the battery.

Rear axle number

The rear axle number is on the bottom surface of the carrier at the cover mounting flange.

Service Parts Identification label

The Service Parts Identification label is located inside the driver's side rear storage compartment. This label will list all of the regular production options (RPO's) installed on the vehicle, as well as standard equipment, mandatory options and paint color codes **(see illustration)**.

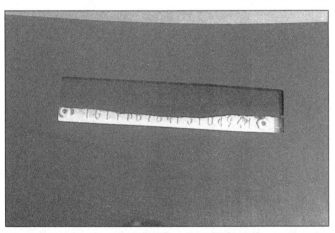

The Vehicle Identification Number is on a plate attached to the left top of the instrument panel where it can be seen through the windshield from outside the vehicle

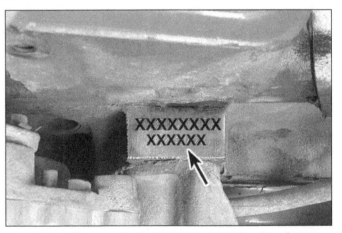

The engine ID number is located on the block head mating surface where it extends in front of the right head, with an optional location just to the rear of the oil filter mount

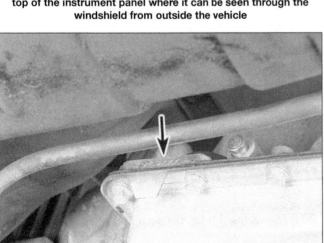

The automatic transmission identification number (arrow) is on the right rear of the pan mounting flange, and the transmission VIN is on the right side of the torque converter housing, with an optional location on the left side of the tailshaft mounting flange

The Service Parts Identification label is mounted inside the driver's side rear storage compartment or under the console lid

Buying parts

Replacement parts are available from many sources, which generally fall into one of two categories - authorized dealer parts departments and independent retail auto parts stores. Our advice concerning these parts is as follows:

Retail auto parts stores: Good auto parts stores will stock frequently needed components which wear out relatively fast, such as clutch components, exhaust systems, brake parts, tune-up parts, etc. These stores often supply new or reconditioned parts on an exchange basis, which can save a considerable amount of money. Discount auto parts stores are often very good places to buy materials and parts needed for general vehicle maintenance such as oil, grease, filters, spark plugs, belts, touch-up paint, bulbs, etc. They also usually sell tools and general accessories, have convenient hours, charge lower prices and can often be found not far from home.

Authorized dealer parts department: This is the best source for parts which are unique to the vehicle and not generally available elsewhere (such as major engine parts, transmission parts, trim pieces, etc.).

Warranty information: If the vehicle is still covered under warranty, be sure that any replacement parts purchased - regardless of the source - do not invalidate the warranty!

To be sure of obtaining the correct parts, have engine and chassis numbers available and, if possible, take the old parts along for positive identification.

Maintenance techniques, tools and working facilities

Maintenance techniques

There are a number of techniques involved in maintenance and repair that will be referred to throughout this manual. Application of these techniques will enable the home mechanic to be more efficient, better organized and capable of performing the various tasks properly, which will ensure that the repair job is thorough and complete.

Fasteners

Fasteners are nuts, bolts, studs and screws used to hold two or more parts together. There are a few things to keep in mind when working with fasteners. Almost all of them use a locking device of some type, either a lockwasher, locknut, locking tab or thread adhesive. All threaded fasteners should be clean and straight, with undamaged threads and undamaged corners on the hex head where the wrench fits. Develop the habit of replacing all damaged nuts and bolts with new ones. Special locknuts with nylon or fiber inserts can only be used once. If they are removed, they lose their locking ability and must be replaced with new ones.

Rusted nuts and bolts should be treated with a penetrating fluid to ease removal and prevent breakage. Some mechanics use turpentine in a spout-type oil can, which works quite well. After applying the rust penetrant, let it work for a few minutes before trying to loosen the nut or bolt. Badly rusted fasteners may have to be chiseled or sawed off or removed with a special nut breaker, available at tool stores.

If a bolt or stud breaks off in an assembly, it can be drilled and removed with a special tool commonly available for this purpose. Most automotive machine shops can perform this task, as well as other repair procedures, such as the repair of threaded holes that have been stripped out.

Flat washers and lockwashers, when removed from an assembly, should always be replaced exactly as removed. Replace any damaged washers with new ones. Never use a lockwasher on any soft metal surface (such as aluminum), thin sheet metal or plastic.

Fastener sizes

For a number of reasons, automobile manufacturers are making wider and wider use of metric fasteners. Therefore, it is important to be able to tell the difference between standard (sometimes called U.S. or SAE) and metric hardware, since they cannot be interchanged.

All bolts, whether standard or metric, are sized according to diameter, thread pitch and length. For example, a standard 1/2 - 13 x 1 bolt is 1/2 inch in diameter, has 13 threads per inch and is 1 inch long. An M12 - 1.75 x 25 metric bolt is 12 mm in diameter, has a thread pitch of 1.75 mm (the distance between threads) and is 25 mm long. The two bolts are nearly identical, and easily confused, but they are not interchangeable.

In addition to the differences in diameter, thread pitch and length, metric and standard bolts can also be distinguished by examining the bolt heads. To begin with, the distance across the flats on a standard bolt head is measured in inches, while the same dimension on a metric bolt is sized in millimeters (the same is true for nuts). As a result, a standard wrench should not be used on a metric bolt and a metric wrench should not be used on a standard bolt. Also, most stan-

dard bolts have slashes radiating out from the center of the head to denote the grade or strength of the bolt, which is an indication of the amount of torque that can be applied to it. The greater the number of slashes, the greater the strength of the bolt. Grades 0 through 5 are commonly used on automobiles. Metric bolts have a property class (grade) number, rather than a slash, molded into their heads to indicate bolt strength. In this case, the higher the number, the stronger the bolt. Property class numbers 8.8, 9.8 and 10.9 are commonly used on automobiles.

Strength markings can also be used to distinguish standard hex nuts from metric hex nuts. Many standard nuts have dots stamped into one side, while metric nuts are marked with a number. The greater the number of dots, or the higher the number, the greater

the strength of the nut.

Metric studs are also marked on their ends according to property class (grade). Larger studs are numbered (the same as metric bolts), while smaller studs carry a geometric code to denote grade.

It should be noted that many fasteners, especially Grades 0 through 2, have no distinguishing marks on them. When such is the case, the only way to determine whether it is standard or metric is to measure the thread pitch or compare it to a known fastener of the same size.

Standard fasteners are often referred to as SAE, as opposed to metric. However, it should be noted that SAE technically refers to a non-metric fine thread fastener only. Coarse thread non-metric fasteners are referred to as USS sizes.

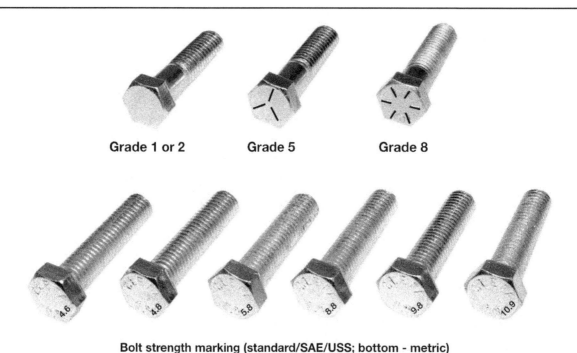

Grade 1 or 2   Grade 5   Grade 8

Bolt strength marking (standard/SAE/USS; bottom - metric)

Grade	Identification
Hex Nut Grade 5	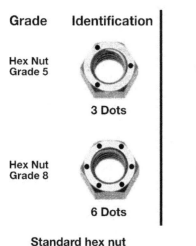 3 Dots
Hex Nut Grade 8	6 Dots

Standard hex nut strength markings

Grade	Identification
Hex Nut Property Class 9	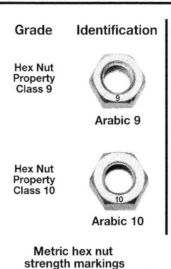 Arabic 9
Hex Nut Property Class 10	Arabic 10

Metric hex nut strength markings

Class 10.9   Class 9.8   Class 8.8

Metric stud strength markings

00-1 HAYNES

Since fasteners of the same size (both standard and metric) may have different strength ratings, be sure to reinstall any bolts, studs or nuts removed from your vehicle in their original locations. Also, when replacing a fastener with a new one, make sure that the new one has a strength rating equal to or greater than the original.

Tightening sequences and procedures

Most threaded fasteners should be tightened to a specific torque value (torque is the twisting force applied to a threaded component such as a nut or bolt). Overtightening the fastener can weaken it and cause it to break, while undertightening can cause it to eventually come loose. Bolts, screws and studs, depending on the material they are made of and their thread diameters, have specific torque values, many of which are noted in the Specifications at the beginning of each Chapter. Be sure to follow the torque recommendations closely. For fasteners not assigned a specific torque, a general torque value chart is presented here as a guide. These torque values are for dry (unlubricated) fasteners threaded into steel or cast iron (not aluminum). As was previously mentioned, the size and grade of a fastener determine the amount of torque that can safely be applied to it. The figures listed here are approximate for Grade 2 and Grade 3 fasteners. Higher grades can tolerate higher torque values.

Fasteners laid out in a pattern, such as cylinder head bolts, oil pan bolts, differential cover bolts, etc., must be loosened or tightened in sequence to avoid warping the component. This sequence will normally be shown in the appropriate Chapter. If a specific pattern is not given, the following procedures can be used to prevent warping.

Metric thread sizes	Ft-lbs	Nm
M-6	6 to 9	9 to 12
M-8	14 to 21	19 to 28
M-10	28 to 40	38 to 54
M-12	50 to 71	68 to 96
M-14	80 to 140	109 to 154
Pipe thread sizes		
1/8	5 to 8	7 to 10
1/4	12 to 18	17 to 24
3/8	22 to 33	30 to 44
1/2	25 to 35	34 to 47
U.S. thread sizes		
1/4 - 20	6 to 9	9 to 12
5/16 - 18	12 to 18	17 to 24
5/16 - 24	14 to 20	19 to 27
3/8 - 16	22 to 32	30 to 43
3/8 - 24	27 to 38	37 to 51
7/16 - 14	40 to 55	55 to 74
7/16 - 20	40 to 60	55 to 81
1/2 - 13	55 to 80	75 to 108

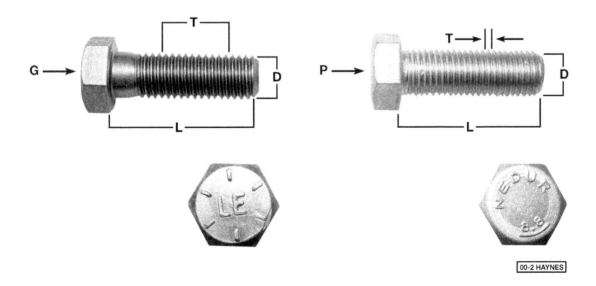

00-2 HAYNES

Standard (SAE and USS) bolt dimensions/grade marks

G   Grade marks (bolt strength)
L   Length (in inches)
T   Thread pitch (number of threads per inch)
D   Nominal diameter (in inches)

Metric bolt dimensions/grade marks

P   Property class (bolt strength)
L   Length (in millimeters)
T   Thread pitch (distance between threads in millimeters)
D   Diameter

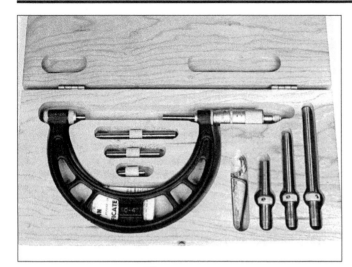

Micrometer set

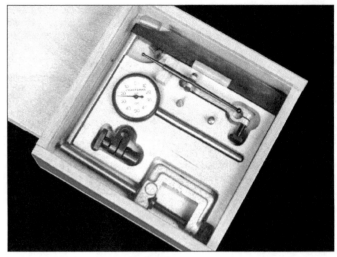

Dial indicator set

Initially, the bolts or nuts should be assembled finger-tight only. Next, they should be tightened one full turn each, in a criss-cross or diagonal pattern. After each one has been tightened one full turn, return to the first one and tighten them all one-half turn, following the same pattern. Finally, tighten each of them one-quarter turn at a time until each fastener has been tightened to the proper torque. To loosen and remove the fasteners, the procedure would be reversed.

Component disassembly

Component disassembly should be done with care and purpose to help ensure that the parts go back together properly. Always keep track of the sequence in which parts are removed. Make note of special characteristics or marks on parts that can be installed more than one way, such as a grooved thrust washer on a shaft. It is a good idea to lay the disassembled parts out on a clean surface in the order that they were removed. It may also be helpful to make sketches or take instant photos of components before removal.

When removing fasteners from a component, keep track of their locations. Sometimes threading a bolt back in a part, or putting the washers and nut back on a stud, can prevent mix-ups later. If nuts and bolts cannot be returned to their original locations, they should be kept in a compartmented box or a series of small boxes. A cupcake or muffin tin is ideal for this purpose, since each cavity can hold the bolts and nuts from a particular area (i.e. oil pan bolts, valve cover bolts, engine mount bolts, etc.). A pan of this type is especially helpful when working on assemblies with very small parts, such as the carburetor, alternator, valve train or interior dash and trim pieces. The cavities can be marked with paint or tape to identify the contents.

Whenever wiring looms, harnesses or connectors are separated, it is a good idea to identify the two halves with numbered pieces of masking tape so they can be easily reconnected.

Gasket sealing surfaces

Throughout any vehicle, gaskets are used to seal the mating surfaces between two parts and keep lubricants, fluids, vacuum or pressure contained in an assembly.

Many times these gaskets are coated with a liquid or paste-type gasket sealing compound before assembly. Age, heat and pressure can sometimes cause the two parts to stick together so tightly that they are very difficult to separate. Often, the assembly can be loosened by striking it with a soft-face hammer near the mating surfaces. A regular hammer can be used if a block of wood is placed between the hammer and the part. Do not hammer on cast parts or parts that could be easily damaged. With any particularly stubborn part, always recheck to make sure that every fastener has been removed.

Avoid using a screwdriver or bar to pry apart an assembly, as they can easily mar the gasket sealing surfaces of the parts, which must remain smooth. If prying is absolutely necessary, use an old broom handle, but keep in mind that extra clean up will be necessary if the wood splinters.

After the parts are separated, the old gasket must be carefully scraped off and the gasket surfaces cleaned. Stubborn gasket material can be soaked with rust penetrant or treated with a special chemical to soften it so it can be easily scraped off. A scraper can be fashioned from a piece of copper tubing by flattening and sharpening one end. Copper is recommended because it is usually softer than the surfaces to be scraped, which reduces the chance of gouging the part. Some gaskets can be removed with a wire brush, but regardless of the method used, the mating surfaces must be left clean and smooth. If for some reason the gasket surface is gouged, then a gasket sealer thick enough to fill scratches will have to be used during reassembly of the components. For most applications, a non-drying (or semi-drying) gasket sealer should be used.

Hose removal tips

Warning: *If the vehicle is equipped with air conditioning, do not disconnect any of the A/C hoses without first having the system depressurized by a dealer service department or a service station.*

Hose removal precautions closely parallel gasket removal precautions. Avoid scratching or gouging the surface that the hose mates against or the connection may leak. This is especially true for radiator hoses. Because of various chemical reactions, the rubber in hoses can bond itself to the metal spigot that the hose fits over. To remove a hose, first loosen the hose clamps that secure it to the spigot. Then, with slip-joint pliers, grab the hose at the clamp and rotate it around the spigot. Work it back and forth until it is completely free, then pull it off. Silicone or other lubricants will ease removal if they can be applied between the hose and the outside of the spigot. Apply the same lubricant to the inside of the hose and the outside of the spigot to simplify installation.

As a last resort (and if the hose is to be replaced with a new one anyway), the rubber can be slit with a knife and the hose peeled from the spigot. If this must be done, be careful that the metal connection is not damaged.

If a hose clamp is broken or damaged, do not reuse it. Wire-type clamps usually weaken with age, so it is a good idea to replace them with screw-type clamps whenever a hose is removed.

Tools

A selection of good tools is a basic requirement for anyone who plans to maintain and repair his or her own vehicle. For the owner who has few tools, the initial investment might seem high, but when compared to the spiraling costs of professional auto maintenance and repair, it is a wise one.

To help the owner decide which tools are needed to perform the tasks detailed in this manual, the following tool lists are offered: *Maintenance and minor repair,*

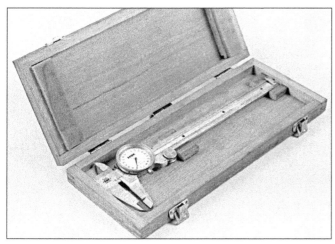

Dial caliper

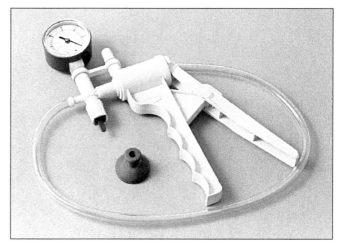

Hand-operated vacuum pump

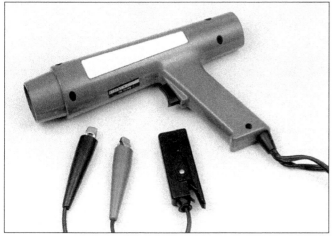

Timing light

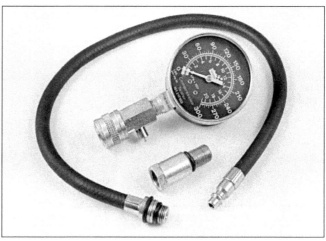

Compression gauge with spark plug hole adapter

Damper/steering wheel puller

General purpose puller

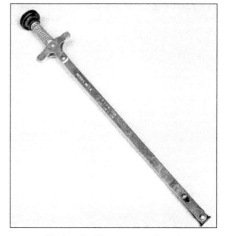

Hydraulic lifter removal tool

Repair/overhaul and *Special.*

The newcomer to practical mechanics should start off with the *maintenance and minor repair* tool kit, which is adequate for the simpler jobs performed on a vehicle. Then, as confidence and experience grow, the owner can tackle more difficult tasks, buying additional tools as they are needed.

Eventually the basic kit will be expanded into the *repair and overhaul* tool set. Over a period of time, the experienced do-it-yourselfer will assemble a tool set complete enough for most repair and overhaul procedures and will add tools from the special category when it is felt that the expense is justified by the frequency of use.

Maintenance and minor repair tool kit

The tools in this list should be considered the minimum required for performance of routine maintenance, servicing and minor repair work. We recommend the purchase of combination wrenches (box-end and open-

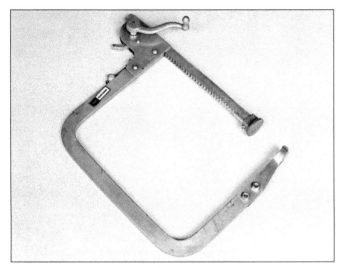

Valve spring compressor

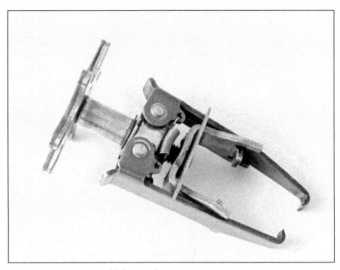

Valve spring compressor

Ridge reamer

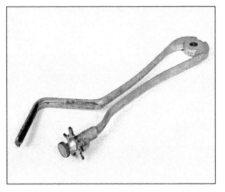

Piston ring groove cleaning tool

Ring removal/installation tool

end combined in one wrench). While more expensive than open end wrenches, they offer the advantages of both types of wrench.

> Combination wrench set (1/4-inch to
> 　1 inch or 6 mm to 19 mm)
> Adjustable wrench, 8 inch
> Spark plug wrench with rubber insert
> Spark plug gap adjusting tool
> Feeler gauge set
> Brake bleeder wrench
> Standard screwdriver (5/16-inch x
> 　6 inch)
> Phillips screwdriver (No. 2 x 6 inch)
> Combination pliers - 6 inch
> Hacksaw and assortment of blades
> Tire pressure gauge
> Grease gun
> Oil can
> Fine emery cloth
> Wire brush
> Battery post and cable cleaning tool
> Oil filter wrench
> Funnel (medium size)
> Safety goggles
> Jackstands (2)
> Drain pan

Note: *If basic tune-ups are going to be part of routine maintenance, it will be necessary to purchase a good quality stroboscopic timing*

light and combination tachometer/dwell meter. Although they are included in the list of special tools, it is mentioned here because they are absolutely necessary for tuning most vehicles properly.

Repair and overhaul tool set

These tools are essential for anyone who plans to perform major repairs and are in addition to those in the maintenance and minor repair tool kit. Included is a comprehensive set of sockets which, though expensive, are invaluable because of their versatility, especially when various extensions and drives are available. We recommend the 1/2-inch drive over the 3/8-inch drive. Although the larger drive is bulky and more expensive, it has the capacity of accepting a very wide range of large sockets. Ideally, however, the mechanic should have a 3/8-inch drive set and a 1/2-inch drive set.

> Socket set(s)
> Reversible ratchet
> Extension - 10 inch
> Universal joint
> Torque wrench (same size drive as
> 　sockets)
> Ball peen hammer - 8 ounce
> Soft-face hammer (plastic/rubber)

Ring compressor

> Standard screwdriver (1/4-inch x 6 inch)
> Standard screwdriver (stubby -
> 　5/16-inch)
> Phillips screwdriver (No. 3 x 8 inch)
> Phillips screwdriver (stubby - No. 2)
> Pliers - vise grip
> Pliers - lineman's
> Pliers - needle nose
> Pliers - snap-ring (internal and external)
> Cold chisel - 1/2-inch

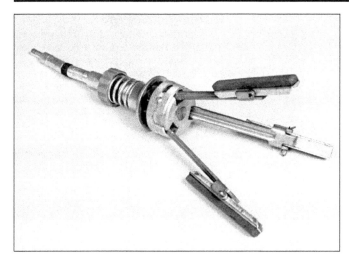

Cylinder hone

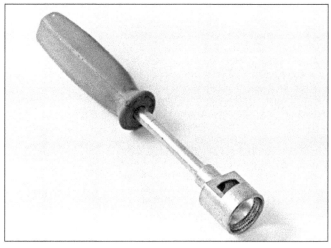

Brake hold-down spring tool

Scribe
Scraper (made from flattened copper
 tubing)
Centerpunch
Pin punches (1/16, 1/8, 3/16-inch)
Steel rule/straightedge - 12 inch
Allen wrench set (1/8 to 3/8-inch or
 4 mm to 10 mm)
A selection of files

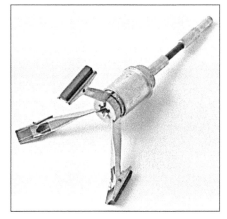

Brake cylinder hone

Wire brush (large)
Jackstands (second set)
Jack (scissor or hydraulic type)

Note: Another tool which is often useful is an electric drill with a chuck capacity of 3/8-inch and a set of good quality drill bits.

Special tools

The tools in this list include those which are not used regularly, are expensive to buy, or which need to be used in accordance with their manufacturer's instructions. Unless these tools will be used frequently, it is not very economical to purchase many of them. A consideration would be to split the cost and use between yourself and a friend or friends. In addition, most of these tools can be obtained from a tool rental shop on a temporary basis.

This list primarily contains only those tools and instruments widely available to the public, and not those special tools produced by the vehicle manufacturer for distribution to dealer service departments. Occasionally, references to the manufacturer's special tools are included in the text of this manual. Generally, an alternative method of doing the job without the special tool is offered. However,

sometimes there is no alternative to their use. Where this is the case, and the tool cannot be purchased or borrowed, the work should be turned over to the dealer service department or an automotive repair shop.

Valve spring compressor
Piston ring groove cleaning tool
Piston ring compressor
Piston ring installation tool
Cylinder compression gauge
Cylinder ridge reamer
Cylinder surfacing hone
Cylinder bore gauge
Micrometers and/or dial calipers
Hydraulic lifter removal tool
Balljoint separator
Universal-type puller
Impact screwdriver
Dial indicator set
Stroboscopic timing light (inductive
 pick-up)
Hand operated vacuum/pressure pump
Tachometer/dwell meter
Universal electrical multimeter
Cable hoist
Brake spring removal and installation
 tools
Floor jack

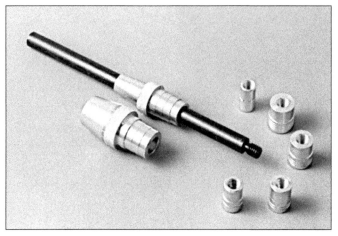

Clutch plate alignment tool

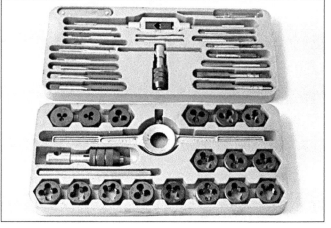

Tap and die set

Buying tools

For the do-it-yourselfer who is just starting to get involved in vehicle maintenance and repair, there are a number of options available when purchasing tools. If maintenance and minor repair is the extent of the work to be done, the purchase of individual tools is satisfactory. If, on the other hand, extensive work is planned, it would be a good idea to purchase a modest tool set from one of the large retail chain stores. A set can usually be bought at a substantial savings over the individual tool prices, and they often come with a tool box. As additional tools are needed, add-on sets, individual tools and a larger tool box can be purchased to expand the tool selection. Building a tool set gradually allows the cost of the tools to be spread over a longer period of time and gives the mechanic the freedom to choose only those tools that will actually be used.

Tool stores will often be the only source of some of the special tools that are needed, but regardless of where tools are bought, try to avoid cheap ones, especially when buying screwdrivers and sockets, because they won't last very long. The expense involved in replacing cheap tools will eventually be greater than the initial cost of quality tools.

Care and maintenance of tools

Good tools are expensive, so it makes sense to treat them with respect. Keep them clean and in usable condition and store them properly when not in use. Always wipe off any dirt, grease or metal chips before putting them away. Never leave tools lying around in the work area. Upon completion of a job, always check closely under the hood for tools that may have been left there so they won't get lost during a test drive.

Some tools, such as screwdrivers, pliers, wrenches and sockets, can be hung on a panel mounted on the garage or workshop wall, while others should be kept in a tool box or tray. Measuring instruments, gauges, meters, etc. must be carefully stored where they cannot be damaged by weather or impact from other tools.

When tools are used with care and stored properly, they will last a very long time. Even with the best of care, though, tools will wear out if used frequently. When a tool is damaged or worn out, replace it. Subsequent jobs will be safer and more enjoyable if you do.

How to repair damaged threads

Sometimes, the internal threads of a nut or bolt hole can become stripped, usually from overtightening. Stripping threads is an all-too-common occurrence, especially when working with aluminum parts, because aluminum is so soft that it easily strips out.

Usually, external or internal threads are only partially stripped. After they've been cleaned up with a tap or die, they'll still work. Sometimes, however, threads are badly damaged. When this happens, you've got three choices:

1) *Drill and tap the hole to the next suitable oversize and install a larger diameter bolt, screw or stud.*

2) *Drill and tap the hole to accept a threaded plug, then drill and tap the plug to the original screw size. You can also buy a plug already threaded to the original size. Then you simply drill a hole to the specified size, then run the threaded plug into the hole with a bolt and jam nut. Once the plug is fully seated, remove the jam nut and bolt.*

3) *The third method uses a patented thread repair kit like Heli-Coil or Slimsert. These easy-to-use kits are designed to repair damaged threads in straight-through holes and blind holes. Both are available as kits which can handle a variety of sizes and thread patterns. Drill the hole, then tap it with the special included tap. Install the Heli-Coil and the hole is back to its original diameter and thread pitch.*

Regardless of which method you use, be sure to proceed calmly and carefully. A little impatience or carelessness during one of these relatively simple procedures can ruin your whole day's work and cost you a bundle if you wreck an expensive part.

Working facilities

Not to be overlooked when discussing tools is the workshop. If anything more than routine maintenance is to be carried out, some sort of suitable work area is essential.

It is understood, and appreciated, that many home mechanics do not have a good workshop or garage available, and end up removing an engine or doing major repairs outside. It is recommended, however, that the overhaul or repair be completed under the cover of a roof.

A clean, flat workbench or table of comfortable working height is an absolute necessity. The workbench should be equipped with a vise that has a jaw opening of at least four inches.

As mentioned previously, some clean, dry storage space is also required for tools, as well as the lubricants, fluids, cleaning solvents, etc. which soon become necessary.

Sometimes waste oil and fluids, drained from the engine or cooling system during normal maintenance or repairs, present a disposal problem. To avoid pouring them on the ground or into a sewage system, pour the used fluids into large containers, seal them with caps and take them to an authorized disposal site or recycling center. Plastic jugs, such as old antifreeze containers, are ideal for this purpose.

Always keep a supply of old newspapers and clean rags available. Old towels are excellent for mopping up spills. Many mechanics use rolls of paper towels for most work because they are readily available and disposable. To help keep the area under the vehicle clean, a large cardboard box can be cut open and flattened to protect the garage or shop floor.

Whenever working over a painted surface, such as when leaning over a fender to service something under the hood, always cover it with an old blanket or bedspread to protect the finish. Vinyl covered pads, made especially for this purpose, are available at auto parts stores.

Booster battery (jump) starting

Observe the following precautions when using a booster battery to start a vehicle:

a) *Before connecting the booster battery, make sure the ignition switch is in the Off position.*
b) *Turn off the lights, heater and other electrical loads.*
c) *Your eyes should be shielded. Safety goggles are a good idea.*
d) *Make sure the booster battery is the same voltage as the dead one in the vehicle.*
e) *The two vehicles MUST NOT TOUCH each other.*
f) *Make sure the transmission is in Neutral (manual transaxle) or Park (automatic transaxle).*
g) *If the booster battery is not a maintenance-free type, remove the vent caps and lay a cloth over the vent holes.*

Connect the red jumper cable to the positive (+) terminals of each battery.

Connect one end of the black cable to the negative (-) terminal of the booster battery. The other end of this cable should be connected to a good ground on the engine block **(see illustration)**. Make sure the cable will not come into contact with the fan, drivebelts or other moving parts of the engine.

Start the engine using the booster battery, then, with the engine running at idle speed, disconnect the jumper cables in the reverse order of connection

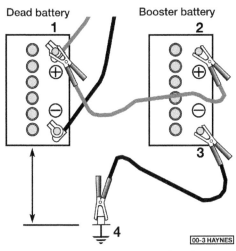

Make the booster battery cable connections in the numerical order shown (note that the negative cable of the booster battery is NOT attached to the negative terminal of the dead battery)

Jacking and towing

Jacking

The jack supplied with the vehicle should only be used for raising the vehicle when changing a tire or placing jackstands under the frame. Caution: Never work under the vehicle or start the engine while this jack is being used as the only means of support.

The vehicle should be on level ground with the wheels blocked and the transmission in Park (automatic) or Reverse (manual). If the wheel is being replaced, loosen the wheel nuts one-half turn but leave them in place until the wheel is raised off the ground.

Block the front and rear of the wheel opposite the one being removed before operating the jack. Place the jack head between the locator triangles on the rocker flange nearest to the wheel being changed **(see illustration)**. Caution: Vehicle damage is likely unless the jack head engages the rocker flange between the locator triangles.

When using a floor type jack against the frame rails to lift the vehicle, use care not to allow the jack to contact the catalytic converter.

When lifting the vehicle by any other means, special care must be exercised to avoid damage to the fuel tank, filler neck, exhaust system or underbody.

Towing

If equipped with an automatic transmission, the vehicle may be towed on all four wheels at speeds less than 35 mph for distances up to 50 miles. These speeds and distances do not apply to vehicles with manual transmissions.

Towing equipment specifically designed for this purpose should be used and should be attached to the main structural members of the vehicle and not the bumper or brackets.

The jack must be mounted between the locator triangles on the rocker flange

Safety is a major consideration when towing and all applicable state and local laws must be obeyed. A safety chain system must be used for all towing.

While towing, the parking brake should be released and the transmission should be in Neutral. The steering must be unlocked (ignition switch in the Off position). Remember that power brakes will not work with the engine off.

Booster battery (jump) starting

Certain precautions must be observed when using a booster battery to jump start a vehicle.

a) *Before connecting the booster battery, make sure that the ignition switch is in the Off position.*

b) *Turn off the lights, heater and other electrical loads.*

c) *The eyes should be shielded. Safety goggles are a good idea.*

d) *Make sure that the booster battery is the same voltage as the dead one in the vehicle.*

e) *The two vehicles must not touch each other.*

f) *Make sure that the transmission is in Neutral (manual transmission) or Park (automatic transmission).*

g) *If the booster battery is not a maintenance-free type, remove the vent caps and lay a cloth over the vent holes.*

Connect the red jumper cable to the Positive (+) terminals of each battery.

Connect one end of the black jumper cable to the negative (-) terminal of the booster battery. The other end of this cable should be connected to a good ground on the vehicle to be started, such as a bolt or bracket on the engine block. Use caution to ensure that the cable will not come into contact with the fan, drivebelt or other moving parts of the engine.

Start the engine using the booster battery, then with the engine running at idle speed, disconnect the jumper cables in the reverse order of connection.

Automotive chemicals and lubricants

A number of automotive chemicals and lubricants are available for use during vehicle maintenance and repair. They include a wide variety of products ranging from cleaning solvents and degreasers to lubricants and protective sprays for rubber, plastic and vinyl.

Cleaners

Carburetor cleaner and choke cleaner is a strong solvent for gum, varnish and carbon. Most carburetor cleaners leave a dry-type lubricant film which will not harden or gum up. Because of this film it is not recommended for use on electrical components.

Brake system cleaner is used to remove grease and brake fluid from the brake system, where clean surfaces are absolutely necessary. It leaves no residue and often eliminates brake squeal caused by contaminants.

Electrical cleaner removes oxidation, corrosion and carbon deposits from electrical contacts, restoring full current flow. It can also be used to clean spark plugs, carburetor jets, voltage regulators and other parts where an oil-free surface is desired.

Demoisturants remove water and moisture from electrical components such as alternators, voltage regulators, electrical connectors and fuse blocks. They are non-conductive, non-corrosive and non-flammable.

Degreasers are heavy-duty solvents used to remove grease from the outside of the engine and from chassis components. They can be sprayed or brushed on and, depending on the type, are rinsed off either with water or solvent.

Lubricants

Motor oil is the lubricant formulated for use in engines. It normally contains a wide variety of additives to prevent corrosion and reduce foaming and wear. Motor oil comes in various weights (viscosity ratings) from 0 to 50. The recommended weight of the oil depends on the season, temperature and the demands on the engine. Light oil is used in cold climates and under light load conditions. Heavy oil is used in hot climates and where high loads are encountered. Multi-viscosity oils are designed to have characteristics of both light and heavy oils and are available in a number of weights from 5W-20 to 20W-50.

Gear oil is designed to be used in differentials, manual transmissions and other areas where high-temperature lubrication is required.

Chassis and wheel bearing grease is a heavy grease used where increased loads and friction are encountered, such as for wheel bearings, balljoints, tie-rod ends and universal joints.

High-temperature wheel bearing grease is designed to withstand the extreme temperatures encountered by wheel bearings in disc brake equipped vehicles. It usually contains molybdenum disulfide (moly), which is a dry-type lubricant.

White grease is a heavy grease for metal-to-metal applications where water is a problem. White grease stays soft under both low and high temperatures (usually from -100 to +190-degrees F), and will not wash off or dilute in the presence of water.

Assembly lube is a special extreme pressure lubricant, usually containing moly, used to lubricate high-load parts (such as main and rod bearings and cam lobes) for initial start-up of a new engine. The assembly lube lubricates the parts without being squeezed out or washed away until the engine oiling system begins to function.

Silicone lubricants are used to protect rubber, plastic, vinyl and nylon parts.

Graphite lubricants are used where oils cannot be used due to contamination problems, such as in locks. The dry graphite will lubricate metal parts while remaining uncontaminated by dirt, water, oil or acids. It is electrically conductive and will not foul electrical contacts in locks such as the ignition switch.

Moly penetrants loosen and lubricate frozen, rusted and corroded fasteners and prevent future rusting or freezing.

Heat-sink grease is a special electrically non-conductive grease that is used for mounting electronic ignition modules where it is essential that heat is transferred away from the module.

Sealants

RTV sealant is one of the most widely used gasket compounds. Made from silicone, RTV is air curing, it seals, bonds, waterproofs, fills surface irregularities, remains flexible, doesn't shrink, is relatively easy to remove, and is used as a supplementary sealer with almost all low and medium temperature gaskets.

Anaerobic sealant is much like RTV in that it can be used either to seal gaskets or to form gaskets by itself. It remains flexible, is solvent resistant and fills surface imperfections. The difference between an anaerobic sealant and an RTV-type sealant is in the curing. RTV cures when exposed to air, while an anaerobic sealant cures only in the absence of air. This means that an anaerobic sealant cures only after the assembly of parts, sealing them together.

Thread and pipe sealant is used for sealing hydraulic and pneumatic fittings and vacuum lines. It is usually made from a Teflon compound, and comes in a spray, a paint-on liquid and as a wrap-around tape.

Chemicals

Anti-seize compound prevents seizing, galling, cold welding, rust and corrosion in fasteners. High-temperature anti-seize, usually made with copper and graphite lubricants, is used for exhaust system and exhaust manifold bolts.

Anaerobic locking compounds are used to keep fasteners from vibrating or working loose and cure only after installation, in the absence of air. Medium strength locking compound is used for small nuts, bolts and screws that may be removed later. High-strength locking compound is for large nuts, bolts and studs which aren't removed on a regular basis.

Oil additives range from viscosity index improvers to chemical treatments that claim to reduce internal engine friction. It should be noted that most oil manufacturers caution against using additives with their oils.

Gas additives perform several functions, depending on their chemical makeup. They usually contain solvents that help dissolve gum and varnish that build up on carburetor, fuel injection and intake parts. They also serve to break down carbon deposits that form on the inside surfaces of the combustion chambers. Some additives contain upper cylinder lubricants for valves and piston rings, and others contain chemicals to remove condensation from the gas tank.

Miscellaneous

Brake fluid is specially formulated hydraulic fluid that can withstand the heat and pressure encountered in brake systems. Care must be taken so this fluid does not come in contact with painted surfaces or plastics. An opened container should always be resealed to prevent contamination by water or dirt.

Weatherstrip adhesive is used to bond weatherstripping around doors, windows and trunk lids. It is sometimes used to attach trim pieces.

Undercoating is a petroleum-based, tar-like substance that is designed to protect metal surfaces on the underside of the vehicle from corrosion. It also acts as a sound-deadening agent by insulating the bottom of the vehicle.

Waxes and polishes are used to help protect painted and plated surfaces from the weather. Different types of paint may require the use of different types of wax and polish. Some polishes utilize a chemical or abrasive cleaner to help remove the top layer of oxidized (dull) paint on older vehicles. In recent years many non-wax polishes that contain a wide variety of chemicals such as polymers and silicones have been introduced. These non-wax polishes are usually easier to apply and last longer than conventional waxes and polishes.

Conversion factors

Length (distance)
Inches (in)	X	25.4	= Millimetres (mm)	X 0.0394	= Inches (in)
Feet (ft)	X	0.305	= Metres (m)	X 3.281	= Feet (ft)
Miles	X	1.609	= Kilometres (km)	X 0.621	= Miles

Volume (capacity)
Cubic inches (cu in; in³)	X	16.387	= Cubic centimetres (cc; cm³)	X 0.061	= Cubic inches (cu in; in³)
Imperial pints (Imp pt)	X	0.568	= Litres (l)	X 1.76	= Imperial pints (Imp pt)
Imperial quarts (Imp qt)	X	1.137	= Litres (l)	X 0.88	= Imperial quarts (Imp qt)
Imperial quarts (Imp qt)	X	1.201	= US quarts (US qt)	X 0.833	= Imperial quarts (Imp qt)
US quarts (US qt)	X	0.946	= Litres (l)	X 1.057	= US quarts (US qt)
Imperial gallons (Imp gal)	X	4.546	= Litres (l)	X 0.22	= Imperial gallons (Imp gal)
Imperial gallons (Imp gal)	X	1.201	= US gallons (US gal)	X 0.833	= Imperial gallons (Imp gal)
US gallons (US gal)	X	3.785	= Litres (l)	X 0.264	= US gallons (US gal)

Mass (weight)
Ounces (oz)	X	28.35	= Grams (g)	X 0.035	= Ounces (oz)
Pounds (lb)	X	0.454	= Kilograms (kg)	X 2.205	= Pounds (lb)

Force
Ounces-force (ozf; oz)	X	0.278	= Newtons (N)	X 3.6	= Ounces-force (ozf; oz)
Pounds-force (lbf; lb)	X	4.448	= Newtons (N)	X 0.225	= Pounds-force (lbf; lb)
Newtons (N)	X	0.1	= Kilograms-force (kgf; kg)	X 9.81	= Newtons (N)

Pressure
Pounds-force per square inch (psi; lbf/in²; lb/in²)	X	0.070	= Kilograms-force per square centimetre (kgf/cm²; kg/cm²)	X 14.223	= Pounds-force per square inch (psi; lbf/in²; lb/in²)
Pounds-force per square inch (psi; lbf/in²; lb/in²)	X	0.068	= Atmospheres (atm)	X 14.696	= Pounds-force per square inch (psi; lbf/in²; lb/in²)
Pounds-force per square inch (psi; lbf/in²; lb/in²)	X	0.069	= Bars	X 14.5	= Pounds-force per square inch (psi; lbf/in²; lb/in²)
Pounds-force per square inch (psi; lbf/in²; lb/in²)	X	6.895	= Kilopascals (kPa)	X 0.145	= Pounds-force per square inch (psi; lbf/in²; lb/in²)
Kilopascals (kPa)	X	0.01	= Kilograms-force per square centimetre (kgf/cm²; kg/cm²)	X 98.1	= Kilopascals (kPa)

Torque (moment of force)
Pounds-force inches (lbf in; lb in)	X	1.152	= Kilograms-force centimetre (kgf cm; kg cm)	X 0.868	= Pounds-force inches (lbf in; lb in)
Pounds-force inches (lbf in; lb in)	X	0.113	= Newton metres (Nm)	X 8.85	= Pounds-force inches (lbf in; lb in)
Pounds-force inches (lbf in; lb in)	X	0.083	= Pounds-force feet (lbf ft; lb ft)	X 12	= Pounds-force inches (lbf in; lb in)
Pounds-force feet (lbf ft; lb ft)	X	0.138	= Kilograms-force metres (kgf m; kg m)	X 7.233	= Pounds-force feet (lbf ft; lb ft)
Pounds-force feet (lbf ft; lb ft)	X	1.356	= Newton metres (Nm)	X 0.738	= Pounds-force feet (lbf ft; lb ft)
Newton metres (Nm)	X	0.102	= Kilograms-force metres (kgf m; kg m)	X 9.804	= Newton metres (Nm)

Vacuum
Inches mercury (in. Hg)	X	3.377	= Kilopascals (kPa)	X 0.2961	= Inches mercury
Inches mercury (in. Hg)	X	25.4	= Millimeters mercury (mm Hg)	X 0.0394	= Inches mercury

Power
Horsepower (hp)	X	745.7	= Watts (W)	X 0.0013	= Horsepower (hp)

Velocity (speed)
Miles per hour (miles/hr; mph)	X	1.609	= Kilometres per hour (km/hr; kph)	X 0.621	= Miles per hour (miles/hr; mph)

Fuel consumption*
Miles per gallon, Imperial (mpg)	X	0.354	= Kilometres per litre (km/l)	X 2.825	= Miles per gallon, Imperial (mpg)
Miles per gallon, US (mpg)	X	0.425	= Kilometres per litre (km/l)	X 2.352	= Miles per gallon, US (mpg)

Temperature
Degrees Fahrenheit = (°C x 1.8) + 32   Degrees Celsius (Degrees Centigrade; °C) = (°F - 32) x 0.56

*It is common practice to convert from miles per gallon (mpg) to litres/100 kilometres (l/100km),
where mpg (Imperial) x l/100 km = 282 and mpg (US) x l/100 km = 235

Safety first!

Regardless of how enthusiastic you may be about getting on with the job at hand, take the time to ensure that your safety is not jeopardized. A moment's lack of attention can result in an accident, as can failure to observe certain simple safety precautions. The possibility of an accident will always exist, and the following points should not be considered a comprehensive list of all dangers. Rather, they are intended to make you aware of the risks and to encourage a safety conscious approach to all work you carry out on your vehicle.

Essential DOs and DON'Ts

DON'T rely on a jack when working under the vehicle. Always use approved jackstands to support the weight of the vehicle and place them under the recommended lift or support points.

DON'T attempt to loosen extremely tight fasteners (i.e. wheel lug nuts) while the vehicle is on a jack - it may fall.

DON'T start the engine without first making sure that the transmission is in Neutral (or Park where applicable) and the parking brake is set.

DON'T remove the radiator cap from a hot cooling system - let it cool or cover it with a cloth and release the pressure gradually.

DON'T attempt to drain the engine oil until you are sure it has cooled to the point that it will not burn you.

DON'T touch any part of the engine or exhaust system until it has cooled sufficiently to avoid burns.

DON'T siphon toxic liquids such as gasoline, antifreeze and brake fluid by mouth, or allow them to remain on your skin.

DON'T inhale brake lining dust - it is potentially hazardous (see *Asbestos* below).

DON'T allow spilled oil or grease to remain on the floor - wipe it up before someone slips on it.

DON'T use loose fitting wrenches or other tools which may slip and cause injury.

DON'T push on wrenches when loosening or tightening nuts or bolts. Always try to pull the wrench toward you. If the situation calls for pushing the wrench away, push with an open hand to avoid scraped knuckles if the wrench should slip.

DON'T attempt to lift a heavy component alone - get someone to help you.

DON'T rush or take unsafe shortcuts to finish a job.

DON'T allow children or animals in or around the vehicle while you are working on it.

DO wear eye protection when using power tools such as a drill, sander, bench grinder, etc. and when working under a vehicle.

DO keep loose clothing and long hair well out of the way of moving parts.

DO make sure that any hoist used has a safe working load rating adequate for the job.

DO get someone to check on you periodically when working alone on a vehicle.

DO carry out work in a logical sequence and make sure that everything is correctly assembled and tightened.

DO keep chemicals and fluids tightly capped and out of the reach of children and pets.

DO remember that your vehicle's safety affects that of yourself and others. If in doubt on any point, get professional advice.

Asbestos

Certain friction, insulating, sealing, and other products - such as brake linings, brake bands, clutch linings, torque converters, gaskets, etc. - may contain asbestos. Extreme care must be taken to avoid inhalation of dust from such products, since it is hazardous to health. If in doubt, assume that they do contain asbestos.

Fire

Remember at all times that gasoline is highly flammable. Never smoke or have any kind of open flame around when working on a vehicle. But the risk does not end there. A spark caused by an electrical short circuit, by two metal surfaces contacting each other, or even by static electricity built up in your body under certain conditions, can ignite gasoline vapors, which in a confined space are highly explosive. Do not, under any circumstances, use gasoline for cleaning parts. Use an approved safety solvent.

Always disconnect the battery ground (-) cable at the battery before working on any part of the fuel system or electrical system. Never risk spilling fuel on a hot engine or exhaust component. It is strongly recommended that a fire extinguisher suitable for use on fuel and electrical fires be kept handy in the garage or workshop at all times. Never try to extinguish a fuel or electrical fire with water.

Fumes

Certain fumes are highly toxic and can quickly cause unconsciousness and even death if inhaled to any extent. Gasoline vapor falls into this category, as do the vapors from some cleaning solvents. Any draining or pouring of such volatile fluids should be done in a well ventilated area.

When using cleaning fluids and solvents, read the instructions on the container carefully. Never use materials from unmarked containers.

Never run the engine in an enclosed space, such as a garage. Exhaust fumes contain carbon monoxide, which is extremely poisonous. If you need to run the engine, always do so in the open air, or at least have the rear of the vehicle outside the work area.

If you are fortunate enough to have the use of an inspection pit, never drain or pour gasoline and never run the engine while the vehicle is over the pit. The fumes, being heavier than air, will concentrate in the pit with possibly lethal results.

The battery

Never create a spark or allow a bare light bulb near a battery. They normally give off a certain amount of hydrogen gas, which is highly explosive.

Always disconnect the battery ground (-) cable at the battery before working on the fuel or electrical systems.

If possible, loosen the filler caps or cover when charging the battery from an external source (this does not apply to sealed or maintenance-free batteries). Do not charge at an excessive rate or the battery may burst.

Take care when adding water to a non maintenance-free battery and when carrying a battery. The electrolyte, even when diluted, is very corrosive and should not be allowed to contact clothing or skin.

Always wear eye protection when cleaning the battery to prevent the caustic deposits from entering your eyes.

Household current

When using an electric power tool, inspection light, etc., which operates on household current, always make sure that the tool is correctly connected to its plug and that, where necessary, it is properly grounded. Do not use such items in damp conditions and, again, do not create a spark or apply excessive heat in the vicinity of fuel or fuel vapor.

Secondary ignition system voltage

A severe electric shock can result from touching certain parts of the ignition system (such as the spark plug wires) when the engine is running or being cranked, particularly if components are damp or the insulation is defective. In the case of an electronic ignition system, the secondary system voltage is much higher and could prove fatal.

Troubleshooting

Contents

This section provides an easy reference guide to the more common problems which may occur during the operation of your vehicle. These problems and their possible causes are grouped under headings denoting various components or systems, such as Engine, Cooling system, etc. They also refer you to the Chapter and/or Section which deals with the problem.

Remember that successful troubleshooting is not a mysterious black art practiced only by professional mechanics. It is simply the result of the right knowledge combined with an intelligent, systematic approach to the problem. Always work by a process of elimination, starting with the simplest solution and working through to the most complex - and never overlook the obvious. Anyone can run the gas tank dry or leave the lights on overnight, so don't assume that you are exempt from such oversights.

Finally, always establish a clear idea of why a problem has occurred and take steps to ensure that it doesn't happen again. If the electrical system fails because of a poor connection, check all other connections in the system to make sure that they don't fail as well. If a particular fuse continues to blow, find out why - don't just replace one fuse after another. Remember, failure of a small component can often be indicative of potential failure or incorrect functioning of a more important component or system.

Engine

1 Engine will not rotate when attempting to start

1   Battery terminal connections loose or corroded (Chapter 1).
2   Battery discharged or faulty (Chapter 1).
3   Automatic transmission not completely engaged in Park (Chapter 7) or clutch not completely depressed (Chapter 8).
4   Broken, loose or disconnected wiring in the starting circuit (Chapters 5 and 12).
5   Starter motor pinion jammed in flywheel ring gear (Chapter 5).
6   Starter solenoid faulty (Chapter 5).
7   Starter motor faulty (Chapter 5).
8   Ignition switch faulty (Chapter 12).
9   Starter pinion or flywheel teeth worn or broken (Chapter 5).

2 Engine rotates but will not start

1   Fuel tank empty.
2   Battery discharged (engine rotates slowly) (Chapter 5).
3   Battery terminal connections loose or corroded (Chapter 1).
4   Leaking fuel injector(s), faulty cold start valve, fuel pump, pressure regulator, etc. (Chapter 4).

5   Fuel not reaching fuel injection system (Chapter 4).
6   Ignition components damp or damaged (Chapter 5).
7   Worn, faulty or incorrectly gapped spark plugs (Chapter 1).
8   Broken, loose or disconnected wiring in the starting circuit (Chapter 5).
9   Loose distributor is changing ignition timing (Chapter 1).
10 Broken, loose or disconnected wires at the ignition coil or faulty coil (Chapter 5).

3 Engine hard to start when cold

1   Battery discharged or low (Chapter 1).
2   Fuel system malfunctioning (Chapter 4).
3   Injector(s) leaking (Chapter 4).
4   Distributor rotor carbon tracked (Chapter 5).
5   Fuel pump relay faulty (Chapter 5).
6   Throttle body dirty (Chapter 4).

4 Engine hard to start when hot

1   Air filter clogged (Chapter 1).
2   Fuel not reaching the fuel injection system (Chapter 4).
3   Corroded battery connections, especially ground (Chapter 1).
4   Fuel pump relay faulty (Chapter 5).
5   Throttle body dirty (Chapter 4).

5 Starter motor noisy or excessively rough in engagement

1   Pinion or flywheel gear teeth worn or broken (Chapter 5).
2   Starter motor mounting bolts loose or missing (Chapter 5).

6 Engine starts but stops immediately

1   Loose or faulty electrical connections at distributor, coil or alternator (Chapter 5).
2   Insufficient fuel reaching the fuel injector(s) (Chapters 1 and 4).
3   Vacuum leak at the gasket between the intake manifold/plenum and throttle body (Chapters 1 and 4).

7 Oil puddle under engine

1   Oil pan gasket and/or oil pan drain bolt seal leaking (Chapter 2).
2   Oil pressure sending unit leaking (Chapter 2).
3   Rocker cover gaskets leaking (Chapter 2).
4   Engine oil seals leaking (Chapter 2).
5   Timing cover sealant or sealing flange leaking (Chapter 2).

8 Engine lopes while idling or idles erratically

1   Vacuum leakage (Chapter 4).
2   Leaking EGR valve or plugged PCV valve (Chapters 1 and 6).
3   Air filter clogged (Chapter 1).
4   Fuel pump not delivering sufficient fuel to the fuel injection system (Chapter 4).
5   Leaking head gasket (Chapter 2).
6   Timing chain and/or gears worn (Chapter 2).
7   Camshaft lobes worn (Chapter 2).

9 Engine misses at idle speed

1   Spark plugs worn or not gapped properly (Chapter 1).
2   Faulty spark plug wires (Chapter 1).
3   Vacuum leaks (Chapter 1).
4   Incorrect ignition timing (Chapter 1).
5   Uneven or low compression (Chapter 2).

10 Engine misses throughout driving speed range

1   Fuel filter clogged and/or impurities in the fuel system (Chapter 1).
2   Low fuel output at the injector (Chapter 4).
3   Faulty or incorrectly gapped spark plugs (Chapter 1).
4   Incorrect ignition timing (Chapter 1).
5   Cracked distributor cap, disconnected distributor wires or damaged distributor components (Chapter 1).
6   Leaking spark plug wires (Chapter 1).
7   Faulty emission system components (Chapter 6).
8   Low or uneven cylinder compression pressures (Chapter 1).
9   Weak or faulty ignition system (Chapter 5).
10 Vacuum leak in fuel injection system, intake manifold or vacuum hoses (Chapter 4).

11 Engine stumbles on acceleration

1   Spark plugs fouled (Chapter 1).
2   Fuel injection system needs adjustment or repair (Chapter 4).
3   Fuel filter clogged (Chapters 1 and 4).
4   Incorrect ignition timing (Chapter 1).
5   Intake manifold air leak (Chapter 4).

12 Engine surges while holding accelerator steady

1   Intake air leak (Chapter 4).
2   Fuel pump faulty (Chapter 4).
3   Loose fuel injector harness connections (Chapters 4 and 6).
4   Defective ECM (Chapter 6).

13 Engine stalls

1 Idle speed incorrect (Chapter 1).
2 Fuel filter clogged and/or water and impurities in the fuel system (Chapter 1).
3 Distributor components damp or damaged (Chapter 5).
4 Faulty emissions system components (Chapter 6).
5 Faulty or incorrectly gapped spark plugs (Chapter 1).
6 Faulty spark plug wires (Chapter 1).
7 Vacuum leak in the fuel injection system, intake manifold or vacuum hoses (Chapter 4).

14 Engine lacks power

1 Incorrect ignition timing (Chapter 1).
2 Excessive play in distributor shaft (Chapter 5).
3 Worn rotor, distributor cap or wires (Chapters 1 and 5).
4 Faulty or incorrectly gapped spark plugs (Chapter 1).
5 Fuel injection system out of adjustment or excessively worn (Chapter 4).
6 Faulty coil (Chapter 5).
7 Brakes binding (Chapter 1).
8 Automatic transmission fluid level incorrect (Chapter 1).
9 Clutch slipping (Chapter 8).
10 Fuel filter clogged and/or impurities in the fuel system (Chapter 1).
11 Emission control system not functioning properly (Chapter 6).
12 Low or uneven cylinder compression pressures (Chapter 2).
13 Exhaust system restricted (Chapter 4).

15 Engine backfires

1 Emissions system not functioning properly (Chapter 6).
2 Ignition timing incorrect (Chapter 1).
3 Faulty secondary ignition system (cracked spark plug insulator, faulty plug wires, distributor cap and/or rotor) (Chapters 1 and 5).
4 Fuel injection system in need of adjustment or worn excessively (Chapter 4).
5 Vacuum leak at fuel injector(s), intake manifold or vacuum hoses (Chapter 4).
6 Valve clearances incorrectly set, and/or valves sticking (Chapter 2).

16 Pinging or knocking engine sounds during acceleration or uphill

1 Incorrect grade of fuel.
2 Ignition timing incorrect (Chapter 1).
3 Fuel injection system in need of adjustment (Chapter 4).
4 Improper or damaged spark plugs or wires (Chapter 1).

5 Worn or damaged distributor components (Chapter 5).
6 Faulty emission system (Chapter 6).
7 Vacuum leak (Chapter 4).

17 Engine runs with oil pressure light on

1 Low oil level (Chapter 1).
2 Idle rpm below specification (Chapter 1).
3 Short in wiring circuit (Chapter 12).
4 Faulty oil pressure sending unit (Chapter 2).
5 Worn engine bearings and/or oil pump (Chapter 2).

18 Engine diesels (continues to run) after switching off

1 Idle speed too high (Chapter 1).
2 Thermo-controlled air cleaner heat valve not operating properly (TBI equipped engines only) (Chapter 6).
3 Excessive engine operating temperature (Chapter 3).
4 Leaking fuel injectors (Chapter 4).

Engine electrical system

19 Battery will not hold a charge

1 Alternator drivebelt defective or not adjusted properly (Chapter 1).
2 Electrolyte level low (Chapter 1).
3 Battery terminals loose or corroded (Chapter 1).
4 Alternator not charging properly (Chapter 5).
5 Loose, broken or faulty wiring in the charging circuit (Chapter 5).
6 Short in vehicle wiring (Chapters 5 and 12).
7 Internally defective battery (Chapters 1 and 5).

20 Voltage warning light fails to go out

1 Faulty alternator or charging circuit (Chapter 5).
2 Alternator drivebelt defective or out of adjustment (Chapter 1).
3 Alternator voltage regulator inoperative (Chapter 5).

21 Voltage warning light fails to come on when key is turned on

1 Warning light bulb defective (Chapter 12).

2 Fault in the printed circuit, dash wiring or bulb holder (Chapter 12).

Fuel system

22 Excessive fuel consumption

1 Dirty or clogged air filter element (Chapter 1).
2 Incorrectly set ignition timing (Chapter 1).
3 Emissions system not functioning properly (Chapter 6).
4 Fuel injection internal parts excessively worn or damaged (Chapter 4).
5 Low tire pressure or incorrect tire size (Chapter 1).

23 Fuel leakage and/or fuel odor

1 Leak in a fuel feed or vent line (Chapter 4).
2 Tank overfilled.
3 Evaporative canister filter clogged (Chapters 1 and 6).
4 Fuel injector internal parts excessively worn (Chapter 4).

Cooling system

24 Overheating

1 Insufficient coolant in system (Chapter 1).
2 Water pump drivebelt defective or out of adjustment (Chapter 1).
3 Radiator core blocked or grille restricted (Chapter 3).
4 Thermostat faulty (Chapter 3).
5 Electric coolant fan blades broken or cracked (Chapter 3).
6 Radiator cap not maintaining proper pressure (Chapter 3).
7 Ignition timing incorrect (Chapter 1).

25 Overcooling

Faulty thermostat (Chapter 3).

26 External coolant leakage

1 Deteriorated/damaged hoses, loose clamps (Chapters 1 and 3).
2 Water pump seal defective (Chapters 1 and 3).
3 Leakage from radiator core or header tank (Chapter 3).
4 Engine drain or water jacket core plugs leaking (Chapter 2).

27 Internal coolant leakage

1 Leaking cylinder head gasket (Chapter 2).
2 Cracked cylinder bore or cylinder head (Chapter 2).
3 Leaking intake manifold gasket (Chapter 2).

28 Coolant loss

1 Too much coolant in system (Chapter 1).
2 Coolant boiling away because of overheating (Chapter 3).
3 Internal or external leakage (Chapter 3).
4 Faulty radiator cap (Chapter 3).

29 Poor coolant circulation

1 Inoperative water pump (Chapter 3).
2 Restriction in cooling system (Chapters 1 and 3).
3 Water pump drivebelt defective/out of adjustment (Chapter 1).
4 Thermostat sticking (Chapter 3).

Clutch

30 Pedal travels to floor - no pressure or very little resistance

1 Master or slave cylinder faulty (Chapter 8).
2 Hose/line ruptured or leaking (Chapter 8).
3 Connections leaking (Chapter 8).
4 No fluid in reservoir (Chapter 8).
5 If fluid is present in master cylinder dust cover, rear master cylinder seal has failed (Chapter 8).
6 If fluid level in reservoir rises as pedal is depressed, master cylinder center valve seal is faulty (Chapter 8).
7 Broken release bearing or fork (Chapter 8).

31 Fluid in area of master cylinder dust cover and on pedal

Rear seal failure in master cylinder (Chapter 8).

32 Fluid on slave cylinder

Slave cylinder plunger seal faulty (Chapter 8).

33 Pedal feels "spongy" when depressed

Air in system (Chapter 8).

34 Unable to select gears

1 Faulty transmission (Chapter 7).
2 Faulty clutch disc (Chapter 8).
3 Fork and bearing not assembled properly (Chapter 8).
4 Faulty pressure plate (Chapter 8).
5 Pressure plate-to-flywheel bolts loose (Chapter 8).

35 Clutch slips (engine speed increases with no increase in vehicle speed)

1 Clutch plate worn (Chapter 8).
2 Clutch plate is oil soaked by leaking rear main seal (Chapter 8).
3 Clutch plate not seated. It may take 30 or 40 normal starts for a new one to seat.
4 Warped pressure plate or flywheel (Chapter 8).
5 Weak diaphragm spring (Chapter 8).
6 Clutch plate overheated. Allow to cool.

36 Grabbing (chattering) as clutch is engaged

1 Oil on clutch plate lining, burned or glazed linings (Chapter 8).
2 Worn or loose engine or transmission mounts (Chapters 2 and 7).
3 Worn splines on clutch plate hub (Chapter 8).
4 Warped pressure plate or flywheel (Chapter 8).

37 Noise in clutch area

1 Fork shaft improperly installed (Chapter 8).
2 Faulty bearing (Chapter 8).

38 Clutch pedal stays on floor

1 Fork shaft binding in housing (Chapter 8).
2 Broken release bearing or fork (Chapter 8).

39 High pedal effort

1 Fork shaft binding in housing (Chapter 8).
2 Pressure plate faulty (Chapter 8).
3 Incorrect size master or slave cylinder installed (Chapter 8).

Manual transmission

40 Vibration

1 Rough wheel bearing (Chapters 1 and 10).
2 Damaged driveaxle (Chapter 8).
3 Out-of-round tires (Chapter 1).
4 Tire out-of-balance (Chapters 1 and 10).
5 Worn U-joint (Chapter 8).

41 Noisy in Neutral with engine running

Damaged clutch release bearing (Chapter 8).

42 Noisy in one particular gear

1 Damaged or worn constant mesh gears (Chapter 7).
2 Damaged or worn synchronizers (Chapter 7).

43 Noisy in all gears

1 Insufficient lubricant (Chapter 1).
2 Damaged or worn bearings (Chapter 7).
3 Worn or damaged input gear shaft and/or output gear shaft (Chapter 7).

44 Slips out of gear

1 Worn or improperly adjusted linkage (Chapter 7).
2 transmission loose on engine (Chapter 7).
3 Shift linkage does not work freely, binds (Chapter 7).
4 Input shaft bearing retainer broken or loose (Chapter 7).
5 Dirt between clutch cover and engine housing (Chapter 7).
6 Worn shift fork (Chapter 7).

45 Leaks lubricant

1 Excessive amount of lubricant in transmission (Chapters 1 and 7).
2 Loose or broken input shaft bearing retainer (Chapter 7).
4 Input shaft bearing retainer O-ring and/or lip seal damaged (Chapter 7).

Automatic transmission

Note: *Due to the complexity of the automatic transmission, it is difficult for the home mechanic to properly diagnose and service this component. For problems other than the following, the vehicle should be taken to a dealer or transmission shop.*

46   Fluid leakage

1   Automatic transmission fluid is a deep red color. Fluid leaks should not be confused with engine oil, which can easily be blown by air flow to the transmission.

2   To pinpoint a leak, first remove all built-up dirt and grime from the transmission housing with degreasing agents and/or steam cleaning. Then drive the vehicle at low speeds so air flow will not blow the leak far from its source. Raise the vehicle and determine where the leak is coming from. Common areas of leakage are:

  a)   **Pan** *(Chapters 1 and 7)*
  b)   **Filler pipe** *(Chapter 7)*
  c)   **Transmission oil lines** *(Chapter 7)*
  d)   **Speedometer sensor** *(Chapter 7)*

47   Transmission fluid brown or has a burned smell

Transmission fluid burned (Chapter 1).

48   General shift mechanism problems

1   Chapter 7 Part B deals with checking and adjusting the shift linkage on automatic transmissions. Common problems which may be attributed to poorly adjusted linkage are:

  a)   *Engine starting in gears other than Park or Neutral.*
  b)   *Indicator on shifter pointing to a gear other than the one actually being used.*
  c)   *Vehicle moves when in Park.*

2   Refer to Chapter 7 Part B for the shift linkage adjustment procedure.

49   Transmission will not downshift with accelerator pedal pressed to the floor

Throttle valve cable out of adjustment (Chapter 7).

50   Engine will start in gears other than Park or Neutral

Neutral start switch malfunctioning (Chapter 7).

51   Transmission slips, shifts roughly, is noisy or has no drive in forward or reverse gears

There are many probable causes for the above problems, but the home mechanic should be concerned with only one possibility - fluid level. Before taking the vehicle to a repair shop, check the level and condition of the fluid as described in Chapter 1. Correct the fluid level as necessary or change the fluid and filter if needed. If the problem persists, have a professional diagnose the probable cause.

Brakes

Note: *Before assuming that a brake problem exists, make sure that:*

  a)   *The tires are in good condition and properly inflated (Chapter 1).*
  b)   *The front end alignment is correct (Chapter 10).*
  c)   *The vehicle is not loaded with weight in an unequal manner.*

52   Vehicle pulls to one side during braking

1   Incorrect tire pressures (Chapter 1).
2   Front end out of line (have the front end aligned).
3   Unmatched tires on same axle.
4   Restricted brake lines or hoses (Chapter 9).
5   Malfunctioning caliper assembly (Chapter 9).
6   Loose suspension parts (Chapter 10).
7   Loose calipers (Chapter 9).

53   Noise (high-pitched squeal when the brakes are applied)

Front and/or rear disc brake pads worn out. The noise comes from the wear sensor rubbing against the disc. Replace pads with new ones immediately (Chapter 9).

54   Brake roughness or chatter (pedal pulsates)

1   Excessive lateral runout (Chapter 9).
2   Parallelism not within specifications (Chapter 9).
3   Uneven pad wear caused by caliper not sliding due to improper clearance or dirt (Chapter 9).
4   Defective rotor (Chapter 9).

55   Excessive pedal effort required to stop vehicle

1   Malfunctioning power brake booster (Chapter 9).
2   Partial system failure (Chapter 9).
3   Excessively worn pads (Chapter 9).
4   Piston in caliper stuck or sluggish (Chapter 9).
5   Brake pads contaminated with oil or grease (Chapter 9).

6   New pads installed and not yet seated. It will take a while for the new material to seat against the rotor.

56   Excessive brake pedal travel

1   Partial brake system failure (Chapter 9).
2   Insufficient fluid in master cylinder (Chapters 1 and 9).
3   Air trapped in system (Chapters 1 and 9).

57   Dragging brakes

1   Master cylinder pistons not returning correctly (Chapter 9).
2   Restricted brakes lines or hoses (Chapters 1 and 9).
3   Incorrect parking brake adjustment (Chapter 9).

58   Grabbing or uneven braking action

1   Malfunction of combination valve (Chapter 9).
2   Malfunction of power brake booster unit (Chapter 9).
3   Binding brake pedal mechanism (Chapter 9).

59   Brake pedal feels spongy when depressed

1   Air in hydraulic lines (Chapter 9).
2   Master cylinder mounting bolts loose (Chapter 9).
3   Master cylinder defective (Chapter 9).

60   Brake pedal travels to the floor with little resistance

Little or no fluid in the master cylinder reservoir caused by leaking caliper piston(s), loose, damaged or disconnected brake lines (Chapter 9).

61   Parking brake does not hold

Parking brake linkage improperly adjusted (Chapters 1 and 9).
Suspension and steering systems
Note: Before attempting to diagnose the suspension and steering systems, perform the following preliminary checks:

  a)   *Tires for wrong pressure and uneven wear.*
  b)   *Steering universal joints from the column to the rack and pinion for loose connections and wear.*

c) Front and rear suspension and the rack and pinion assembly for loose and damaged parts.

d) Out-of-round or out-of-balance tires, bent rims and loose and/or rough wheel bearings.

62 Vehicle pulls to one side

1   Mismatched or uneven tires (Chapter 10).
2   Broken or sagging springs (Chapter 10).
3   Front wheel or rear wheel alignment incorrect (Chapter 10).
4   Front brakes dragging (Chapter 9).

63 Abnormal or excessive tire wear

1   Front wheel or rear wheel alignment incorrect (Chapter 10).
2   Sagging or broken springs (Chapter 10).
3   Tire out-of-balance (Chapter 10).
4   Worn shock absorber (Chapter 10).
5   Overloaded vehicle.
6   Tires not rotated regularly.

64 Wheel makes a "thumping" noise

1   Blister or bump on tire (Chapter 10).
2   Improper shock absorber action (Chapter 10).

65 Shimmy, shake or vibration

1   Tire or wheel out-of-balance or out-of-round (Chapter 10).
2   Loose, worn or out-of-adjustment wheel bearings (Chapters 8 and 10).
3   Worn tie-rod ends (Chapter 10).
4   Worn lower balljoints (Chapter 10).
5   Excessive wheel runout (Chapter 10).
6   Blister or bump on tire (Chapter 10).

66 Hard steering

1   Lack of lubrication at balljoints, tie-rod ends and rack and pinion assembly (Chapter 10).
2   Front wheel alignment (Chapter 10).
3   Low tire pressure(s) (Chapters 1 and 10).

67 Steering wheel does not return to center position correctly

1   Lack of lubrication at balljoints and tie-rod ends (Chapter 10).

2   Binding in balljoints (Chapter 10).
3   Binding in steering column (Chapter 10).
4   Lack of lubricant in rack and pinion assembly (Chapter 10).
5   Front wheel alignment incorrect (Chapter 10).

68 Abnormal noise at the front end

1   Lack of lubrication at balljoints and tie-rod ends (Chapters 1 and 10).
2   Damaged shock absorber mount (Chapter 10).
3   Worn control arm bushings or tie-rod ends (Chapter 10).
4   Loose stabilizer bar (Chapter 10).
5   Loose wheel nuts (Chapters 1 and 10).
6   Loose suspension bolts (Chapter 10).

69 Wander or poor steering stability

1   Mismatched or uneven tires (Chapter 10).
2   Lack of lubrication at balljoints and tie-rod ends (Chapters 1 and 10).
3   Worn shock absorbers (Chapter 10).
4   Loose stabilizer bar (Chapter 10).
5   Broken or sagging springs (Chapter 10).
6   Front or rear wheel alignment incorrect (Chapter 10).

70 Erratic steering when braking

1   Wheel bearings worn (Chapters 8 and 10).
2   Broken or sagging springs (Chapter 10).
3   Leaking wheel cylinder or caliper (Chapter 9).
4   Warped rotors (Chapter 9).

71 Excessive pitching and/or rolling around corners or during braking

1   Loose stabilizer bar (Chapter 10).
2   Worn shock absorbers or mounts (Chapter 10).
3   Broken or sagging springs (Chapter 10).
4   Overloaded vehicle.

72 Suspension bottoms

1   Overloaded vehicle.
2   Worn shock absorbers (Chapter 10).
3   Incorrect, broken or sagging springs (Chapter 10).

73 Cupped tires

1   Front wheel or rear wheel alignment incorrect (Chapter 10).
2   Worn shock absorbers (Chapter 10).
3   Wheel bearings worn (Chapters 8 and 10).
4   Excessive tire or wheel runout (Chapter 10).
5   Worn balljoints (Chapter 10).

74 Excessive tire wear on outside edge

1   Inflation pressures incorrect (Chapter 1).
2   Excessive speed in turns.
3   Front end alignment incorrect (excessive toe-in). Have professionally aligned.
4   Suspension arm bent or twisted (Chapter 10).

75 Excessive tire wear on inside edge

1   Inflation pressures incorrect (Chapter 1).
2   Front end alignment incorrect (toe-out). Have the front end professionally aligned.
3   Loose or damaged steering components (Chapter 10).

76 Tire tread worn in one place

1   Tires out-of-balance.
2   Damaged or buckled wheel. Inspect and replace if necessary.
3   Defective tire (Chapter 1).

77 Excessive play or looseness in steering system

1   Wheel bearing(s) worn (Chapter 10).
2   Tie-rod end loose or worn (Chapter 10).
3   Rack and pinion loose (Chapter 10).

78 Rattling or clicking noise in rack and pinion

1   Insufficient or improper lubricant in rack and pinion assembly (Chapter 10).
2   Rack and pinion mounts loose (Chapter 10).

Notes

Chapter 1
Tune-up and routine maintenance

Contents

Specifications

Recommended lubricants and fluids

Note: *Listed here are manufacturer recommendations at the time this manual was written. Manufacturers occasionally upgrade their fluid and lubricant specifications, so check with your local auto parts store for current recommendations.*

Engine oil type
    1984 through 1991 .... SF, SF/CC or SF/CD
    1992 and later .... SG or SH synthetic
Engine oil viscosity .... See accompanying chart

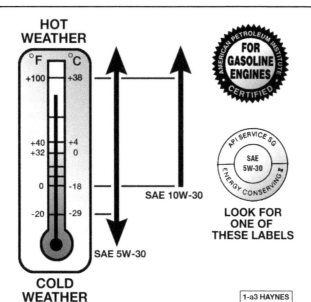

Engine oil viscosity chart - for best fuel economy and cold starting, select the lowest SAE viscosity grade for the expected temperature range

1-a3 HAYNES

Recommended lubricants and fluids

Automatic transmission fluid type	Dexron II, IIE or III Automatic Transmission Fluid (ATF)
Manual transmission fluid type	
Four-speed..	SAE 80W gear lubricant
Six-speed	SAE 5W-30 transmission lubricant
Overdrive unit	Dexron II ATF
Differential (rear axle)	SAE 80W or SAE 80W-90 GL-5 gear lubricant with special limited-slip additive
Engine coolant	
1995 and earlier..	Mixture of water and ethylene glycol-based antifreeze
1996	Mixture of water and DEX-COOL silicate free antifreeze
Brake fluid...	DOT 3 brake fluid
Clutch fluid..	DOT 3 brake fluid
Power steering fluid	GM power steering fluid, or equivalent
Chassis lubrication	Multi-purpose lithium-base chassis grease

Capacities*

Engine oil	5 quarts
Cooling system	
1990 and earlier..	14 quarts
1991 and later	17.8 quarts
Automatic transmission (drain and refill)	4.5 quarts
Manual transmission	
Four-speed..	Not available
Six-speed	4.4 pints

All capacities approximate, add as necessary to fill to appropriate level.

General

Ignition timing	Refer to the *Vehicle Emission Control Information* label located in the engine compartment
Engine idle speed	Not owner adjustable (see Section 26)
Spark plug type	
1984	AC R45TS
1985 through 1987 (cast-iron cylinder heads)	AC R43CTS
1987 through 1991 (aluminum cylinder heads)	AC FR5LS
1992 and later	AC 41-906
Spark plug gap	
1984	0.045 inch
1985 through 1991	0.035 inch
1992 and later	0.050 inch
Firing order	1-8-4-3-6-5-7-2

Filters

Oil filter type	
1984 through 1991	AC type PF25
1992 and later	AC type PF51
Air filter type	
1984	AC type A862C
1985 through 1991	AC type A917C
1992 and later	AC type A1097C
PCV valve	
1984	AC type CV853C
1985 through 1991	AC type CV774C
1992 and later	AC type CV8956

Brakes

Minimum disc refinish thickness	0.724 inch
Brake pedal travel..	2-3/4 inches
Brake pad wear limit	1/8 inch

Cylinder location and distributor rotation - 1991 and earlier

The blackened terminal shown on the distributor cap indicates the Number One spark plug wire position

FRONT

24017-1-B HAYNES

Cylinder and spark plug wire terminal locations - 1992 and later

Torque specifications

	Ft-lbs (unless otherwise indicated)
Differential (rear axle) fill plug	30
Spark plugs	
Cast-iron cylinder heads	22
Aluminum cylinder heads	12
Oil pan drain plug	20
Wheel lug nuts	100
Manual transmission fill/drain plug	26
Overdrive unit oil pan bolts	96 in-lbs
Automatic transmission oil pan bolts	96 in-lbs

Right-side view of the engine compartment (typical 1991 and earlier)

1	Engine oil dipstick	6	Radiator cap	11	Hood latch and cable
2	Engine oil fill cap	7	Air conditioning compressor	12	Windshield wiper arm
3	Fuel injection throttle body	8	Windshield washer reservoir	13	Windshield wiper blade and rubber
4	Fuel injection air intake	9	Heater hoses		element
5	Air filter	10	Distributor		

Left-side view of the engine compartment (typical 1991 and earlier)

1	Throttle linkage	7	Battery	12	Alternator
2	Fuel injectors	8	Battery cables	13	Drivebelt
3	PVC valve	9	Brake caliper	14	Power steering fluid reservoir
4	Brake master cylinder fluid reservoirs	10	Brake hose	15	Vehicle Emissions Control
5	Vehicle identification number (VIN)	11	Brake disc (rotor)	16	Radiator hose
6	Hood latch and cable				

Right-side view of the engine compartment (typical 1992 and later models)

1	Fuel injection throttle body	5	Power steering reservoir	9	Transmission fluid dipstick
2	Air intake duct	6	Engine oil fill cap	10	Fuel lines
3	Air filter	7	Engine oil dipstick	11	Fuel injectors (under cover)
4	Coolant reservoir	8	Cooling system pressure cap		

Left-side view of the engine compartment (Typical 1992 and later models)

1	Alternator	5	Underhood fuse box	9	Vehicle Emissions Control
2	PCV valve (under cover)	6	Brake system reservoir		Information (VECI) label
3	Electronic Control Module (ECM)	7	Air conditioning compressor	10	Drivebelt
4	Battery	8	Upper radiator hose		

Underside view of engine/transmission

1 Oil filter
2 Oil pan drain plug
3 Manual transmission check and fill plug

4 Overdrive unit check plug
5 Overdrive unit oil pan

1 Chevrolet Corvette Maintenance schedule

The following recommendations are given with the assumption that the vehicle owner will be doing the maintenance or service work, as opposed to having a dealer service department do the work. The following are factory maintenance recommendations. However, the owner, interested in keeping his or her vehicle in peak condition at all times and with the vehicle's ultimate resale in mind, may want to perform many of these operations more often. Specifically, we would encourage the shortening of fluid and filter replacement intervals.

When the vehicle is new it may be wise to have the vehicle serviced initially by a factory authorized dealer service department to protect the factory warranty. In many cases the initial maintenance check is done at no cost to the owner. Check with your local dealer for additional information.

Every 250 miles or weekly, whichever comes first

Check the engine oil level (Section 4)
Check the engine coolant level (Section 4)
Check the windshield washer fluid level (Section 4)
Check the brake fluid and clutch fluid levels (Section 4)
Check the tires and tire pressures (Section 5)

Every 3000 miles or 3 months, whichever comes first

All items listed above plus:
Check the automatic transmission fluid level (Section 6)*
Check the power steering fluid level (Section 7)*
Check and service the battery (Section 8)
Check the cooling system (Section 9)
Inspect and replace if necessary all underhood hoses (Section 10)
Inspect and replace if necessary the windshield wiper blades (Section 11)

Every 7500 miles or 12 months, whichever comes first

All items listed above plus:
Change the engine oil and oil filter (Section 12)*
Lubricate the chassis components (Section 13)
Inspect the suspension and steering components (Section 14)*
Inspect the exhaust system (Section 15)*
Check the manual transmission oil level (Section 16)*
Check the overdrive unit fluid level (Section 17)*
Check the differential (rear axle) oil level (Section 18)*
Rotate the tires (Section 19)
Check the brakes (Section 20)*
Inspect the fuel system (Section 21)
Replace the air filter and PCV filter (Section 22)

Check the TBI mounting torque (Section 23)
Check the throttle linkage (Section 24)
Check the thermostatically-controlled air cleaner (Section 25)
Check the engine drivebelt (Section 27)
Check the seat belts (Section 28)
Check the starter safety switch (Section 29)
Check the seat back latch (Section 30)
Check the spare tire and jack (Section 31)

Every 30,000 miles or 24 months, whichever comes first

All items listed above plus:
Change the overdrive fluid (Section 32)*
Change the automatic transmission fluid and filter (Section 33)
Service the cooling system (drain, flush and refill) (1995 and earlier models) (Section 34)
Inspect and replace if necessary the PCV valve (Section 35)
Inspect the evaporative emissions control system (Section 36)
Replace the spark plugs (1991 and earlier models) (Section 37)
Inspect the spark plug wires, distributor cap and rotor (Section 38)
Check and adjust if necessary the ignition timing (Section 39)

Every 100,000 miles or 72 months, whichever comes first

Replace the spark plugs (1992 and later models) (Section 37)
Service the cooling system (drain, flush and refill) (1996 models equipped with DEX-COOL coolant) (Section 34)

* *These items are affected by severe operating conditions as described below. If your vehicle is operated under severe conditions, perform all maintenance indicated with a * at 3000 mile/3 month intervals.*

Severe conditions are indicated if you mainly operate your car under one or more of the following conditions:
Operating in dusty areas
Towing a trailer
Idling for extended periods and/or low speed operation
Operating when outside temperatures remain below freezing and when most trips are less than four miles long

If operated under one or more of the following conditions, change the automatic transmission fluid every 15,000 miles:
In heavy city traffic where the outside temperature regularly reaches 90-degrees F (32-degrees C) or higher
In hilly or mountainous terrain
Frequent trailer pulling

2   Introduction

This Chapter is designed to help the home mechanic maintain his or her vehicle for peak performance, economy, safety and long life.

On the following pages you will find a maintenance schedule along with Sections which deal specifically with each item on the schedule. Included are visual checks, adjustments and item replacements.

Servicing your vehicle using the time/mileage maintenance schedule and the sequenced Sections will give you a planned program of maintenance. Keep in mind that it is a full plan, and maintaining only a few items at the specified intervals will not give you the same results.

In many cases the manufacturer will recommend additional owner checks such as warning lamp operation, defroster operation, condition of window glass, etc. We assume these to be obvious and thus have not included such items in our maintenance plan. Consult your owner's manual for additional information.

You will find as you service your vehicle that many of the procedures can, and should, be grouped together, due to the nature of the job at hand. Examples of this are as follows:

If the vehicle is raised for a chassis lubrication, for example, it is an ideal time for the following checks: exhaust system, suspension, steering and fuel system.

If the tires and wheels are removed, as during a routine tire rotation, check the brakes at the same time.

If you must borrow or rent a torque wrench, service the spark plugs and check the TBI mounting torque all in the same day to save time and money.

The first step of the maintenance plan is to prepare yourself before the actual work begins. Read through the appropriate Sections of this Chapter for all work that is to be performed before you begin. Gather together all the necessary parts and tools. If it appears that you could have a problem during a particular job, don't hesitate to seek advice from your local parts man or dealer service department.

3   Tune-up general information

The term tune-up is used in this manual to represent a combination of individual operations rather than one specific procedure.

If, from the time the vehicle is new, the routine maintenance schedule is followed closely and frequent checks are made of fluid levels and high wear items, as suggested throughout this manual, the engine will be kept in relatively good running condition and the need for additional work will be minimized.

More likely than not, however, there will be times when the engine is running poorly due to lack of regular maintenance. This is even more likely if a used vehicle, which has not received regular and frequent maintenance checks, is purchased. In such cases, an engine tune-up will be needed outside of the regular routine maintenance intervals.

The first step in any tune-up or engine diagnosis to help correct a poor running engine would be a cylinder compression check. A check of the engine compression will give valuable information regarding the overall condition of many internal components and should be used as a basis for tune-up and repair procedures. If, for instance, a compression check indicates serious internal engine wear, a conventional tune-up will not help the running condition of the engine and would be a waste of time and money. Due to its importance, compression checking should be performed by someone who has the proper compression testing gauge and who is knowledgeable with its use. Further information on compression testing can be found in Chapter 2 of this manual.

The following series of operations are those most often needed to bring a generally poor running engine back into a proper state of tune.

Minor tune-up

Clean, inspect and test the battery
Check all engine related fluids
Check and adjust the drivebelt
Replace the spark plugs
Inspect the distributor cap and rotor
*   (1991 and earlier models)*
Inspect the spark plug wires
Check the idle speed
Check and adjust the ignition timing
*   (1991 and earlier models)*
Check the PCV valve
Check the air and PCV filters
Check the cooling system
Check all underhood hoses

Major tune-up

(the above operations and those listed below)
Check the EGR system
Check the ignition system
Check the charging system
Check the fuel system
Replace the air and PCV filters
Replace the distributor cap and rotor
*   (1991 and earlier models)*
Replace the spark plug wires

4   Fluid level checks

Refer to illustrations 4.2, 4.4, 4.9, 4.14, 4.17 and 4.18

Note: *The following are fluid level checks to be done on a 250 mile or weekly basis. Additional fluid level checks can be found in specific Sections that follow. Regardless of intervals, be alert to fluid leaks under the vehicle that would indicate a fault to be corrected immediately.*

1   There are a number of components on a vehicle which rely on the use of fluids to perform their job. During normal operation of the vehicle, these fluids are used up and must be replenished before damage occurs. See Recommended lubricants and fluids at the front of this Chapter for the specific fluid to be used when addition is required. When checking fluid levels, it is important to have the vehicle on a level surface.

Engine oil

2   The engine oil level is checked with a dipstick **(see illustration)**. The dipstick travels through a tube and into the bottom of the engine.

3   The oil level should be checked before the vehicle has been driven, or about 15 minutes after the engine has been shut off. If the oil is checked immediately after driving the vehicle, some of the oil will remain in the upper engine components, producing an inaccurate reading on the dipstick.

4   Pull the dipstick from the tube and wipe all the oil from the end with a clean rag or paper towel. Insert the clean dipstick all the

4.2 The oil dipstick T-handle is clearly marked "engine oil" (arrow), as is the oil cap which threads into the rocker arm cover (arrow)

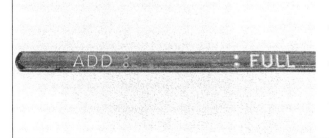

4.4 The marks at the end of the oil dipstick. The oil level must be maintained between the marks at all times. It takes one quart of oil to raise the level from the ADD mark to the FULL mark

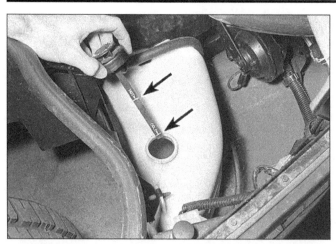

4.9 The radiator coolant reservoir is located just ahead of the right front tire and has a built-in dipstick. The HOT and COLD marks on the dipstick allow you to check the coolant level regardless of engine temperature. Always maintain the level between the marks

4.14 The reservoir for the windshield washer is located just behind the right front tire. How often you use the washers will dictate how often you need to check this reservoir

way back into the oil pan and pull it out again. Observe the oil at the end of the dipstick. Add oil as necessary to keep the level between the ADD mark and the FULL mark on the dipstick **(see illustration)**.

5   Do not overfill the engine by adding too much oil since this may result in oil fouled spark plugs, oil leaks or oil seal failures.

6   Oil is added to the engine after removing a twist off cap located on the rocker arm cover **(see illustration 4.2)**. An oil can spout or funnel may help to reduce spills.

7   Checking the oil can also be an important preventive maintenance step. If you find the oil level dropping abnormally, it is an indication of oil leakage or internal engine wear that should be corrected. If there are water droplets in the oil, or if it is milky looking, component failure is indicated and the engine should be checked immediately.

Engine coolant

8   All vehicles covered by this manual are equipped with a pressurized coolant recovery system. A white coolant reservoir at the right front side of the engine compartment is connected by a hose to the base of the radiator cap. As the engine heats up during operation, coolant is forced from the radiator, through the connecting tube and into the reservoir. As the engine cools, the coolant is automatically drawn back into the radiator to keep the level correct.

9   You usually do not need to remove the radiator cap to check the coolant level. After locating the coolant reservoir, twist off the cap and use the attached dipstick to check the coolant level **(see illustration)**. When the engine is cold the level should be at or slightly above the COLD mark on the dipstick. With the engine fully warm, the level should be at or slightly above the HOT mark on the dipstick. On a periodic basis, or if the coolant in the reservoir runs completely dry, check the coolant level in the radiator as well (see Section 9). **Warning:** *Under no circum-*

stances should the radiator cap be removed when the system is hot, because escaping steam and scalding liquid could cause serious personal injury.

10   If only a small amount of coolant is required to bring the system up to the proper level, plain water can be used. However, to maintain the proper antifreeze/water mixture in the system, both should be mixed together to replenish a low level. High quality antifreeze offering protection to -34-degrees F should be mixed with water in the proportion specified on the container. Do not allow antifreeze to come in contact with your skin or painted surfaces of the vehicle. Flush contacted areas immediately with plenty of water.

11   Coolant should be added to the reservoir until it reaches the proper mark on the dipstick.

12   As the coolant level is checked, note the condition of the coolant. It should be relatively transparent. If it is brown or a rust color, the system should be drained, flushed and refilled (Section 34).

13   If the cooling system requires repeated additions to maintain the proper level, have the radiator cap checked for proper sealing ability. Also check for leaks in the system from cracked hoses, loose hose connections, leaking gaskets, etc.

Windshield washer fluid

14   Fluid for the windshield washer system is located in a plastic reservoir just behind the right front tire **(see illustration)**. The reservoir should be kept no more than 2/3 full to allow for expansion should the fluid freeze. Windshield washer fluid, available at auto parts stores, will not freeze and will result in better cleaning of the windshield surface. Do not use antifreeze in the reservoir because it will cause damage to the vehicle's paint.

15   To help prevent icing in cold weather, warm the windshield with the defroster before using the washer.

Battery electrolyte

16   All vehicles with which this manual is concerned are equipped with a battery which is permanently sealed (except for vent holes) and has no filler caps. Water does not have to be added to these batteries at any time.

Brake and clutch fluid

17   The brake master cylinder is mounted on the front of the power booster unit in the engine compartment. The clutch cylinder used on manual transmissions is mounted adjacent to it **(see illustration)**.

18   The fluid inside is readily visible. The level should be between the MIN and MAX marks on the reservoirs **(see illustration)**. If a low level is indicated, be sure to wipe the top of the reservoir cover with a clean rag to prevent contamination of the brake and/or clutch system before removing the cover.

19   When adding fluid, pour it carefully into the reservoir, taking care not to spill any onto surrounding painted surfaces. Be sure the specified fluid is used, since mixing different types of brake fluid can cause damage to the system. See Recommended lubricants and fluids at the front of this Chapter or your owner's manual.

20   At this time the fluid and cylinder can be inspected for contamination. The system should be drained and refilled if deposits, dirt particles or water droplets are seen in the fluid.

21   After filling the reservoir to the proper level, make sure the lid is on tight to prevent fluid leakage and/or system pressure loss.

22   The brake fluid in the master cylinder will drop slightly as the pads at each wheel wear down during normal operation. If the master cylinder requires repeated replenishing to keep it at the proper level, this is an indication of leakage in the brake system, which should be corrected immediately. Check all brake lines and connections (see Section 20 for more information).

4.17 On early models, the brake system has two fluid reservoirs, which are mounted in tandem (arrows). If equipped with a manual transmission, there will also be a clutch fluid reservoir attached to the firewall (arrow)

4.18 The fluid level inside the brake reservoirs is easily checked by observing the level from outside - later models have a single reservoir feeding both circuits

23   If, upon checking the master cylinder fluid level, you discover one or both reservoirs empty or nearly empty, the brake system should be bled (Chapter 9).

5   Tire and tire pressure checks

Refer to illustrations 5.6a and 5.6b

1   Periodically inspecting the tires may not only prevent you from being stranded with a flat tire, but can also give you clues as to possible problems with the steering and suspension systems before major damage occurs.

2   Proper tire inflation adds miles to the lifespan of the tires, allows the vehicle to achieve maximum miles per gallon of gas and contributes to the overall quality of the ride. Offered as an option on some models is a low tire pressure warning system to alert the driver if air pressure drops more than one pound per square inch below a pre-set level. Other than this early warning system, all inspection and maintenance procedures are the same as for standard models.

3   When inspecting the tires, first check the wear of the tread. Irregularities in the tread pattern (cupping, flat spots, more wear on one side than the other) are indications of front end alignment and/or balance problems. If any of these conditions are noted, take the vehicle to a repair shop to correct the problem.

4   Check the tread area for cuts and punctures. Many times a nail or tack will embed itself into the tire tread and yet the tire may hold air pressure for a period of time. In most cases, a repair shop or gas station can repair the punctured tire.

5   It is important to check the sidewalls of the tires, both inside and outside. Check for deteriorated rubber, cuts, and punctures. Inspect the inboard side of the tire for signs of brake fluid leakage, indicating that a thorough brake inspection is needed immediately.

6   Incorrect tire pressure cannot be determined merely by looking at the tire **(see illustration)**. This is especially true for radial tires.

UNDERINFLATION

CUPPING

Cupping may be caused by:
- Underinflation and/or mechanical irregularities such as out-of-balance condition of wheel and/or tire, and bent or damaged wheel.
- Loose or worn steering tie-rod or steering idler arm.
- Loose, damaged or worn front suspension parts.

OVERINFLATION

5.6a This chart will help you determine the condition of the tires, the probable cause(s) of abnormal wear and the corrective action necessary

INCORRECT TOE-IN OR EXTREME CAMBER

FEATHERING DUE TO MISALIGNMENT

5.6b Tire pressure gauges are available in a variety of styles. Because service station gauges are often inaccurate, keep a gauge in your glove compartment and occasionally check it for accuracy against a "master" air pressure gauge

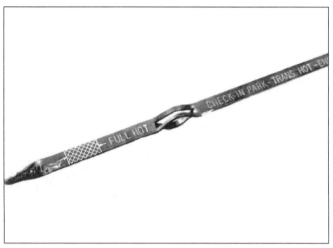

6.6 When checking the automatic transmission fluid level, it is important to note the temperature of the fluid

7.2 The power steering fluid reservoir is located near the front of the engine (arrow). Note the directions on the cap which indicate that it must be turned clockwise for removal

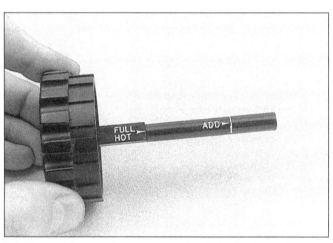

7.6 The markings on the power steering fluid dipstick indicate the safe range. The COLD level mark is on the other side of the dipstick

A tire pressure gauge must be used (see illustration). If you do not already have a reliable gauge, it is a good idea to purchase one and keep it in the glovebox. Built-in pressure gauges at gas stations are often inaccurate.

7 Always check tire inflation when the tires are cold. Cold, in this case, means the vehicle has not been driven more than one mile after sitting for three hours or more. It is normal for the pressure to increase four to eight pounds when the tires are hot.

8 Unscrew the valve cap protruding from the wheel and press the gauge firmly onto the valve. Observe the reading on the gauge and compare the figure to the recommended tire pressure listed on the tire placard. The tire placard is usually attached to the driver's door jamb.

9 Check all tires and add air as necessary to bring them up to the recommended pressure levels. Do not forget the spare tire. Be sure to reinstall the valve caps, which will keep dirt and moisture out of the valve stem mechanism.

6 Automatic transmission fluid level check

Refer to illustration 6.6

1 The level of the automatic transmission fluid should be carefully maintained. Low fluid level can lead to slipping or loss of drive, while overfilling can cause foaming and loss of fluid.

2 With the parking brake set, start the engine, then move the shift lever through all the gear ranges, ending in Park. The fluid level must be checked with the vehicle level and the engine running at idle. **Note:** *Incorrect fluid level readings will result if the vehicle has just been driven at high speeds for an extended period, in hot weather in city traffic, or if it has been pulling a trailer. If any of these conditions apply, wait until the fluid has cooled (about 30 minutes).*

3 With the transmission at normal operating temperature, remove the dipstick from the filler tube. The dipstick is located at the

rear of the engine compartment on the passenger's side.

4 Carefully touch the fluid at the end of the dipstick to determine if the fluid is cool, warm or hot. Wipe the fluid from the dipstick with a clean rag and push it back into the filler tube until the cap seats.

5 Pull the dipstick out again and note the fluid level.

6 If the fluid felt cool, the level should be about 1/8 to 3/8-inch below the ADD mark (see illustration). If it felt warm, the level should be close to the ADD mark. If the fluid was hot, the level should be at the FULL mark. If additional fluid is required, add the recommended fluid directly into the tube by using a funnel. It takes about one pint to raise the level from the ADD mark to the FULL mark with a hot transmission, so add the fluid a little at a time and keep checking the level until it is correct.

7 The condition of the fluid should also be checked along with the level. If the fluid at the end of the dipstick is a dark reddish-brown

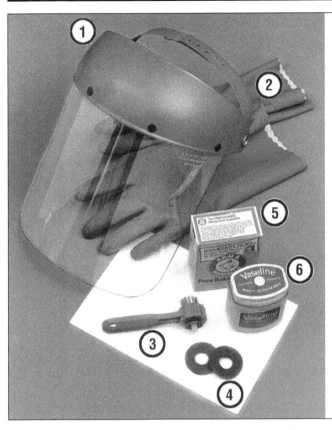

8.1 Tools and materials required for battery maintenance

1   **Face shield/safety goggles** - *When removing corrosion with a brush, the acidic particles can easily fly up into your eyes*
2   **Rubber gloves** - *Another safety item to consider when servicing the battery - remember that's acid inside the battery!*
3   **Battery terminal/cable cleaner** - *This wire brush cleaning tool will remove all traces of corrosion from the battery and cable*
4   **Treated felt washers** - *Placing one of these on each terminal, directly under the cable end, will help prevent corrosion (be sure to get the correct type for side-terminal batteries)*
5   **Baking soda** - *A solution of baking soda and water can be used to neutralize corrosion*
6   **Petroleum jelly** - *A layer of this on the battery terminal bolts will help prevent corrosion*

---

color, or if the fluid has a burned smell, the fluid should be changed. If you are in doubt about the condition of the fluid, purchase some new fluid and compare the two for color and smell.

7   Power steering fluid level check

Refer to illustrations 7.2 and 7.6

1   Unlike manual steering, the power steering system relies on fluid which may, over a period of time, require replenishing.
2   The fluid reservoir for the power steering pump is located behind the radiator near the front of the engine **(see illustration)**.
3   For the check, the front wheels should

be pointed straight ahead and the engine should be off.
4   Use a clean rag to wipe off the reservoir cap and the area around the cap. This will help prevent any foreign matter from entering the reservoir during the check. **Note:** *The power steering reservoir cap has left-hand threads. To remove it, turn it clockwise instead of the normal counterclockwise.*
5   Twist off the cap and check the temperature of the fluid at the end of the dipstick with your finger.
6   Wipe off the fluid with a clean rag, reinsert it, then withdraw it and read the fluid level. The level should be at the FULL HOT mark if the fluid was hot to the touch **(see illustration)**. It should be at the COLD mark if

the fluid was cool to the touch. Note that the marks (Full Hot and Cold) are on opposite sides of the dipstick. At no time should the fluid level drop below the ADD mark.
7   If additional fluid is required, pour the specified type directly into the reservoir, using a funnel to prevent spills.
8   If the reservoir requires frequent fluid additions, all power steering hoses, hose connections, the power steering pump and the rack and pinion assembly should be carefully checked for leaks.

8   Battery check and maintenance

Refer to illustrations 8.1 and 8.3

1   Battery maintenance is an important preventive step in ensuring your car's engine starts quickly and consistently. Refer to the accompanying **illustration** for the tools and materials necessary for battery servicing.
2   A sealed battery is standard equipment on all vehicles with which this manual is concerned. Although this type of battery has many advantages over the older, capped cell type, and never requires the addition of water, it should nevertheless be routinely maintained according to the procedures that follow. **Warning:** *Hydrogen gas in small quantities is present in the area of the side vents on sealed batteries, so keep lighted tobacco and open flames or sparks away from them.*
3   The battery is located on the left side of the engine compartment, behind the tire **(see illustration)**. The external condition of the

8.3 The battery is located just behind the left front tire and is of the side post design. Facing the battery from the front of the car, the positive cable is on the left (usually red in color and indicated on the battery case with a +) and the negative cable is on the right indicated with a -) (arrows)

Check for a soft area indicating
the hose has deteriorated inside.

Check for a chafed area that
could fail prematurely.

Overtightening the clamp on a
hardened hose will damage the
hose and cause a leak.

Check each hose for swelling and
oil-soaked ends. Cracks and breaks
can be located by squeezing the hose.

9.4 Hoses, like drivebelts, have a habit of failing at the worst possible time - to prevent the inconvenience of a blown radiator or heater hose, inspect them carefully as shown here

battery should be inspected periodically for damage such as a cracked case or cover.

4   Check the tightness of the battery cable clamps to ensure good electrical connections and check the entire length of each cable for cracks and frayed conductors.

5   If corrosion (visible as white, fluffy deposits) is evident, remove the cables from the terminals, clean them with a battery brush and reinstall the cables.

6   Make sure that the rubber protector (if so equipped) over the positive terminal is not torn or missing. It should completely cover the terminal.

7   Make sure that the battery tray is in good condition and that the hold-down clamp bolts are tight. If the battery is removed from the tray, make sure that no parts remain in the bottom of the tray when the battery is reinstalled. When reinstalling the hold-down clamp bolts, do not over-tighten them.

8   Corrosion on the hold-down compo-nents, battery case and surrounding areas may be removed with a solution of water and baking soda, but take care to prevent any solution from coming in contact with your eyes, skin or clothes. Protective gloves should be worn. Thoroughly wash all cleaned areas with plain water.

9   Any metal parts of the vehicle damaged by corrosion should be covered with a zinc-based primer then painted.

10   Further information on the battery, charging and jump-starting can be found in Chapter 5 and at the front of this manual.

9   Cooling system check

Refer to illustration 9.4

1   Many major engine failures can be attributed to a faulty cooling system. If the vehicle is equipped with an automatic trans-mission, the cooling system also cools the

transmission fluid and thus plays an impor-tant role in prolonging transmission life.

2   The cooling system should be checked with the engine cold. Do this before the vehi-cle is driven for the day or after it has been shut off for at least three hours.

3   Remove the radiator cap by turning it to the left until it reaches a stop. If you hear any hissing sound (indicating there is still pres-sure in the system), wait until this stops. Now press downward on the cap with the palm of your hand and continue turning to the left until the cap can be removed. Thoroughly clean the cap, inside and out, with clean water. Also clean the filler neck on the radia-tor. All traces of corrosion should be removed. The coolant inside the radiator should be relatively transparent. If it is rust colored, the system should be drained and refilled (Section 34). If the coolant level is not up to the top, add additional anti-freeze/coolant mixture (see Section 4).

4   Carefully check the large upper and lower radiator hoses along with the smaller diameter heater hoses that run from the engine to the firewall. Inspect each hose along its entire length, replacing any hose that is cracked, swollen or shows signs of deterioration. Cracks may become more apparent if the hose is squeezed **(see illus-tration)**.

5   Make sure that all hose connections are tight. A leak in the cooling system will usually show up as white or rust colored deposits on the areas adjoining the leak. If wire-type clamps are used at the ends of the hoses, it may be wise to replace them with more secure screw-type clamps.

6   Use compressed air or a soft brush to remove bugs, leaves, etc. from the front of the radiator or air conditioning condenser. Be careful not to damage the delicate cooling fins or cut yourself on them.

7   Every other inspection, or at the first indication of cooling system problems, have

the cap and system pressure tested. If you do not have a pressure tester, most gas sta-tions and repair shops will do this for a mini-mal charge.

10   Underhood hose check and replacement

Refer to illustration 10.0
Caution: *Replacement of air conditioning hoses must be left to a dealer or air condi-tioning specialist who has the proper equip-ment to depressurize the system safely. Never remove air conditioning components or hoses* **(see illustration)** *until the system has been depressurized.*

General

1   High temperatures under the hood can cause the deterioration of the rubber and plastic hoses used for engine, accessory and emission systems operation. Periodic inspec-tion should be made for cracks, loose clamps, material hardening and leaks.

2   Information specific to the cooling sys-tem hoses can be found in Section 9.

3   Some, but not all, hoses use clamps to secure the hoses to fittings. Where clamps are used, check to be sure they haven't lost their tension, allowing the hose to leak. Where clamps are not used, make sure the hose has not expanded and/or hardened where it slips over the fitting, allowing it to leak.

Vacuum hoses

4   It is quite common for vacuum hoses, especially those in the emissions system, to be color coded or identified by colored stripes molded into the hose. Various sys-tems require hoses with different wall thick-ness, collapse resistance and temperature resistance. When replacing hoses, be sure to use the same hose material on the new hose.

10.0 Air conditioning hoses are best identified by the metal tubes used at all bends (arrows). Do not disconnect or accidentally damage the air conditioning hoses as the system is under great pressure

5   Often the only effective way to check a hose is to remove it completely from the vehicle. Where more than one hose is removed, be sure to label the hoses and their attaching points to insure proper reattachment.

6   When checking vacuum hoses, be sure to include any plastic T-fittings in the check. Check the fittings for cracks and the hose where it fits over the fitting for enlargement, which could cause leakage.

7   A small piece of vacuum hose (1/4-inch inside diameter) can be used as a stethoscope to detect vacuum leaks. Hold one end of the hose to your ear and probe around vacuum hoses and fittings, listening for the "hissing" sound characteristic of a vacuum leak. **Warning:** *When probing with the vacuum hose stethoscope, be careful not to allow your body or the hose to come into contact with moving engine components such as the drivebelt, cooling fan, etc.*

Fuel hose

Warning: *There are certain precautions that must be taken when inspecting or servicing fuel system components. Work in a well ventilated area and do not allow open flames*

(cigarettes, appliance pilot lights, etc.) or bare light bulbs near the work area. Mop up any spills immediately and do not store fuel soaked rags where they could ignite. The fuel system is under pressure, so if any fuel lines are to be disconnected, the pressure in the system must be relieved (see Chapter 4 for more information).

8   Check all rubber fuel lines for deterioration and chafing. Check especially for cracking in areas where the hose bends and just before clamping points, such as where a hose attaches to the fuel filter and fuel injection unit.

9   High quality fuel line, specifically designed for fuel injection systems should be used for fuel line replacement. Under no circumstances should unreinforced vacuum line, clear plastic tubing or water hose be used for fuel line replacement.

10   Spring-type clamps are commonly used on fuel lines. These clamps often lose their tension over a period of time, and can be sprung during the removal process. Therefore it is recommended that all spring-type clamps be replaced with screw clamps whenever a hose is replaced.

Metal lines

11   Sections of metal line are often used for fuel line between the fuel pump and fuel injection unit. Check carefully to be sure the line has not been bent and crimped and that cracks have not started in the line.

12   If a section of metal fuel line must be replaced, only seamless steel tubing should be used, since copper and aluminum tubing do not have the strength necessary to withstand normal engine operating vibration.

13   Check the metal brake lines where they enter the master cylinder and brake proportioning unit (if used) for cracks in the lines or loose fittings. Any sign of brake fluid leakage calls for an immediate thorough inspection of the brake system.

11   Wiper blade inspection and replacement

Refer to illustrations 11.5a, 11.5b and 11.7

1   The windshield wiper and blade assembly should be inspected periodically for damage, loose components and cracked or worn blade elements.

2   Road film can build up on the wiper blades and affect their efficiency, so they should be washed regularly with a mild detergent solution.

3   The action of the wiping mechanism can loosen the bolts, nuts and fasteners, so they should be checked and tightened, as necessary, at the same time the wiper blades are checked.

4   If the wiper blade elements (sometimes called inserts) are cracked, worn or warped, they should be replaced with new ones.

5   Remove the wiper blade assembly from the wiper arm by inserting a small screwdriver in the opening and gently prying on the spring while pulling on the blade to release it **(see illustrations)**.

6   With the blade removed from the vehicle, you can remove the rubber element from the blade.

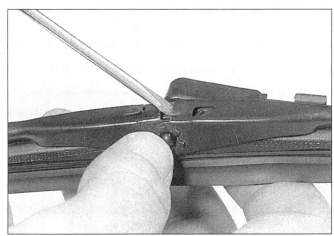

11.5a Using a small screwdriver, gently pry on the spring at the center of the windshield wiper arm

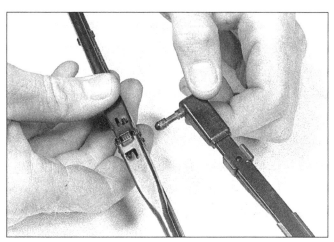

11.5b While prying on the spring, pull the blade assembly away from the arm

7   Using pliers, pinch the metal backing of the element **(see illustration)** and then slide the element out of the blade assembly.

8   Compare the new element with the old for length, design, etc.

9   Slide the new element into place. It will automatically lock at the correct location.

10   Reinstall the blade assembly on the arm, wet the windshield glass and test for proper operation.

12   Oil and oil filter change

Refer to illustrations 12.3, 12.9 and 12.14

1   Frequent oil changes may be the best form of preventive maintenance available to the home mechanic. When engine oil ages, it becomes diluted and contaminated, which leads to premature engine wear.

11.7 The rubber element is retained to the blade by little clips. At one end, the metal backing to the rubber element can be compressed with pliers, allowing the element to slide out of the clips

2   Although some sources recommend oil filter changes every other oil change, we feel that the minimal cost of an oil filter and the relative ease with which it is installed dictate that a new filter be used whenever the oil is changed.

3   Gather together all necessary tools and materials before beginning the procedure **(see illustration)**.

4   In addition, you should have plenty of clean rags and newspapers handy to mop up any spills. Access to the underside of the

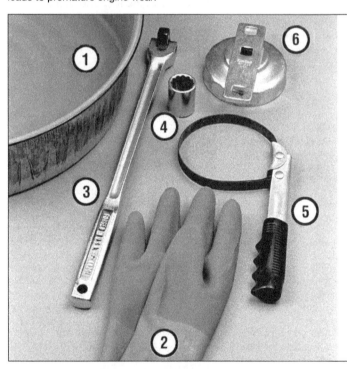

12.3 These tools are required when changing the engine oil and filter

1   **Drain pan** - *It should be fairly shallow in depth, but wide to prevent spills*

2   **Rubber gloves** - *When removing the drain plug and filter, you will get oil on your hands (the gloves will prevent burns)*

3   **Breaker bar** - *Sometimes the oil drain plug is tight, and a long breaker bar is needed to loosen it*

4   **Socket** - *To be used with the breaker bar or a ratchet (must be the correct size to fit the drain plug - six-point preferred)*

5   **Filter wrench** - *This is a metal band-type wrench, which requires clearance around the filter to be effective*

6   **Filter wrench** - *This type fits on the bottom of the filter and can be turned with a ratchet or breaker bar (different-size wrenches are available for different types of filters)*

12.9 The oil drain plug is located at the rear of the pan and should be removed using either a socket or a box-end wrench like the one shown. Do not use an open-end wrench, as the bolt can be easily rounded-off

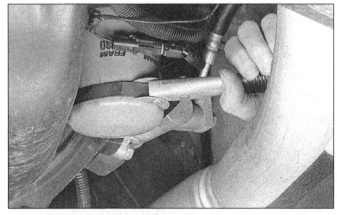

12.14 A metal band-type oil filter wrench is used here to loosen the oil filter. Note that the wrench is positioned at the bottom of the filter where the filter has the most strength. If access makes removal difficult, other types of filter wrenches are available

vehicle is greatly improved if the vehicle can be lifted on a hoist, driven onto ramps or supported by jackstands. **Warning:** *Do not work under a vehicle which is supported only by a bumper, hydraulic or scissors-type jack.*

5   If this is your first oil change, get under the vehicle and familiarize yourself with the locations of the oil drain plug and the oil filter. The engine and exhaust components will be warm during the actual work, so figure out any potential problems before the engine and accessories are hot.

6   Warm the engine to normal operating temperature. If the new oil or any tools are needed, use this warm-up time to gather everything necessary for the job. The correct type of oil for your application can be found in Recommended lubricants and fluids at the beginning of this Chapter.

7   With the engine oil warm (warm engine oil will drain better and more built-up sludge will be removed with the oil), raise and support the vehicle. Make sure it is safely supported.

8   Move all necessary tools, rags and newspapers under the vehicle. Position the drain pan under the drain plug. Keep in mind that the oil will initially flow from the pan with some force, so place the pan accordingly.

9   Being careful not to touch any of the hot exhaust components, use the wrench to remove the drain plug near the bottom of the oil pan **(see illustration)**. Depending on how hot the oil has become, you may want to wear gloves while unscrewing the plug the final few turns.

10   Allow the old oil to drain into the pan. It may be necessary to move the pan farther under the engine as the oil flow slows to a trickle.

11   After all the oil has drained, wipe off the drain plug with a clean rag. Small metal particles may cling to the plug, which would immediately contaminate the new oil.

12   Clean the area around the drain plug opening and reinstall the plug. Tighten the plug securely with the wrench. If a torque wrench is available, use it to tighten the plug.

13   Move the drain pan into position under the oil filter.

14   Use the filter wrench to loosen the oil filter **(see illustration)**. Chain or metal band filter wrenches may distort the filter canister, but this is of no concern as the filter will be discarded anyway.

15   Completely unscrew the old filter. Be careful, as it is full of oil. Empty the oil inside the filter into the drain pan.

16   Compare the old filter with the new one to make sure they are the same type.

17   Use a clean rag to remove all oil, dirt and sludge from the area where the oil filter mounts to the engine. Check the old filter to make sure the rubber gasket is not stuck to the engine mounting surface. If the gasket is stuck to the engine (use a flashlight if necessary), remove it.

18   Apply a light coat of oil around the full circumference of the rubber gasket of the new oil filter.

19   Attach the new filter to the engine, following the tightening directions printed on the filter canister or packing box. Most filter manufacturers recommend against using a filter wrench due to the possibility of overtightening and damage to the seal.

20   Remove all tools, rags, etc. from under the vehicle, being careful not to spill the oil in the drain pan, then lower the vehicle.

21   Move to the engine compartment and locate the oil filler cap.

22   If an oil can spout is used, push the spout into the top of the oil can and pour the fresh oil through the filler opening. A funnel may also be used.

23   Pour three quarts of fresh oil into the engine. Wait a few minutes to allow the oil to drain into the pan, then check the level on the oil dipstick (see Section 4 if necessary). If the oil level is above the ADD mark, start the engine and allow the new oil to circulate.

24   Run the engine for only about a minute and then shut it off. Immediately look under the vehicle and check for leaks at the oil pan drain plug and around the oil filter. If either is leaking, tighten with a bit more force.

25   With the new oil circulated and the filter now completely full, recheck the level on the dipstick and add more oil as necessary.

26   During the first few trips after an oil change, make it a point to check frequently for leaks and proper oil level.

27   The old oil drained from the engine cannot be reused in its present state and should be disposed of. Oil reclamation centers, auto repair shops and gas stations will normally accept the oil, which can be refined and used again. After the oil has cooled it can be drained into a suitable container (capped plastic jugs, topped bottles, milk cartons, etc.) for transport to one of these disposal sites.

28   On vehicles equipped with an oil life monitor (CHANGE OIL light), it will be necessary to reset the light after the oil change is complete. Proceed as follows:

29   Turn the ignition key to on - the engine should not be running.

30   Press the ENG/MET button on the trip monitor. Then within five seconds press and release the ENG/MET button again.

31   Within five seconds of Step 30, press and hold the RANGE button (1991) or the GAUGES button (1992 through 1996).

32   The CHANGE OIL light should flash. Hold the GAUGES or RANGE button until the light stops flashing and goes out.

33   If you make a mistake during this procedure, turn the ignition off and retry the procedure starting with Step 29.

13   Chassis lubrication

Refer to illustrations 13.1, 13.6 and 13.9

1   Refer to Recommended lubricants and fluids at the front of this Chapter to obtain the necessary grease, etc. You will also need a grease gun **(see illustration)**. Occasionally

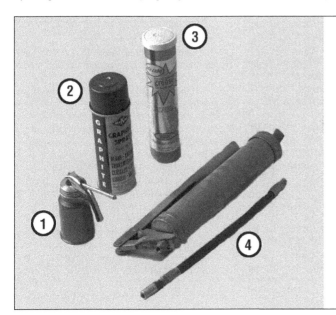

13.1 Materials required for chassis and body lubrication

1   ***Engine oil*** - *Light engine oil in a can like this can be used for door and hood hinges*

2   ***Graphite spray*** - *Used to lubricate lock cylinders*

3   ***Grease*** - *Grease, in a variety of types and weights, is available for use in a grease gun. Check the Specifications for your requirements*

4   ***Grease gun*** - *A common grease gun, shown here with a detachable hose and nozzle, is needed for chassis lubrication. After use, clean it thoroughly*

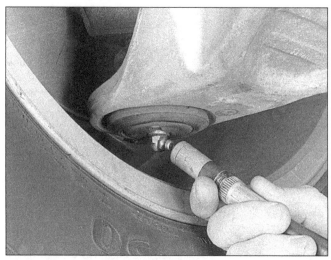

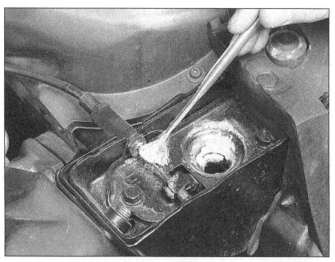

13.6 After wiping the grease fitting clean, push the nozzle firmly into place and pump the grease into the component. Usually about two pumps of the gun will be sufficient

13.9 A little grease on the hood release cable and the latch mechanism will keep the release operating properly. The cables and latches are on both side of the hood

plugs will be installed rather than grease fittings. If so, grease fittings will have to be purchased and installed.

2   Look under the vehicle and see if grease fittings or plugs are installed. If there are plugs, remove them and buy grease fittings, which will thread into the component. A dealer or auto parts store will be able to supply the correct fittings. Straight, as well as angled, fittings are available. **Note:** *There are grease fittings located at three points near each front wheel and also at one point at each rear wheel (at each end of the rear tie-rod).*

3   For easier access under the vehicle, raise it with a jack and place jackstands under the frame. Make sure it is securely supported by the stands. If the wheels are to be removed at this interval for rotation or brake inspection, loosen the lug nuts slightly while the vehicle is still on the ground.

4   Before beginning, force a little grease out of the nozzle to remove any dirt from the end of the gun. Wipe the nozzle clean with a rag.

5   With the grease gun and plenty of clean rags, crawl under the vehicle and begin lubricating the components.

6   Wipe the balljoint grease fitting nipple clean and push the nozzle firmly over it **(see illustration)**. Squeeze the trigger on the grease gun to force grease into the component. The balljoints should be lubricated until the rubber seal is firm to the touch. Do not pump too much grease into the fittings as it could rupture the seal. For all other suspension and steering components, continue pumping grease into the fitting until it oozes out of the joint between the two components. If it escapes around the grease gun nozzle, the nipple is clogged or the nozzle is not completely seated on the fitting. Resecure the gun nozzle to the fitting and try again. If necessary, replace the fitting with a new one.

7   Wipe the excess grease from the com-

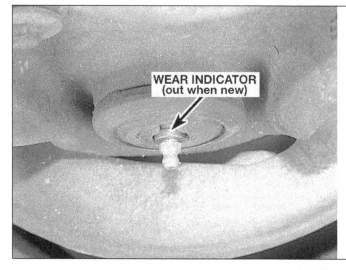

WEAR INDICATOR
(out when new)

14.5 Wear indicators are built into the lower balljoints to aid in their inspection

ponents and the grease fitting. Repeat the procedure for the remaining fittings.

8   If equipped with a manual transmission, lubricate the shift linkage with a little multi-purpose grease. While you are under the vehicle, clean and lubricate the parking brake cable along with the cable guides and levers. This can be done by smearing some of the chassis grease onto the cable and its related parts with your fingers.

9   Open the hood and smear a little chassis grease on the hood latch mechanism **(see illustration)**. Have an assistant pull the release lever from inside the vehicle as you lubricate the cable at the latch.

10   Lubricate all the hinges (door, hood, etc.) with engine oil to keep them in proper working order.

11   The key lock cylinders can be lubricated with spray-on graphite or silicone lubricant, which is available at auto parts stores.

12   Lubricate the door weatherstripping with silicone spray. This will reduce chafing and retard wear.

14 Suspension and steering check

Refer to illustrations 14.5 and 14.7

1   Raise the front of the vehicle periodically and visually check the suspension and steering components for wear.

2   Indications of a fault in these systems are excessive play in the steering wheel before the front wheels react, excessive sway around corners, body movement over rough roads or binding at some point as the steering wheel is turned.

3   Raise the front end of the vehicle and support it securely on jackstands placed under the frame rails. Because of the work to be done, make sure the vehicle cannot fall from the stands.

4   Check the wheel bearings. Do this by spinning the front wheels. Listen for any abnormal noises and watch to make sure the wheel spins true (does not wobble). Grabbing the top and bottom of the tire, pull inward and then outward on the tire, noticing any

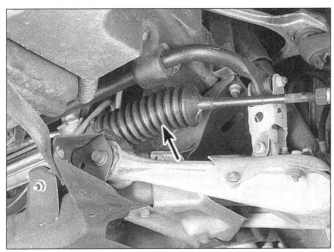

14.7 The steering system uses rubber boots to protect key components. Check the boots (one on each side) for tears, cracks or other damage

15.4a The exhaust system should be inspected from the engine back to the tailpipes, this is the forward portion

15.4b Rear section of the exhaust system

16.1 The manual transmission has two plugs, one for the checking and filling (A) and the lower for draining (B)

movement that would indicate a loose wheel bearing assembly. If the bearings are suspect, refer to Chapter 10 for more information or see a dealer for a more thorough inspection.

5   From under the vehicle check for loose bolts, broken or disconnected parts and deteriorated rubber bushings on all suspension and steering components. Look for grease or fluid leaking from the steering assembly. Check the power steering hoses and connections for leaks. **Note:** *The bottom of the balljoint grease fitting normally protrudes about 0.050-inch from the surface of the balljoint housing* **(see illustration)**. *If the unit is worn enough to allow the bottom of the fitting to move below this surface, then the balljoint must be replaced.*

6   Have an assistant turn the steering wheel from side-to-side and check the steering components for free movement, chafing and binding. If the steering does not react with the movement of the steering wheel, try to determine where the slack is located.

7   The steering system uses flexible rubber boots that should be carefully checked for tears, oil contamination or damage **(see illustration)**.

15   Exhaust system check

Refer to illustrations 15.4a and 15.4b

1   With the engine cold (at least three hours after the vehicle has been driven), check the complete exhaust system from its starting point at the engine to the end of the tailpipe. Be careful around the area of the catalytic converter, which may be hot even after three hours. The inspection should be done on a hoist where unrestricted access is available.

2   Check the pipes and connections for signs of leakage and/or corrosion indicating a potential failure. Make sure that all brackets and hangers are in good condition and tight.

3   At the same time, inspect the underside of the body for holes, corrosion, open seams, etc. which may allow exhaust gases to enter the passenger compartment. Seal all body openings with silicone or body putty.

4   Rattles and other noises can often be traced to the exhaust system, especially the hangers, mounts and heat shields **(see illustrations)**. Try to move the pipes, mufflers and catalytic converter. If the components can come in contact with the body or suspension parts, secure the exhaust system with new brackets and hangers.

16   Manual transmission oil level check

Refer to illustration 16.1

1   Manual transmissions do not have a dipstick. The oil level is checked by removing the plug in the side of the transmission case **(see illustration)**.

2   If the oil level is not at the bottom of the

17.1 The overdrive unit has a large square-head plug (arrow) which must be removed for checking the fluid level

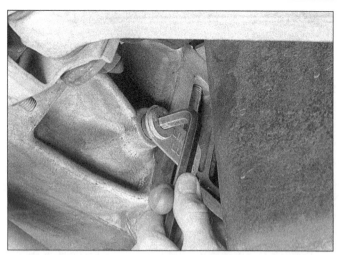

18.1 An Allen wrench must be used to remove the check/filter plug on the rear axle. Note also that a tag is attached to the plug. This tag gives specific lubricant information

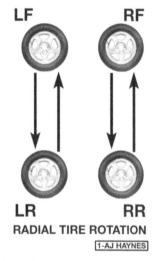

19.2a On this vehicle the wheels must remain on the same side during tire rotation

19.2b The inside of the wheel indicates on which side of the vehicle the wheel must be installed and also the size of the wheel (arrow). Only wheels that are the same size can be rotated front-to-rear

plug opening, use a syringe to squeeze the appropriate lubricant into the opening until it just starts to run out of the hole.

3   Install the plug and tighten it securely. Drive the vehicle a short distance, then check for leaks.

17   Overdrive unit fluid level check (manual transmission only)

Refer to illustration 17.1

1   The overdrive unit (located just behind the manual transmission) fluid is checked by removing a square-head plug on the left side of the overdrive unit **(see illustration)**.

2   Remove the plug and reach inside the hole. The fluid level should be just at the bottom of the hole. If not, add the appropriate fluid through the opening. Install the plug and tighten it securely.

18   Differential (rear axle) oil level check

Refer to illustration 18.1

1   On the side of the differential, remove the lubricant check/filler plug **(see illustration)**.

2   The oil level should be at the bottom of the plug opening. If not, use a syringe to add the proper lubricant until the oil just starts to run out of the opening. Note that a tag is attached to the plug, which gives information regarding lubricant type.

19   Tire rotation

Refer to illustrations 19.2a and 19.2b
Note: *Offered as options with this vehicle are certain handling and suspension packages,*

which include different width wheels on the front and rear. If it is determined that different width wheels are installed, conventional tire rotation is not permitted. The only option is to have the tires removed from the wheels and remounted to their new locations. If this is done, the tires must stay on the same side of the vehicle (left or right) and directional rotation must be retained. Rebalancing must also be done.

1   The tires should be rotated at the specified intervals and whenever uneven wear is noticed.

2   Refer to the accompanying **illustration** for the preferred tire rotation pattern for these special "directional" tires. The direction in which the tires must roll is indicated on the sidewall with an arrow. The wheels are marked "left side only" - "right side only" on the back of the wheel **(see illustration)**. Due to this, the front-to-rear rotation pattern is the only arrangement you should use.

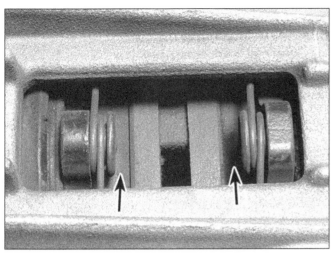

20.5 Looking through the cutout on the back of the caliper, the brake pads (arrows) can be inspected. The pad lining, which rubs against the disc, can also be inspected by looking at each end of the caliper

20.10 The rear brakes look much the same as the front, with the caliper (A) and disc or rotor (B) easily accessible for inspection. The parking brake components can be inspected by looking through one of the two inspection holes (C). Turn the assembly until the linings come into view

3   Refer to the information in Jacking and towing at the front of this manual for the proper procedures to follow when raising the vehicle and changing a tire. If the brakes are to be checked, do not apply the parking brake as stated. Make sure the tires are blocked to prevent the vehicle from rolling as it is raised.

4   Preferably, the entire vehicle should be raised at the same time. This can be done on a hoist or by jacking up each corner and then lowering the vehicle onto jackstands placed under the frame rails. Always use four jackstands and make sure the vehicle is safely supported.

5   After rotation, check and adjust the tire pressures as necessary and be sure to check the lug nut tightness.

6   For further information on the wheels and tires, refer to Chapter 10.

20   Brake check

Refer to illustrations 20.5, 20.10 and 20.14

Note: *For detailed photographs of the brake system, refer to Chapter 9.*

1   In addition to the specified intervals, the brakes should be inspected every time the wheels are removed or whenever a defect is suspected.

Brake pads

2   Disc brakes are used on both the front and the rear on this vehicle. Extensive rotor damage can occur if the pads are not replaced when needed.

3   Raise the vehicle and place it securely on jackstands. Remove the wheels (see Jacking and towing at the front of the manual, if necessary).

4   The disc brake calipers, which contain the pads, are now visible. There is an outer pad and an inner pad in each caliper. All pads

should be inspected.

5   The caliper has a window to inspect the pads **(see illustration)**. If the pad material has worn to about 1/8-inch or less, the pads require replacement.

6   If you are unsure about the exact thickness of the remaining lining material, remove the pads for further inspection or replacement (refer to Chapter 9).

7   Before installing the wheels, check for leakage and/or damage (cracking, splitting, etc.) around the brake hose connections. Replace the hose or fittings as necessary, referring to Chapter 9.

8   Check the condition of the rotor. Look for scoring, gouging and burned spots. If these conditions exist, the hub/rotor assembly should be removed for servicing (Chapter 9).

Parking brake

9   The parking brake operates from a hand lever and locks the rear brake system. The easiest, and perhaps most obvious, method of periodically checking the operation of the parking brake assembly is to park the vehicle on a steep hill with the parking brake set and the transmission in Neutral. If the parking brake cannot prevent the vehicle from rolling,

it is in need of adjustment or the linings are in need of replacement (see Chapter 9).

10   With the wheels off the vehicle, the parking brake linings can be inspected by looking through the holes in the rear rotors **(see illustration)**. If necessary, rotate the rotor until the linings are visible. If the linings are less than 1/8-inch in thickness, replace the shoe and lining components. Refer to Chapter 9 for additional information.

Brake pedal travel

11   The brakes should be periodically checked for correct pedal travel, which is the distance the brake pedal moves toward the floor from a fully released position. The brakes must be cold while performing this test.

12   Using a ruler, measure the distance from the floor to the brake pedal.

13   Now pump the brakes at least three times (without starting the engine). Press firmly on the brake pedal and measure the distance between the floor and the pedal in this position.

14   The distance the pedal travels between these two positions should not exceed 2 3/4-inch (70 mm) **(see illustration)**. If adjustment is required, refer to Section 10 in Chapter 9.

20.14 Using a ruler held firmly in place, the brake pedal travel can be checked

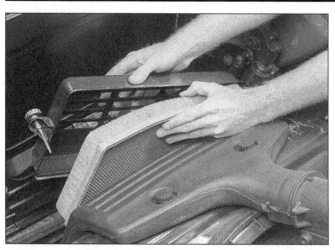

22.2 On models with Tuned Port Injection (TPI), the air filter is located inside a housing at the extreme front of the engine compartment

24.1 The throttle linkage should be periodically checked for free movement

21 Fuel system check

Warning: *There are certain precautions to take when inspecting or servicing the fuel system components. Work in a well ventilated area and do not allow open flames (cigarettes, appliance pilot lights, etc.) to get near the work area. Mop up spills immediately and do not store fuel soaked rags where they could ignite. The fuel system is under pressure and no component should be disconnected without first relieving the pressure (see Chapter 4).*

1   The fuel system is most easily checked with the vehicle raised on a hoist so the components underneath the vehicle are readily visible and accessible.

2   If the smell of gasoline is noticed while driving or after the vehicle has been in the sun, the system should be thoroughly inspected immediately.

3   Remove the gas filler cap and check for damage, corrosion and an unbroken sealing imprint on the gasket. Replace the cap with a new one if necessary.

4   With the vehicle raised, inspect the gas tank and filler neck for punctures, cracks and other damage. The connection between the filler neck and the tank is especially critical. Sometimes a rubber filler neck will leak due to loose clamps or deteriorated rubber, problems a home mechanic can usually rectify.

Warning: *Do not, under any circumstances, try to repair a fuel tank yourself (except rubber components) unless you have had considerable experience. A welding torch or any open flame can easily cause the fuel vapors to explode if the proper precautions are not taken.*

5   Carefully check all rubber hoses and metal lines leading away from the fuel tank. Check for loose connections, deteriorated hoses, crimped lines and other damage. Follow the lines to the front of the vehicle, carefully inspecting them all the way. Repair or replace damaged sections as necessary.

6   If a fuel odor is still evident after the inspection, refer to Section 36.

22 Air filter and PCV filter replacement

Refer to illustration 22.2

1   At the specified intervals, the air filter and PCV filter should be replaced with new ones. A thorough program of preventive maintenance would call for the two filters to be inspected between changes. **Note:** *Because of its advanced design, the Tuned Port Injection (TPI) system does not use an individual PCV filter. The engine air cleaner supplies filtered air to the PCV system.*

2   The air filter is located inside the air cleaner housing located at the front of the engine on TPI models and is replaced by unscrewing the plastic knobs on the sides of the cleaner assembly and lifting off the top plate **(see illustration)**. The filter is located on top of the "Crossfire" Throttle Body Injection (TBI) units on those models and is replaced by unscrewing the nuts from the top of the filter housing and lifting the cover.

3   While the top plate is off, be careful not to drop anything down into the TBI or air cleaner assembly.

4   Lift the air filter element out of the housing.

5   Wipe out the inside of the air cleaner housing with a clean rag.

6   Place the new filter into the air cleaner housing. Make sure it seats properly in the bottom of the housing.

7   The PCV filter on the TBI types is located in the right (passenger) side air intake snorkel, and can be removed through the air cleaner housing when the air cleaner is removed. Remove the top plate and air filter as described previously, then locate the PCV filter on the inside of the housing.

8   Use a screwdriver to pry the filter up and then pull it from the housing.

9   Install a new PCV filter and the air filter.

10   Install the top plate.

23 TBI mounting torque check (1984 models only)

1   The "Crossfire" fuel injection throttle bodies are attached to the top of the intake manifold by four nuts. These fasteners can sometimes work loose from vibration and temperature changes during normal engine operation and cause a vacuum leak.

2   If you suspect that a vacuum leak exists at the bottom of the throttle bodies, obtain a length of hose about the diameter of fuel hose. Start the engine and place one end of the hose next to your ear as you probe around the base with the other end. You will hear a hissing sound if a leak exists (be careful of hot or moving engine components).

3   Remove the air cleaner assembly, tagging each hose to be disconnected with a piece of numbered tape to make reassembly easier.

4   Locate the mounting nuts at the base of the throttle body. Decide what special tools or adapters will be necessary, if any, to tighten the fasteners.

5   Tighten the nuts securely and evenly. Do not overtighten them, as the threads could strip.

6   If, after the nuts are properly tightened, a vacuum leak still exists, the throttle body must be removed and a new gasket installed. See Chapter 4 for more information.

7   After tightening the fasteners, reinstall the air cleaner and return all hoses to their original positions.

24 Throttle linkage inspection

Refer to illustration 24.1

1   Inspect the throttle linkage for damaged or missing parts and for binding and interfer-

27.5 The drivebelt tensioner automatically keeps the proper tension on the drivebelt, but does have limits. The indicator mark (left arrow) should remain within the operating range (right arrows) of the tensioner

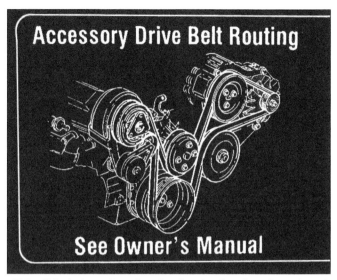

Accessory Drive Belt Routing

See Owner's Manual

27.9a Drivebelt routing, also showing the direction of rotation of the various pulleys - 1984 model shown

ence when the accelerator pedal is operated **(see illustration)**.

2   Lubricate the various linkage pivot points with a drop of engine oil.

25   Thermostatic air cleaner check (1984 models only)

1   On models equipped with "Crossfire" throttle body fuel injection, a heated air system is used to give good driveability under a variety of climatic conditions. By having a relatively constant inlet air temperature, the fuel system can be calibrated to reduce exhaust emissions and eliminate throttle plate icing.

2   The thermostatic air cleaner, Thermac as the system is sometimes called, operates on intake manifold vacuum and heated air. Air can enter from outside the engine compartment or through a stove built around the exhaust manifold.

3   A vacuum diaphragm motor, built into the air cleaner snorkel, operates a damper door to admit hot air from the stove, air from outside, or a combination of both. A temperature sensor inside the air cleaner housing that is sensitive to intake air temperature controls how much vacuum goes to the motor.

4   The air cleaner has two air intake snorkels, both having damper doors. the snorkel on the left (driver) side has both outside and exhaust manifold air inlets and the right (passenger) side has an outside air inlet only. Full manifold vacuum is supplied to this (passenger side) door at all times to keep the door closed. At wide open throttle, vacuum drops and the damper door opens letting in outside air for additional power.

5   No Thermac device is used with the Tuned Port Injection system. A hot water passage is built into the bottom of the throttle assembly to prevent throttle plate icing.

6   The first step in checking the Thermac

air cleaner is to make sure that the hoses and heat stove tube are connected, and not kinked, plugged or damaged.

7   Check the gasket between the air cleaner and the throttle bodies and the air cleaner cover gasket for wear or damage.

8   With the air cleaner installed, the damper door should be open to outside air.

9   Watch the damper door and start the engine. The damper door should close off the outside air right away. **Warning:** *Keep away from the drivebelt and cooling fan when running the engine.*

10   If the door does not close, some simple checks can be made.

  a) *First, with the engine off, disconnect the vacuum hose at the diaphragm motor.*
  b) *With a hand-held vacuum pump, apply at least 7 in-Hg (23 kPa) of vacuum to the motor. The door should completely block off outside air. If not, check for binding linkage or a corroded snorkel.*
  c) *With vacuum still applied, trap the vacuum in the motor by bending the hose. The door should remain closed. If not, replace the vacuum motor assembly.*
  d) *If the vacuum motor, linkage and snorkel check out all right, the temperature sensor is most likely defective and should be replaced with a new one.*

11   Additional information on this and other emissions systems can be found in Chapter 6.

26   Idle speed check and adjustment

Note: *The engine idle speed on these models is controlled by the ECM and thus is not a regular maintenance item. In fact, most models are equipped with tamper-proof plugs to discourage changing the factory pre-set idle speeds. For more information, refer to Chapter 4 and Chapter 6.*

27   Drivebelt check and replacement

Refer to illustrations 27.5, 27.9a and 27.9b

Check

1   A single drivebelt, or serpentine belt as it is more often called, is located at the front of the engine and plays an important role in the overall operation of the engine and its components. Due to its function and material make up, the belt is prone to failure as it wears and should be periodically inspected.

2   A single belt is be used to drive the alternator, power steering pump, water pump and air conditioning compressor.

3   With the engine off, open the hood and use your fingers (and a flashlight, if necessary), to move along the belt checking for cracks and separation of the belt plies. Also check for fraying and glazing, which gives the belt a shiny appearance.

4   Check the ribs on the underside of the belt. They should all be the same depth, with none of the surface uneven.

5   The tension of the belt is checked visually. Locate the belt tensioner at the front of the engine under the air conditioning compressor on the right (passenger) side, and then locate the tensioner operating marks **(see illustration)** located on the side of the tensioner.

6   If the indicator mark is outside of the operating range, the belt should be replaced.

Replacement

7   Using a 1/2-inch breaker bar, rotate the tensioner clockwise to allow the belt to be removed or replaced.

8   Remove the belt from the auxiliary components and carefully release the tensioner.

9   Route the new belt over the various pulleys **(see illustrations)**, again rotating the tensioner to allow the belt to be positioned,

then release the belt tensioner. **Note:** *Most models have a drivebelt routing decal on the radiator to help when installing the drivebelt.*

28   Seat belt check

1   Check the seat belts, buckles, latch plates and guide loops for any obvious damage or signs of wear.
2   Check that the seat belt reminder light comes on when the key is turned to the Run or Start positions. A chime should also sound.
3   The seat belts are designed to lock up during a sudden stop or impact, yet allow free movement during normal driving. Check that the retractors return the belt against your chest while driving and rewind the belt fully when the buckle is unlatched.
4   A special cinch button is integrated into this seat belt system which allows the user to tighten the belt securely and bypass the free movement feature. With the belt in the desired position, push in the cinch button, which should lock the belt in place.
5   If any of the above checks reveal problems with the seat belt system, replace parts as necessary.

29   Starter safety switch check

Caution: *During the following checks there is a chance that the vehicle could lunge forward, possibly causing damage or injuries. Allow plenty of room around the vehicle, firmly apply the parking brake and hold down the regular brake pedal during the following checks.*
1   On automatic transmission vehicles, try to start the vehicle in each gear. The engine should crank only in Park or Neutral.
2   If equipped with a manual transmission, place the shift lever in Neutral and push the clutch pedal down about halfway. The engine should crank only with the clutch pedal fully depressed.
3   Check that the steering column lock allows the key to go into the Lock position only when the shift lever is in Park (automatic transmissions) or Reverse (manual transmissions).
4   The ignition key should come out only in the Lock position.

30   Seat back latch check

1   It is important to periodically check the seat back latch mechanism to prevent the seat back from falling forward during a sudden stop or an accident.
2   Grasping the top of the seat, attempt to tilt the seat back forward. It should tilt only when the latch on the rear of the seat is pulled upward. Note that there is a certain amount of free play built into the latch mechanism.

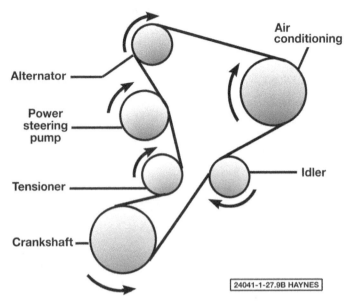

27.9b Drivebelt routing - 1992 and later models

3   When returned to the upright position, the seat back should latch securely.

31   Spare tire and jack check

1   Periodically checking the security and condition of the spare tire and jack will help to familiarize you with the procedures necessary for emergency tire replacement and also help to insure that no components work loose during normal vehicle operation.
2   Following the instructions in your owner's manual or under Jacking and towing near the front of this manual, lower the spare tire and jack.
3   Using a reliable air pressure gauge, check the pressure in the spare tire. It should be kept at 60 psi.
4   Check that the jack operates freely and all components are undamaged.
5   Upon completion, check that the spring holds the jack firmly to the wheel and that the tire tray is securely fastened to the vehicle.

32   Overdrive unit fluid change

Refer to illustrations 32.9 and 32.12
1   Raise the vehicle.
2   On models equipped with a convertible top, body bracing that runs under the overdrive oil pan will have to be removed.
3   With a drain pan placed under the overdrive unit, remove the oil pan attaching bolts from the front and sides of the pan. Loosen the bolts a little at a time to prevent warping of the pan.
4   Loosen the rear pan attaching bolts approximately four turns.
5   Carefully pry the oil pan loose allowing the oil to drain.
6   Remove the remaining bolts and the pan.
7   Drain the fluid from the pan, clean the pan and round magnet with solvent and dry with compressed air.
8   Remove the filter from the overdrive unit.
9   Using a gasket scraper or putty knife,

32.9 The gasket surfaces of the overdrive unit and the oil pan must be carefully cleaned before reassembly

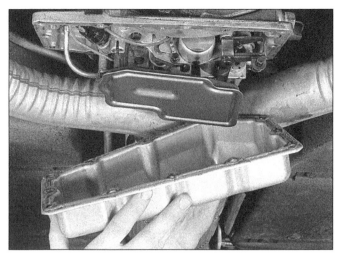

32.12 With the new filter screen pushed into place and a bead of sealant on the oil pan flange, the pan is ready for reassembly. Remember to tighten each of the bolts only a little at a time to prevent warping

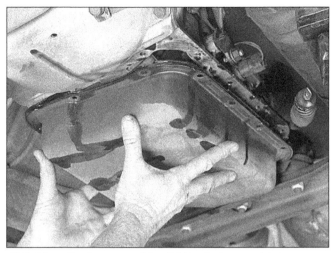

33.8 After allowing some of the fluid to drain, completely remove the remaining bolts and lower the pan. Be careful, there is still plenty of fluid in the pan

scrape away all traces of the old sealant from the oil pan and the overdrive unit pan surface **(see illustration)**. Note that no gasket is used - only sealant.

10   Carefully install a new filter, making sure it is the correct type and in the correct position.

11   Place the magnet in the narrow end of the oil pan towards the front.

12   Apply a bead of RTV sealant to the oil pan flange, and place the pan into position **(see illustration)**.

13   Fill the overdrive unit with automatic transmission fluid to the bottom of the plug hole in the left side of the unit.

33   Automatic transmission fluid and filter change

Refer to illustrations 33.8 and 33.10

1   At the specified time intervals, the transmission fluid should be drained and replaced. Since the fluid will remain hot long after driving, perform this procedure only after sufficient cooling.

2   Before beginning work, purchase the specified transmission fluid (see Recommended lubricants and fluids at the front of this Chapter) and filter.

3   Other tools necessary for this job include jackstands to support the vehicle in a raised position, a drain pan capable of holding at least 8 pints, newspapers and clean rags.

4   Raise and support the vehicle on jackstands.

5   With a drain pan in place, remove the front and side oil pan attaching bolts.

6   Loosen the rear pan attaching bolts approximately four turns.

7   Carefully pry the transmission oil pan loose with a screwdriver allowing the fluid to drain.

8   Remove the remaining bolts, pan and

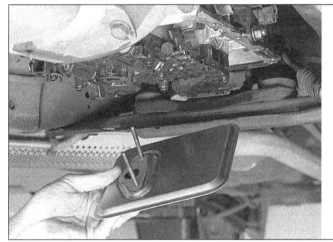

33.10 The new filter will attach to the transmission with either one or two bolts. Some models will have a gasket that must be placed between the filter and the transmission valve body

gasket **(see illustration)**. Carefully clean the gasket surface of the transmission to remove all traces of the old gasket and sealant.

9   Drain the fluid from the transmission oil pan, clean it with solvent and dry it well with compressed air.

10   Remove the filter screen from its mount inside the transmission **(see illustration)**.

11   Install a new filter screen and O-ring. Be sure to lubricate the O-ring with petroleum jelly before installation.

12   Make sure the gasket surface on the transmission oil pan is clean, then install a new gasket. Put the pan in place against the transmission and, working around the pan, tighten each pan bolt a little at a time until the final torque figure is reached.

13   Lower the vehicle and add the specified amount of automatic transmission fluid through the filler tube (Section 6).

14   With the selector in Park and the parking brake set, run the engine at a fast idle, but do not race the engine.

15   Move the gear selector through each range and back to Park. Check the fluid level.

16   Check under the vehicle for leaks during the first few trips.

34   Cooling system servicing (draining, flushing and refilling)

Refer to illustration 34.6

Note: *1996 models should only be serviced with DEX-COOL orange-colored silicate-free coolant. The use of other types of coolant can damage cooling system components.*

1   Periodically, the cooling system should be drained, flushed and refilled to replenish the antifreeze mixture and prevent formation of rust and corrosion, which can impair the performance of the cooling system and cause engine damage.

2   At the same time the cooling system is serviced, all hoses and the radiator cap should be inspected and replaced if defective (see Section 9).

3   Since antifreeze is a corrosive and poisonous solution, be careful not to spill any of the coolant mixture on the vehicle's paint or your skin. If this happens, rinse immediately with plenty of clean water. Consult your local authorities about the dumping of antifreeze before draining the cooling system. In many areas, reclamation centers have been set up

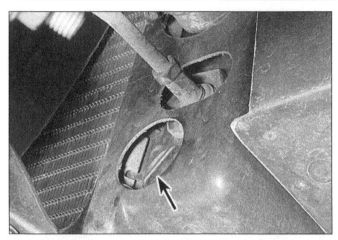

34.6 The drain for the radiator is recessed inside a round cutout (arrow) and is best loosened with a pair of pliers

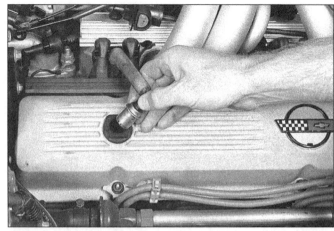

35.2 The PVC valve simply pushes into the valve cover. Feel for suction at the end of the valve and shake the valve, listening for a clicking sound

to collect automobile oil and drained antifreeze/water mixtures, rather than allowing them to be added to the sewage system.
4   With the engine cold, remove the radiator cap.
5   Move a large container under the radiator to catch the coolant as it is drained.
6   Drain the radiator by removing the drain plug at the bottom **(see illustration)**. If this drain has excessive corrosion and cannot be turned easily, or if the radiator is not equipped with a drain, disconnect the lower radiator hose to allow the coolant to drain. Be careful that none of the solution is splashed on your skin or into your eyes.
7   Disconnect the hose from the coolant reservoir and remove the reservoir. Flush it out with clean water.
8   Place a garden hose in the radiator filler neck and flush the system until the water runs clear at all drain points.
9   In severe cases of contamination or clogging of the radiator, remove it (see Chapter 3) and reverse flush it. This involves inserting the hose in the bottom radiator outlet to allow the water to run against the normal flow, draining through the top. A radiator repair shop should be consulted if further cleaning or repair is necessary.
10   When the coolant is regularly drained and the system refilled with the correct antifreeze/water mixture, there should be no need to use chemical cleaners or descalers.
11   To refill the system, reconnect the radiator hoses and install the reservoir and the overflow hose.
12   Open the bleed screws on the throttle body and bypass hose (1992 and later models).
13   Fill the radiator or radiator surge tank with a 50/50 mixture of coolant and water to the base of the filler neck or, on 1992 and later models, until a steady stream of coolant comes out of the bleed screws. On 1992 and later models, after a steady stream of coolant begins flowing from the bleed screws, close the bleed screws and fill to the base of the filler neck.

14   With the radiator cap still removed, start the engine and run until normal operating temperature is reached. With the engine idling, add additional coolant to the radiator and the reservoir to bring up to the proper levels. Install the radiator and reservoir caps.
15   Keep a close watch on the coolant level and the cooling system hoses during the first few miles of driving. Tighten the hose clamps and/or add more coolant as necessary.

35 Positive Crankcase Ventilation (PCV) valve check and replacement

Refer to illustration 35.2
1   On 1995 and earlier models, the PCV valve is located in the passenger side rocker arm cover. On 1996 models, the PCV valve is located on the driver's side of the intake manifold.
2   With the engine idling at normal operating temperature, pull the valve (with hose attached) from the rubber grommet in the cover **(see illustration)**.
3   Place your finger over the end of the valve. If there is no vacuum at the valve,

check for a plugged hose, manifold port, or the valve itself. Replace any plugged or deteriorated hoses.
4   Turn off the engine and shake the PCV valve, listening for a rattle. If the valve does not rattle, replace it with a new one.
5   To replace the valve, pull it from the end of the hose, noting its installed position and direction.
6   When purchasing a replacement PCV valve, make sure it is for your particular vehicle, model year and engine size. Compare the old valve with the new one to make sure they are the same.
7   Push the valve into the end of the hose until it is seated.
8   Inspect the rubber grommet for damage and replace it with a new one if necessary.
9   Push the PCV valve and hose securely into position in the rocker arm cover.

36 Evaporative emissions control system check

Refer to illustration 36.2
1   The function of the evaporative emissions control system is to draw fuel vapors

36.2 The evaporative system canister is located at the left front corner of the engine compartment (arrow). Inspect the various hoses attached to it and the canister itself for any damage

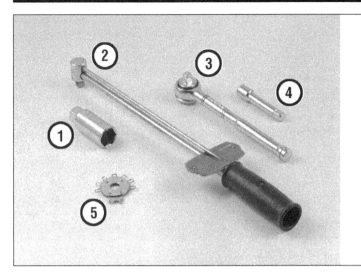

37.1 Tools required for changing spark plugs

1   **Spark plug socket** - *This will have special padding inside to protect the spark plug's porcelain insulator*
2   **Torque wrench** - *Although not mandatory, using this tool is the best way to ensure the plugs are tightened properly*
3   **Ratchet** - *Standard hand tool to fit the spark plug socket*
4   **Extension** - *Depending on model and accessories, you may need special extensions and universal joints to reach one or more of the plugs*
5   **Spark plug gap gauge** - *This gauge for checking the gap comes in a variety of styles. Make sure the gap for your engine is included*

37.4a Spark plug manufacturers recommend using a wire-type gauge when checking the gap - if the wire does not slide between the electrodes with a slight drag, adjustment is required

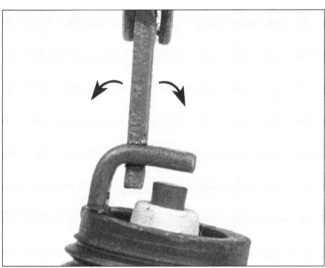

37.4b To change the gap, bend the side electrode only, as indicated by the arrows, and be very careful not to crack or chip the porcelain insulator surrounding the center electrode

from the gas tank and fuel system, store them in a charcoal canister and then burn them during normal engine operation.

2   The most common symptom of a fault in the evaporative emissions system is a strong fuel odor in the engine compartment. If a fuel odor is detected, inspect the charcoal canister **(see illustration)**. The charcoal canister is located at the left front corner of the engine compartment on 1984 through 1991 models or at the right rear quarter panel, above the muffler on 1992 and later models.

3   The evaporative emissions control system is explained in more detail in Chapter 6.

---

37   Spark plug replacement

Refer to illustrations 37.1, 37.4a, 37.4b and 37.5
Note: *On models equipped with aluminum*

cylinder heads, it is essential that the spark plug threads be coated with anti-seize compound. Failure to coat the threads with anti-seize will cause galling of the threads or electrolytic action, corroding the threads, which can make future spark plug removal difficult and may damage cylinder head threads.

1   Before beginning, obtain the necessary tools which will include a special spark plug wrench or socket and a gap measuring tool **(see illustration)**.

2   The best procedure to follow when replacing the spark plugs is to purchase the new spark plugs beforehand, adjust them to the proper gap, and then replace each plug one at a time. When buying the new spark plugs it is important to obtain the correct plugs for your specific engine. This information can be found on the Vehicle Emissions Control Information label located under the hood or in the owner's manual. If differences exist between these sources, purchase the

spark plug type specified on the Emissions Control label, because the information was printed for your specific engine.

3   With the new spark plugs at hand, allow the engine to cool completely before attempting plug removal. During this time, each of the new spark plugs can be inspected for defects and the gaps can be checked.

4   The gap is checked by inserting the proper thickness gauge between the electrodes at the tip of the plug **(see illustration)**. The gap between the electrodes should be the same as that given in the Specifications or on the Emissions Control label. The wire should just touch each of the electrodes. If the gap is incorrect, use the notched adjuster on the feeler gauge body to bend the curved side electrode slightly until the proper gap is achieved **(see illustration)**. If the side electrode is not exactly over the center electrode, use the notched adjuster to align the two.

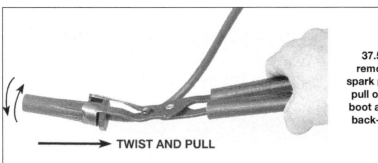

37.5 When removing the spark plug wires, pull only on the boot and twist it back-and-forth

TWIST AND PULL

Check for cracks in the porcelain insulator, indicating the spark plug should not be used.

5   With the engine cool, remove the spark plug wire from one spark plug. Do this by grabbing the boot at the end of the wire, not the wire itself **(see illustration)**. Sometimes it is necessary to use a twisting motion while the boot and plug wire are pulled free.

6   If compressed air is available, use it to blow any dirt or foreign material away from the spark plug area. The idea here is to eliminate the possibility of debris falling into the cylinder as the spark plug is removed.

7   Place the spark plug socket over the plug and remove it from the engine by turning in a counterclockwise direction.

8   Compare the spark plug with those shown in the accompanying color photos to get an indication of the overall running condition of the engine. **Caution:** *Some models are equipped with aluminum cylinder heads which makes it critical that the spark plugs are not cross-threaded or overtightened during installation. A small amount of anti-seize lubricant should be used on the threads of spark plugs before inserting them into aluminum heads.*

9   Thread the new plug into the engine until you can no longer turn it with your fingers, then tighten it with the socket. Where there might be difficulty in inserting the spark plugs into the spark plug holes, or the possibility of cross-threading them into the head, a short piece of 3/16-inch rubber tubing can be fitted over the end of the spark plug. The flexible tubing will act as a universal joint to help align the plug with the plug hole, and should the plug begin to cross-thread, the hose will slip on the spark plug, preventing thread damage. If one is available, use a torque wrench to tighten the plug to ensure that it is seated correctly. The correct torque figure is included in the Specifications at the front of this Chapter.

10   Before pushing the spark plug wire onto the end of the plug, inspect the wire following the procedures outlined in Section 38.

11   Attach the plug wire to the new spark plug, again using a twisting motion on the boot until it is firmly seated on the spark plug. Make sure the wire is routed away from the exhaust manifold.

12   Follow the above procedure for the remaining spark plugs, replacing them one at a time to prevent mixing up the spark plug wires.

38   Spark plug wire, distributor cap and rotor check and replacement

Refer to illustrations 38.6, 38.14 and 38.17

Spark plug wires

Note: *Often the spark plug wire holders will be too brittle to be reused. In many cases they will break as you remove them. It is essential the spark plug wires be routed as they were from the factory. Be certain to replace any wire holders that cannot be reused.*

1   The spark plug wires should be checked at the recommended intervals and whenever new spark plugs are installed in the engine.

2   The wires should be inspected one at a time to prevent mixing up the order, which is essential for proper engine operation.

3   Disconnect the plug wire from the spark plug. To do this, grab the rubber boot, twist slightly and pull the wire free. Do not pull on the wire itself, only on the rubber boot.

4   Inspect inside the boot for corrosion, which will look like a white crusty powder. Push the wire and boot back onto the end of the spark plug. It should be a tight fit on the plug. If it is not, remove the wire and use pliers to carefully crimp the metal connector inside the boot until it fits securely on the end of the spark plug.

5   Using a clean rag, wipe the entire length of the wire to remove any built-up dirt and grease. Once the wire is clean, check for burns, cracks and other damage. Do not

bend the wire excessively or pull the wire since the conductor inside might break.

6   To disconnect the wires from the distributor cap, remove the cover plate (later models), push the latches on the spark plug retaining ring out (away from the coil) and lift the retaining ring up. The wires are held in the retaining ring by rubber plugs at the top of each distributor boot. Press the plugs out of the retaining ring, remove the ring, then remove the wires from the distributor cap. Note that both the wires and retaining ring are numbered to ensure that the wires go back on to the distributor cap in the proper order **(see illustration)**.

7   Check the wires for corrosion and a tight fit on the distributor cap nipples. When the check is complete replace the spark plug wires and retaining ring on the distributor cap.

8   If new spark plug wires are required, purchase a new set for your specific engine model. Wire sets are available pre-cut, with the rubber boots already installed. Remove and replace the wires one at a time to avoid mix-ups in the firing order.

Distributor cap and rotor (1991 and earlier models only)

9   It is common practice to install a new cap and rotor whenever new spark plug wires are installed. Although the breakerless distributor used on this car requires much less maintenance than conventional distributors, periodic inspections should be performed when the plug wires are inspected.

10   To gain access to the distributor cap, remove either the air cleaner assembly (1984 models) or the top cover plate (1985 and later).

11   To remove the distributor cap, begin by removing the retaining ring on the top of the distributor. This ring will have to be pried from the top of each plug wire, and more than likely each of the spark plug wires will come off the cap at this time. Note that each spark plug wire is numbered, as is the retaining ring.

12   Fully loosen the four screws that hold the distributor cap to the distributor body. Note that these screws have shouldered

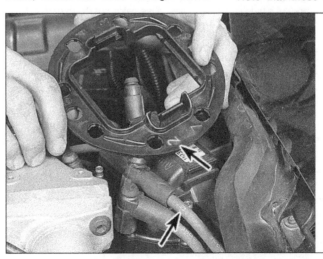

38.6 The spark plug wire retainer ring at the top of the distributor is marked with each of the eight cylinder numbers, as are each of the spark plug wires (arrows)

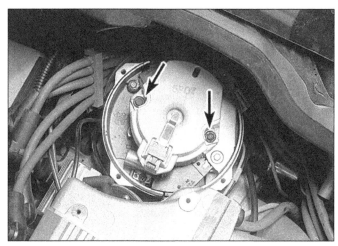

38.14 Visible with the distributor cap removed are the internal components. The rotor is held in place with two screws (arrows). Be careful not to drop anything down into the distributor

38.17 Inspect the inside of the cap, especially the metal contacts. If in doubt, replace it with a new one

39.1 The Vehicle Emission Control Information label is located on the side of the radiator and will give important information regarding ignition timing and other tune-up-related procedures

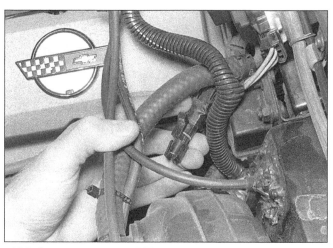

39.2 On most models an EST wire leading to the distributor will have to be disconnected before the ignition timing adjustment can be performed. Specific instructions regarding the location and color of this wire will be given on the VECI label

areas so they do not come completely out.

13 Release the wires on the side of the distributor cap and remove the cap.

14 The rotor is now visible at the center of the distributor. It is held in place by two screws **(see illustration)**.

15 Visually inspect the rotor for cracks or damage. Carefully check the condition of the metal contact at the top of the rotor for excessive burning or pitting. If in doubt as to its condition, replace it with a new one.

16 Install the rotor onto the distributor shaft. It is keyed to fit properly only one way.

17 Before installing the distributor cap, inspect the cap for cracks or other damage. Closely examine the contacts on the inside of the cap for excessive corrosion or damage **(see illustration)**. Slight scoring is normal. Again, if in doubt as to the condition of the cap, replace it with a new one.

18 Replace the cap and secure with the screws.

19 Place the spark plug wires onto the cap in their proper positions and secure with the

retaining ring at the top. Note that the retaining ring fits properly in only one position, so if the numbers on the ring match the numbers on the spark plug wires, everything is in the proper position.

---

39 Ignition timing check and adjustment (1991 and earlier models)

Refer to illustrations 39.1, 39.2 and 39.5

Note: *It is imperative that the procedures included on the Vehicle Emissions Control Information (VECI) label be followed when adjusting the ignition timing. The label will include all information concerning preliminary steps to be performed before adjusting the timing, as well as the timing specifications. The ignition timing on 1992 and later models is non-adjustable.*

Caution: *Always disconnect and reconnect electrical connections with the ignition key*

Off to prevent damaging "voltage spikes" from being sent to the ECM.

1 With the ignition Off, locate the VECI label under the hood **(see illustration)** and read through and perform all preliminary instructions concerning ignition timing.

2 Vehicles covered in this manual are equipped with Electric Spark Timing (EST). It will be necessary to disconnect the EST connector near the brake booster in order to bypass the computer timing mode **(see illustration)**.

3 With the ignition Off, connect the pick up lead of the timing light to the number one spark plug. Use a jumper lead, if necessary, between the plug and wire, or an inductive pick up. **Note:** *Do not pierce the plug wire or force anything between the wire and boot. This will ruin the wire.*

4 Connect the power leads of the timing light according to the manufacturer's instructions.

5 Start the engine and point the timing light at the timing mark on the harmonic bal-

ancer and the timing scale on the front cover of the engine **(see illustration)**.

6   The line on the harmonic balancer will line up with the scale. If an adjustment is required, loosen the distributor hold-down clamp bolt and rotate the distributor slightly until the line indicates the correct timing. Because access to the distributor hold-down bolt is tight, a special distributor wrench made for this purpose should be used.

7   Tighten the hold-down bolt and re-check the timing.

8   Turn off the engine and remove the timing light, reconnecting the number one plug wire (if removed).

9   Reconnect the EST connector near the brake booster. On some models you will have to clear a stored trouble code from the ECM memory (indicated by a Check engine light on the dashboard). If you get a Check engine light, refer to Chapter 6 for instructions regarding removing this code.

39.5 To properly view the timing indicator, hold the timing light in this position. Be very careful of moving or hot engine parts during this procedure

Chapter 2 Part A Engine

Contents

Specifications

General

Firing order 1-8-4-3-6-5-7-2

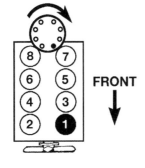

The blackened terminal shown on the distributor cap indicates the Number One spark plug wire position

FRONT

Cylinder location and distributor rotation - 1991 and earlier

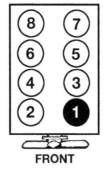

FRONT

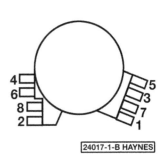

24017-1-B HAYNES

Cylinder and spark plug terminal locations - 1992 and later

Camshaft

Bearing journal

Diameter	1.8682 to 1.8692 inches
Out-of-round limit	0.001 inch

Lobe Lift

1984 thru 1991

Intake	0.2733 inch
Exhaust	0.2820 inch

1992 thru 1993

Intake	0.300 inch
Exhaust	0.300 inch

1994 thru 1995

Intake	0.300 inch
Exhaust	0.307 inch

1996

LT1

Intake	0.298 inch
Exhaust	0.306 inch

LT4

Intake	0.298 inch
Exhaust	0.299 inch
End play	0.004 to 0.012 inch

Torque specifications

Ft-lbs (unless otherwise indicated)

Rocker arm cover

Nuts

with rocker arm cover gasket	80 to 120 in-lb
without rocker arm cover gasket	50 in-lb
Bolts	55 to 125 in-lb
Intake manifold bolts	35
Exhaust manifold bolts	20
Rocker arm nut (LT-4 only)	18

Cylinder head bolts*

1984 thru 1991

Cast iron heads	65

Aluminum heads

short bolts	60
medium and long bolts	65
1992 thru 1995 (all)	65

1996

Step 1	22

Step 2

short bolts	Tighten an additional 67-degrees
medium and long bolts	Tighten an additional 80-degrees

Timing chain cover bolts

1984 thru 1991	80 in-lbs
1992 and later	106 in-lbs

Camshaft sprocket bolts

1984 through 1991	20
1992 and later	216 in-lbs
Vibration damper bolt	60

Oil pan

Bolts/studs	80 in-lb
Nuts	165 in-lb
Oil pump bolt	65
Rear main bearing cap bolts	80

*Use Permatex no. 2 on the bolt threads

1 General information

This Part of Chapter 2 is devoted to in-vehicle repair procedures for the engine. All information concerning engine removal and installation and engine block and cylinder head overhaul can be found in Part B of this Chapter.

Since the repair procedures included in this Part are based on the assumption that the engine is still installed in the vehicle, if they are being used during a complete engine overhaul (with the engine already out of the vehicle and on a stand) many of the steps included here will not apply.

The specifications included in this Part of Chapter 2 apply only to the procedures found here. The specifications necessary for rebuilding the block and cylinder heads are included in Part B.

2 Repair operations possible with the engine in the vehicle

Many major repair operations can be accomplished without removing the engine from the vehicle.

Clean the engine compartment and the exterior of the engine with some type of pressure washer before any work is done. A clean engine will make the job easier and will help keep dirt out of the internal areas of the engine.

Depending on the components involved, it may be a good idea to remove the hood to improve access to the engine as repairs are performed (refer to Chapter 11 if necessary).

If oil or coolant leaks develop, indicating a need for gasket or seal replacement, the repairs can generally be made with the engine in the vehicle. The oil pan gasket, the cylinder head gaskets, intake and exhaust manifold gaskets, timing cover gaskets and the crankshaft oil seals are accessible with the engine in place.

Exterior engine components, such as the water pump, the starter motor, the alternator, the distributor and the EFI components, as well as the intake and exhaust manifolds, can be removed for repair with the engine in place.

Since the cylinder heads can be removed without pulling the engine, valve component servicing can also be accomplished with the engine in the vehicle.

Replacement of, repairs to or inspection of the timing chain and sprockets and the oil pump are all possible with the engine in place.

In extreme cases caused by a lack of necessary equipment, repair or replacement of piston rings, pistons, connecting rods and rod bearings is possible with the engine in the vehicle. However, this practice is not recommended because of the cleaning and preparation work that must be done to the components involved.

3 Rocker arm covers - removal and installation

Removal - 1984 models

1   Disconnect the negative battery cable from the battery, then remove the air cleaner assembly and the heat stove ducts. Cover the TBI units to prevent tools and parts from falling into the intake manifold.

Right side

2   Detach the hose from the AIR system exhaust check valve.

3   Disconnect the fuel lines from the TBI units (see Chapter 4).

4   Remove the drivebelt from the pulleys (see Chapter 1).

5   Refer to Chapter 3, remove the A/C compressor from the bracket and position it out of the way (do not loosen or disconnect the compressor lines).

6   Remove the four rocker arm cover mounting nuts. Slip the spark plug wire clip brackets and washers off the lower rocker arm cover studs and position the brackets/wires out of the way. Remove the washers from the upper studs.

7   Raise the vehicle and support it securely on jackstands, then remove the bolt holding the rear spark plug wire clip bracket to the end of the cylinder head (the bolt is visible just to the rear of the last exhaust manifold tube). Lower the vehicle (if the left side cover is also being removed, remove the bolt on the left side before lowering the vehicle).

8   Detach the spark plug wires from the plugs, then lay the spark plug wire harness up over the distributor and remove the rocker arm cover. **Note:** *If the cover is stuck to the head, bump the front end with a block of wood and a hammer to release it. If it still will not come loose, try to slip a flexible putty knife between the head and cover to break the seal. Do not pry at the cover-to-head joint or damage to the sealing surface and cover flange will result and oil leaks will develop.*

Left side

9   Pull the PCV valve out of the cover and detach the hose from the intake manifold fitting.

10   Detach the vacuum brake line from the manifold fitting.

11   Remove the drivebelt from the pulleys (see Chapter 1).

12   Refer to Chapter 5 and detach the alternator - lay it aside, out of the way, without disconnecting the wires.

13   Refer to steps 6 through 8 above for the remainder of the procedure.

Removal - 1985 and later models

14   Disconnect the negative battery cable from the battery.

15   Detach the PCV pipe and elbow assembly from the right side rocker arm cover and pipe fitting. Remove the PCV valve and hose from the left side rocker arm cover and fitting.

16   Remove the screws and detach the plastic EFI wire harness shroud.

17   Remove the four rocker arm cover mounting nuts. Detach the spark plug wire clip brackets and washers from the lower rocker arm cover studs and position the brackets/wires out of the way.

18   Slip the PCV pipe/hose assembly off the upper studs and manipulate it out of the wire harness and hose maze on the side of the intake manifold. Remove the washers from the upper studs. **Note:** *Some later model engines with aluminum heads have rocker arm covers that are held in place with four bolts arranged in a row down the center of the cover rather than with nuts along the sides.*

19   Refer to steps 7 and 8 above for the remainder of the procedure. If your engine is equipped with rocker arm cover gaskets, remove and discard them.

Installation - all models

20   The mating surfaces of each cylinder head and rocker arm cover must be perfectly clean when the covers are installed. Use a gasket scraper to remove all traces of sealant or old gasket, then wipe the mating surfaces with a cloth saturated with lacquer thinner or acetone (if there is sealant or oil on the mating surfaces when the cover is installed, oil leaks may develop). The rocker arm covers (and the heads on some engines) are made of aluminum, so be extra careful not to nick or gouge the mating surfaces with the scraper.

21   Clean the mounting nut stud threads with a die to remove any corrosion and restore damaged threads. On models that utilize bolts, make sure the threaded holes in the head are clean; run a tap into them to remove corrosion and restore damaged threads.

22   Apply an unbroken 1/8-inch bead of RTV sealant to the rocker arm cover flange. Make sure it is applied to the inside of all stud holes or oil may be forced out around the studs when the engine is running. **Note:** *If gaskets are being used, they should be mated to the covers before the covers are installed. Apply a thin coat of RTV sealant to the cover flange, then position the gasket inside the cover lip and allow the sealant to set up so the gasket adheres to the cover (if the sealant is not allowed to set, the gasket may fall out of the cover as it is installed on the engine).*

23   Carefully position the cover on the head and install the nuts. If a gasket is not being used, this must be done while the sealant is still wet. Don' t forget to slip the spark plug wire clip brackets and washers over the studs before the nuts are threaded on.

24   Tighten the nuts or bolts in three or four steps to the specified torque.

25   The remaining installation steps are the reverse of removal.

26   Start the engine and check carefully for oil leaks as the engine warms up.

4 Valve spring, retainer and seals - replacement on vehicle

Refer to illustrations 4.8, 4.9, 4.16 and 4.17
Note: *Broken valve springs and defective valve stem seals can be replaced without removing the cylinder head. Two special tools and a compressed air source are normally required to perform this operation, so read through this Section carefully and rent or buy the tools before beginning the job. If compressed air is not available, a length of nylon rope can be used to keep the valves from falling into the cylinder during this procedure.*

1   Refer to Section 3 and remove the rocker arm cover from the affected cylinder head. If all of the valve stem seals are being replaced, remove both rocker arm covers.

2   Remove the spark plug from the cylinder that has the defective component. If all of the valve stem seals are being replaced, all of the

4.8 Once the spring is depressed, the keepers can be removed with a small magnet or needle-nose pliers (a magnet is preferred to prevent dropping the keepers)

4.9 The O-ring seal (arrow) should be replaced with a new one each time the keepers and retainer are removed

spark plugs should be removed.

3   Turn the crankshaft until the piston in the affected cylinder is at top dead center on the compression stroke (refer to Section 9 for instructions). If you are replacing all of the valve stem seals, begin with cylinder number one and work on the valves for one cylinder at a time. Move from cylinder-to-cylinder following the firing order sequence (1-8-4-3-6-5-7-2).

4   Thread an adapter into the spark plug hole and connect an air hose from a compressed air source to it. Most auto parts stores can supply the air hose adapter. **Note:** *Many cylinder compression gauges utilize a screw-in fitting that may work with your air hose quick-disconnect fitting.*

5   Remove the nut, pivot ball and rocker arm for the valve with the defective part and pull out the pushrod. If all of the valve stem seals are being replaced, all of the rocker arms and pushrods should be removed (refer to Section 7).

6   Apply compressed air to the cylinder. The valves should be held in place by the air pressure. If the valve faces or seats are in poor condition, leaks may prevent the air pressure from retaining the valves - refer to the alternative procedure below.

7   If you do not have access to compressed air, an alternative method can be used. Position the piston at a point just before TDC on the compression stroke, then feed a long piece of nylon rope through the spark plug hole until it fills the combustion chamber. Be sure to leave the end of the rope hanging out of the engine so it can be removed easily. Use a large breaker bar and socket to rotate the crankshaft in the normal direction of rotation until slight resistance is felt.

8   Stuff shop rags into the cylinder head holes above and below the valves to prevent parts and tools from falling into the engine, then use a valve spring compressor to compress the spring/damper assembly. Remove the keepers with small needle-nose pliers or

4.16 Make sure the O-ring seal under the retainer is seated in the groove and not twisted before installing the keepers

a magnet **(see illustration)**. **Note:** *A couple of different types of tools are available for compressing the valve springs with the head in place. One type grips the lower spring coils and presses on the retainer as the knob is turned, while the other type, shown here, utilizes the rocker arm stud and nut for leverage. Both types work very well, although the lever type is usually less expensive.*

9   Remove the spring retainer or rotator, oil shield and valve spring assembly, then remove the valve stem O-ring seal and the umbrella-type guide seal (the O-ring seal will most likely be hardened and will probably break when removed, so plan on installing a new one each time the original is removed) **(see illustration)**. Note: If air pressure fails to hold the valve in the closed position during this operation, the valve face or seat is probably damaged. If so, the cylinder head will have to be removed for additional repair operations.

10   Wrap a rubber band or tape around the top of the valve stem so the valve will not fall into the combustion chamber, then release the air pressure. **Note:** *If a rope was used instead of air pressure, turn the crankshaft slightly in the direction opposite normal rotation.*

11   Inspect the valve stem for damage. Rotate the valve in the guide and check the end for eccentric movement, which would indicate that the valve is bent.

12   Move the valve up-and-down in the guide and make sure it doesn't bind. If the valve stem binds, either the valve is bent or the guide is damaged. In either case, the head will have to be removed for repair.

13   Reapply air pressure to the cylinder to retain the valve in the closed position, then remove the tape or rubber band from the valve stem. If a rope was used instead of air pressure, rotate the crankshaft in the normal direction of rotation until slight resistance is felt.

14   Lubricate the valve stem with engine oil and install a new umbrella-type guide seal.

15   Install the spring/damper assembly and shield in position over the valve.

16   Install the valve spring retainer or rotator. Compress the valve spring assembly and carefully install the new O-ring seal in the lower groove of the valve stem. Make sure the seal is not twisted - it must lie perfectly flat in the groove **(see illustration)**.

17   Position the keepers in the upper groove. Apply a small dab of grease to the inside of each keeper to hold it in place if

4.17 Apply a small dab of grease to each keeper as shown here before installation - it will hold them in place on the valve stem as the spring is released

5.11 After covering the lifter valley, use a gasket scraper to remove all traces of sealant and old gasket material from the head and manifold mating surfaces

5.12a The bolt hole threads must be clean and dry to ensure accurate torque readings when the manifold mounting bolts are installed

necessary **(see illustration)**. Remove the pressure from the spring tool and make sure the keepers are seated. Refer to Chapter 2, Part B, Section 11 and check the seals with a vacuum pump.

18   Disconnect the air hose and remove the adapter from the spark plug hole. If a rope was used in place of air pressure, pull it out of the cylinder.

19   Refer to Section 7 and install the rocker arm(s) and pushrod(s) and adjust the valves.

20   Install the spark plug(s) and hook up the wire(s).

21   Refer to Section 3 and install the rocker arm cover(s).

22   Start and run the engine, then check for oil leaks and unusual sounds coming from the rocker arm cover area.

5   Intake manifold - removal and installation

Refer to illustrations 5.11, 5.12a, 5.12b, 5.13, 5.15, 5.17a, 5.17b and 5.17c

Removal

1   Disconnect the negative battery cable from the battery, then refer to Chapter 1, drain the cooling system and remove the serpentine drivebelt from the pulleys.

2   Remove the air cleaner assembly.

3   Refer to the appropriate Sections in Chapter 4 and remove the fuel injection components to expose the intake manifold mounting bolts. On TBI equipped models, the fuel lines, vacuum hoses, wiring harnesses and control cables must be detached to remove the intake cover/TBI assemblies. On MPFI equipped models, remove the mass air flow sensor, the plenum and the runners, but leave the fuel rail assembly in place.

4   On 1991 and earlier models, refer to Chapter 3 and detach the upper radiator hose from the thermostat housing and the alternator brace (if not already done). The thermostat housing must also be detached to provide room for removal of the left-front manifold bolt.

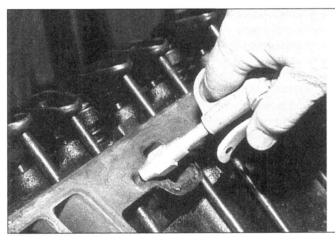

5   On 1991 and earlier models, refer to Chapter 5 and remove the distributor.

6   Detach the heater hose from the rear of the intake manifold. If the hose is stuck to the fitting, grasp the hose with a pair of water pump pliers and twist it to break the seal, then pull it off the fitting.

7   On 1991 and earlier models, refer to Chapter 6 and detach the AIR pump.

8   Disconnect the wire harness from the coolant temperature sensor (the sensor is threaded into the front of the intake manifold).

9   Loosen the manifold mounting bolts in 1/4-turn increments until they can be removed by hand. Note that the bolts have Torx-type heads and require a special driver for removal and installation.

10   The manifold will probably be stuck to the cylinder heads and force may be required to break the gasket seal. A large pry bar can be positioned under the cast-in lug near the thermostat housing to pry up the front of the manifold. **Caution:** *Do not pry between the block and manifold or the heads and manifold or damage to the gasket sealing surfaces will result and vacuum leaks could develop.*

Installation

Note: *The mating surfaces of the cylinder*

5.12b Clean the bolt holes with compressed air, but be careful - wear safety goggles!

heads, block and manifold must be perfectly clean when the manifold is installed. Gasket removal solvents in aerosol cans are available at most auto parts stores and may be helpful when removing old gasket material that is stuck to the heads and manifold (since the manifold and some heads are made of aluminum, aggressive scraping can cause damage). Be sure to follow the directions printed on the container.

11   Use a gasket scraper to remove all traces of sealant and old gasket material, then wipe the mating surfaces with a cloth saturated with lacquer thinner or acetone. If there is old sealant or oil on the mating surfaces when the manifold is installed, oil or vacuum leaks may develop. When working on the heads and block, cover the lifter valley with shop rags to keep debris out of the engine **(see illustration)**. **Note:** *The heads on some engines are made of aluminum, so be extra careful not to nick or gouge the mating surfaces with the scraper. Use a vacuum cleaner to remove any gasket material that falls into the intake ports in the heads.*

12   Use a tap of the correct size to chase the threads in the bolt holes, then use compressed air (if available) to remove the debris from the holes **(see illustrations)**. **Warning:**

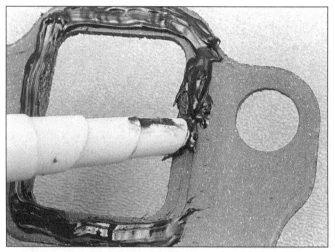

5.13 RTV sealant should be used around the coolant passage holes in the new intake manifold gaskets

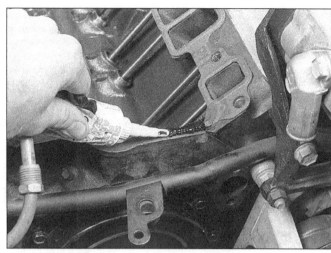

5.15 The ends of the manifold are sealed with RTV - make sure the beads overlap the gaskets slightly

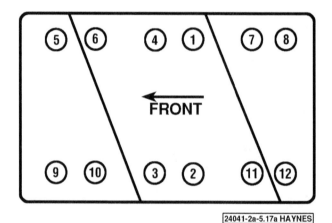

5.17a Intake manifold bolt tightening sequence - 1984 models only

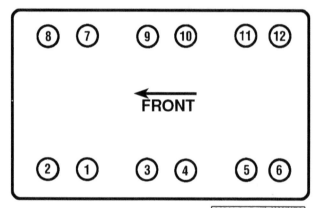

5.17b Intake manifold bolt tightening sequence - 1985 through 1991

Wear safety glasses or a face shield to protect your eyes when using compressed air.

13   Apply a thin coat of RTV sealant around the coolant passage holes on the cylinder head side of the new intake manifold gaskets (there is normally one hole at each end and one in the center) **(see illustration)**.

14   Position the gaskets on the cylinder heads. Make sure all intake port openings, coolant passage holes and bolt holes are aligned correctly. Note that some gaskets may have small tabs, which must be bent over until they are flush with the rear surface of each head.

15   Apply a 3/16-inch wide bead of RTV sealant to the front and rear manifold mating surfaces of the block **(see illustration)**. Make sure the beads overlap the gaskets slightly at each end. Coat the areas around the coolant passage holes on the exposed sides of the gaskets as well.

16   Carefully set the manifold in place while the sealant is still wet. **Note:** *Do not disturb the gaskets and do not move the manifold fore-and-aft after it contacts the sealant on*

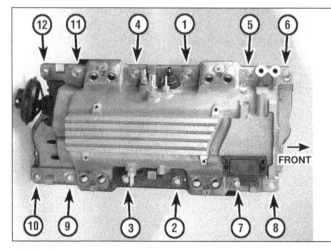

5.17c Intake manifold bolt tightening sequence - 1992 and later

the block.

17   Apply a non-hardening sealant (such as Permatex Number 2) to the manifold bolt threads, then install the bolts. While the sealant is still wet, tighten the bolts to the specified torque following the recommended sequence **(see illustrations)**. Work up to the

final torque in three steps. Note that 1984 models require a different tightening sequence than later models.

18   The remaining installation steps are the reverse of removal. Start the engine and check carefully for oil and coolant leaks at the intake manifold joints.

6.10 Use a backup wrench when removing the air injection system check valve from the exhaust manifold fitting

6.13 The manifold should be positioned with the rear end lower than the front as it is withdrawn

6   Exhaust manifolds - removal and installation

Refer to illustrations 6.10 and 6.13
Note: *New exhaust manifold gaskets are very expensive - a fact you may want to consider before proceeding with manifold removal (since new gaskets will be required for installation). Unless removal of the manifold(s) is absolutely necessary, you may want to consider alternative approaches.*

Removal

1   Disconnect the negative battery cable from the battery. If your vehicle is a 1984 model with Cross-Fire Injection, remove the intake ducts attached to the air cleaner assembly.
2   Refer to Chapter 1 and drain the cooling system.

Right manifold

3   Raise the vehicle and support it securely on jackstands, then remove the three exhaust pipe-to-manifold nuts. These nuts are usually rusted in place, so penetrating oil should be applied to the stud threads before attempting to remove them.
4   Separate the exhaust pipe flange from the manifold studs, then pull the pipe down slightly to break the seal at the manifold joint.
5   The catalytic converter air injection pipe is attached to the manifold bracket with a hose clamp, which must be loosened at this point. A small wrench may be more useful than a screwdriver, since access to the clamp is very restricted.
6   If both manifolds are being removed, repeat Steps 3 and 4 for the left manifold before lowering the vehicle. If only the right side manifold is being removed, lower the vehicle and perform the remaining steps in the engine compartment.
7   Remove the nut and detach the air conditioning compressor bracket from the front exhaust manifold stud. The bolt in the com-

pressor must be loosened, but leave it in place and allow the bracket to hang from the compressor. Refer to Chapter 3 and remove the air conditioning compressor. Do not detach the hoses. Move the compressor ahead to provide room for exhaust manifold removal.
8   Remove the engine oil dipstick from the guide tube. Remove the nut and pull the guide tube bracket off the manifold mounting stud.
9   Detach the spark plug wires from the plugs, then remove the two lower rocker arm cover mounting nuts. Slip the spark plug wire clip brackets and washers off the lower rocker arm cover studs and position the brackets/wires out of the way. **Note:** *It would be a good idea, although not absolutely necessary, to remove the spark plugs at this time to prevent damage to them as the manifold is withdrawn.*
10   Loosen the air injection system check valve hose clamps, then unthread the valve from the manifold fitting. Use a backup wrench on the fitting to avoid damaging the manifold **(see illustration)**. The air injection line for the catalytic converter, particularly on 1984 models, may interfere with manifold removal. If it does, remove the check valve and hose.
11   Disconnect the wire from the cooling fan temperature switch (located just to the rear of the oil dipstick guide tube).
12   Loosen the exhaust manifold mounting bolts/nuts 1/4-turn at a time each, working from the inside out, until they can be removed by hand.
13   Separate the manifold from the head and withdraw it by dropping the rear end and lifting the front end up and out **(see illustration)**.

Left manifold

14   The left manifold removal procedure is essentially the same as the one for the right manifold with several minor differences.
15   Ignore any references to the oil dipstick

and guide tube, the catalytic converter air injection pipe and the air conditioning compressor.
16   Refer to Chapter 5 and remove the alternator to provide room for exhaust manifold removal. The lower rear alternator bracket is held in place with the exhaust manifold fasteners and will be detached when the manifold is removed.
17   Disconnect the wire from the coolant temperature sending unit (located between the two front spark plugs).

Installation

18   The manifold and cylinder head mating surfaces must be clean before the manifolds are reinstalled. Use a gasket scraper to remove all traces of old gasket material and carbon deposits.
19   It may be difficult to hold the manifold in place, position the gaskets behind it and install the bolts without assistance, but it can be done. An alternative would be to attach the gaskets to the manifold with sewing thread before attempting to install the manifold. Do not use sealant to hold the gaskets in place.
20   When tightening the mounting bolts, work from the center to the ends and be sure to use a torque wrench. Tighten the bolts in three equal steps until the specified torque is reached.
21   The remaining installation steps are the reverse of removal.

7   Rocker arms and pushrods - removal, inspection and installation

Refer to illustrations 7.4, 7.10, 7.11 and 7.13

Removal

1   Refer to Section 3 and detach the rocker arm covers from the cylinder heads.
2   Beginning at the front of one cylinder

7.4 A perforated cardboard box can be used to store the pushrods to ensure that they are reinstalled in their original locations - note the label indicating the front of the engine

7.10 The ends of the pushrods and the valve stems should be lubricated with moly-base grease prior to installation of the rocker arms

7.11 Moly-base grease applied to the pivot balls will ensure adequate lubrication as oil pressure builds up when the engine is started

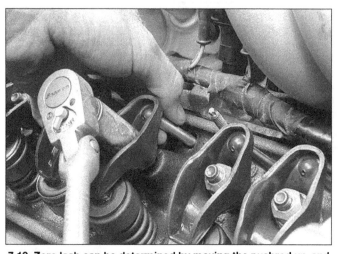

7.13 Zero lash can be determined by moving the pushrod up-and-down and rotating the pushrod between your thumb and index finger as the nut is tightened. At the point where all clearance between the pushrod and rocker arm is removed and a slight drag is just felt as you spin the pushrod, all lash has been removed.

head, loosen and remove the rocker arm stud nuts. Store them separately in marked containers to ensure that they will be reinstalled in their original locations. **Note:** *If the pushrods are the only items being removed, loosen each nut just enough to allow the rocker arms to be rotated to the side so the pushrods can be lifted out.*

3   Lift off the rocker arms and pivot balls and store them in the marked containers with the nuts (they must be reinstalled in their original locations).

4   Remove the pushrods and store them separately to make sure they will not get mixed up during installation **(see illustration)**.

Inspection

5   Check each rocker arm for wear, cracks and other damage, especially where the pushrods and valve stems contact the rocker arm faces.

6   Make sure the hole at the pushrod end of each rocker arm is open.

7   Check each rocker arm pivot area for wear, cracks and galling. If the rocker arms are worn or damaged, replace them with new ones and use new pivot balls as well.

8   Inspect the pushrods for cracks and excessive wear at the ends. Roll each pushrod across a piece of plate glass to see if it is bent (if it wobbles, it is bent).

Installation

9   Lubricate the lower end of each pushrod with clean engine oil or moly-base grease and install them in their original locations. Make sure each pushrod seats completely in the lifter.

10   Apply moly-base grease to the ends of the valve stems and the upper ends of the pushrods before positioning the rocker arms over the studs **(see illustration)**.

11   Set the rocker arms in place, then install

the pivot balls and nuts. Apply moly-base grease to the pivot balls to prevent damage to the mating surfaces before engine oil pressure builds up **(see illustration)**. Be sure to install each nut with the flat side against the pivot ball.

Valve adjustment

12   Refer to Section 9 and bring the number one piston to top dead center on the compression stroke.

13   Tighten the rocker arm nuts (number one cylinder only) until all play is removed at the pushrods (zero lash). This can be determined by moving the pushrod up-and down and rotating the pushrod between your thumb and index finger as the nut is tightened **(see illustration)**. At the point where all clearance between the pushrod and rocker arm is removed and a slight drag is just felt as you spin the pushrod, all lash has been removed.

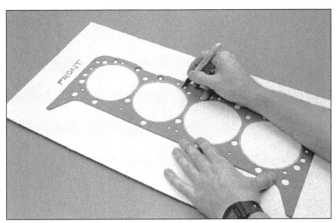

8.5 To avoid mixing up the head bolts, use a new gasket to transfer the bolt hole pattern to a piece of cardboard, then punch holes to accept the bolts

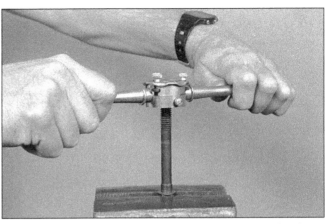

8.12 A die should be used to remove sealant and corrosion from the head bolt threads prior to installation

14   On all except LT4 models, tighten each nut an additional one full turn (360-degrees) to center the lifters. If equipped with the LT4 engine, instead of tightening the rocker nut 360-degrees after zero lash, tighten the rocker nut to 18 ft-lbs. Valve adjustment for cylinder number one is now complete.

15   Turn the crankshaft 90-degrees in the normal direction of rotation until the next piston in the firing order (number eight) is at TDC on the compression stroke. The distributor rotor should be pointing in the direction of terminal number eight on the cap (see Section 9 for additional information).

16   Repeat the procedure described in Paragraphs 13 and 14 for the number eight cylinder valves.

17   Turn the crankshaft another 90-degrees and adjust the number four cylinder valves. Continue turning the crankshaft 90-degrees at a time and adjust both valves for each cylinder before proceeding. Follow the firing order sequence - a cylinder number illustration is also included - in the Specifications.

18   Refer to Section 3 and install the rocker arm covers. Start the engine, listen for unusual valvetrain noises and check for oil leaks at the rocker arm cover joints.

8   Cylinder heads - removal and installation

Refer to illustrations 8.5, 8.12, 8.13a, 8.13b, 8.15 and 8.16

Removal

1   Refer to Section 3 and remove the rocker arm covers.

2   Refer to Section 5 and remove the intake manifold. Note that the cooling system must be drained to prevent coolant from getting into internal areas of the engine when the manifold and heads are removed. The alternator/power steering pump, air pump and air conditioner compressor brackets must be removed as well.

3   Refer to Section 6 and detach both exhaust manifolds.

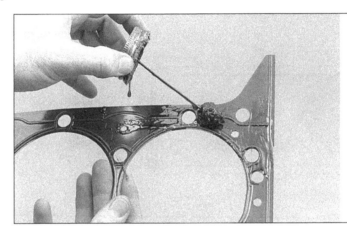

8.13a Steel gaskets - used with cast iron heads - should be coated with a sealant such as KW Copper Coat before installation

4   Refer to Section 7 and remove the pushrods.

5   Using a new head gasket, outline the cylinders and bolt pattern on a piece of cardboard **(see illustration)**. Be sure to indicate the front of the engine for reference. Punch holes at the bolt locations.

6   Loosen the head bolts in 1/4-turn increments until they can be removed by hand. Work from bolt-to-bolt in a pattern that is the reverse of the tightening sequence shown in **illustration 8.16**. **Note:** *Don't overlook the row of bolts on the lower edge of each head, near the spark plug holes. Store the bolts in the cardboard holder as they are removed; this will ensure that the bolts are reinstalled in their original holes.*

7   Lift the heads off the engine. If resistance is felt, do not pry between the head and block as damage to the mating surfaces will result. To dislodge the head, place a block of wood against the end of it and strike the wood block with a hammer. Store the heads on blocks of wood to prevent damage to the gasket sealing surfaces.

8   Cylinder head disassembly and inspection procedures are covered in detail in Chapter 2, Part B.

Installation

9   The mating surfaces of the cylinder heads and block must be perfectly clean

when the heads are installed.

10   Use a gasket scraper to remove all traces of carbon and old gasket material, then wipe the mating surfaces with a cloth saturated with lacquer thinner or acetone. If there is oil on the mating surfaces when the heads are installed, the gaskets may not seal correctly and leaks may develop. When working on the block, cover the lifter valley with shop rags to keep debris out of the engine. **Note:** *Since the heads on some engines are made of aluminum, aggressive scraping can cause damage. Be extra careful not to nick or gouge the mating surfaces with the scraper. Use a vacuum cleaner to remove any debris that falls into the cylinders.*

11   Check the block and head mating surfaces for nicks, deep scratches and other damage. If damage is slight, it can be removed with a file; if it is excessive, machining may be the only alternative.

12   Use a tap of the correct size to chase the threads in the head bolt holes. Mount each bolt in a vise and run a die down the threads to remove corrosion and restore the threads **(see illustration)**. Dirt, corrosion, sealant and damaged threads will affect torque readings.

13   Position the new gaskets over the dowel pins in the block. **Note:** *If a steel gasket is used, apply a thin, even coat of sealant such as KW Copper Coat to both sides prior to*

8.13b Steel gaskets must be installed with the raised bead (arrow) facing UP

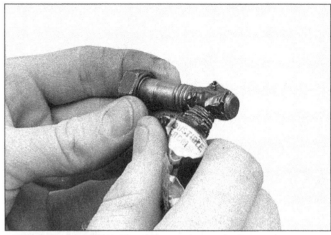

8.15 The head bolts MUST be coated with a non-hardening sealant (such as Permatex no. 2) before they are installed - coolant will leak past the bolts if this is not done

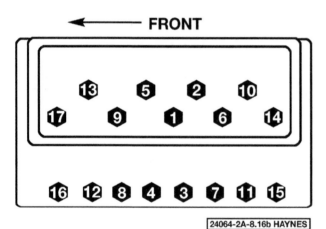

← **FRONT**

⑬　⑤　②　⑩
⑰　⑨　①　⑥　⑭

⑯ ⑫ ⑧ ④ ③ ⑦ ⑪ ⑮

24064-2A-8.16b HAYNES

8.16 Cylinder head bolt tightening sequence - note that the bolts along the lower edge of the head (arrow) are shorter than the others

9.5 On later models, make a mark on the plenum directly opposite the number one spark plug wire terminal in the distributor cap (in this photo the number one on the wire retainer has been highlighted with chalk for clarity)

---

installation **(see illustration)**. *Steel gaskets must be installed with the raised bead UP* **(see illustration)***. Engines with aluminum heads require a composition gasket - do not use a steel gasket with aluminum heads. The composition gasket must be installed dry; do not use sealant.*

14　Carefully position the heads on the block without disturbing the gaskets.

15　Before installing the head bolts, coat the threads with a non-hardening sealant such as Permatex number 2 **(see illustration)**.

16　Install the bolts in their original locations and tighten them finger tight. Following the recommended sequence, tighten the bolts in several steps to the specified torque **(see illustration)**. **Note:** *On engines with aluminum heads, the short bolts installed along the lower edge do not require the same torque as the rest of the bolts (see the Specifications).*

17　The remaining installation steps are the reverse of removal.

9　Top Dead Center (TDC) for number 1 piston - locating

Refer to illustrations 9.5, 9.7 and 9.8

1991 and earlier models

1　Top Dead Center (TDC) is the highest point in the cylinder that each piston reaches as it travels up-and-down when the crankshaft turns. Each piston reaches TDC on the compression stroke and again on the exhaust stroke, but TDC generally refers to piston position on the compression stroke. The timing marks on the vibration damper installed on the front of the crankshaft are referenced to the number one piston at TDC on the compression stroke.

2　Positioning the piston(s) at TDC is an essential part of many procedures such as rocker arm removal, valve adjustment, timing chain and sprocket replacement and distributor removal.

3　In order to bring any piston to TDC, the crankshaft must be turned using one of the methods outlined below. When looking at the front of the engine, normal crankshaft rotation is clockwise. **Warning:** *Before beginning this procedure, be sure to place the transmission in Neutral and disconnect the BAT wire at the distributor cap to disable the ignition system.*

　a)　*The preferred method is to turn the crankshaft with a large socket and breaker bar attached to the vibration damper bolt that is threaded into the front of the crankshaft.*

　b)　*A remote starter switch, which may save some time, can also be used. Attach the switch leads to the S (switch) and B (battery) terminals on the starter motor. Once the piston is close to TDC, use a socket and breaker bar as described in the previous paragraph.*

　c)　*If an assistant is available to turn the ignition switch to the Start position in*

9.7 Turn the crankshaft until the line on the vibration damper is directly opposite the zero mark on the timing plate as shown here

9.8 If the rotor is pointing directly at the mark on the plenum or air cleaner housing, as shown here, the number one piston is at TDC on the compression stroke

short bursts, you can get the piston close to TDC without a remote starter switch. Use a socket and breaker bar as described in Paragraph a to complete the procedure.

4   On 1984 models, make a mark on the air cleaner assembly housing directly across from the number one spark plug wire terminal on the distributor. **Note:** *The terminal numbers are marked on the spark plug wire retainer attached to the top of the distributor.*

5   On later models, remove the screws and detach the distributor cover from the plenum. Using a felt tip pen, make a mark on the plenum directly across from the number one spark plug wire terminal on the distributor **(see illustration)**. The terminal numbers are marked on the spark plug wire retainer attached to the top of the distributor.

6   Remove the distributor cap as described in Chapter 1.

7   Turn the crankshaft (see Paragraph 3 above) until the line on the vibration damper is aligned with the zero mark on the timing plate **(see illustration)**. The timing plate and vibration damper are located low on the front of the engine, near the pulley that turns the drivebelt.

8   The rotor should now be pointing directly at the mark on the air cleaner housing or plenum **(see illustration)**. If it isn't, the piston is at TDC on the exhaust stroke.

9   To get the piston to TDC on the compression stroke, turn the crankshaft one complete turn (360-degrees) clockwise. The rotor should now be pointing at the mark. When the rotor is pointing at the number one spark plug wire terminal in the distributor cap (which is indicated by the mark on the plenum or housing) and the timing marks are aligned, the number one piston is at TDC on the compression stroke.

10   After the number one piston has been positioned at TDC on the compression stroke, TDC for any of the remaining cylinders can be located by turning the crankshaft

90-degrees at a time and following the firing order (refer to the Specifications).

1992 and later models

11   Insert a compression gauge (screw-in type with a hose) in the number 1 spark plug hole and zero it. Place the gauge dial where you can see it while turning the balancer hub bolt.

12   Turn the crankshaft until the arrow on the balancer approaches straight up (or 12 o' clock) **(see illustration 14.9)**. If you see compression building up on the gauge, you are on the compression stroke for number one. Stop turning when the arrow is straight up. If you did not see compression build up, continue rotating one more complete revolution to achieve TDC for number one.

13   After the number one piston has been positioned at TDC on the compression stroke, TDC for any of the remaining cylinders can be located by turning the crankshaft 90-degrees (1/4-turn) at a time and following the firing order (refer to the Specifications). For example, turning 90-degrees past number one TDC would give you TDC for number eight cylinder, the next in the firing order.

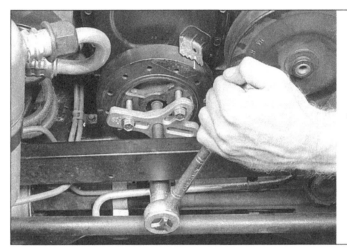

10.4 Use the recommended puller to remove the vibration damper - if a puller that applies force to the outer edge is used, the damper will be damaged

10   Timing cover, chain and sprockets - removal and installation

1991 and earlier models

Refer to illustrations 10.4, 10.5, 10.10, 10.11, 10.12 and 10.13

Removal

1   Refer to Chapter 3 and remove the water pump.

2   Remove the bolts and separate the crankshaft drivebelt pulley from the vibration damper.

3   Refer to Section 9 and position the number one piston at TDC on the compression stroke. **Caution:** *Once this has been done, do not turn the crankshaft until the timing chain and sprockets have been reinstalled.*

4   Remove the bolt from the front of the crankshaft, then use a puller to detach the vibration damper **(see illustration)**. **Caution:** *Do not use a puller with jaws that grip the outer edge of the damper. The puller must be the type shown in the illustration that utilizes*

10.5 A putty knife can be used to break the timing chain cover-to-block seal, but do not pry it off as damage to the cover will result

10.10 With the number one piston at TDC on the compression stroke and the timing marks on the cam and crankshaft sprockets in the 12 o' clock positions, a straight line should pass through the camshaft timing mark, the center of the camshaft, the crankshaft timing mark and the center of the crankshaft as shown here (the camshaft dowel pin, which is in the 9 o' clock position, is hidden by the ruler)

10.11 Be sure to apply Loctite to the camshaft sprocket bolts before tightening them to the specified torque

10.12 Old sealant at the block-to-oil pan junctions will prevent the timing chain cover from forming an oil tight seal

bolts to apply force to the damper hub only.

5   Remove the bolts and separate the timing chain cover from the block. It may be stuck; if so, use a putty knife to break the gasket seal **(see illustration)**. The cover is easily distorted, so do not attempt to pry it off.

6   Remove the three bolts from the end of the camshaft, then detach the camshaft sprocket and chain as an assembly. The sprocket on the crankshaft can be removed with a two or three jaw puller, but be careful not to damage the threads in the end of the crankshaft. **Note:** *If the timing chain cover oil seal has been leaking, refer to Section 14 and install a new one.*

Installation

7   Use a gasket scraper to remove all traces of old gasket material and sealant from the cover and engine block. Stuff a shop rag into the opening at the front of the oil pan to keep debris out of the engine. Wipe the cover

and block sealing surfaces with a cloth saturated with lacquer thinner or acetone.

8   Check the cover flange for distortion, particularly around the bolt holes. If necessary, place the cover on a block of wood and use a hammer to flatten and restore the gasket surface.

9   If new parts are being installed, be sure to align the keyway in the crankshaft sprocket with the Woodruff key in the end of the crankshaft. Press the sprocket onto the crankshaft with the vibration damper bolt, a large socket and some washers or tap it gently into place until it is completely seated. **Caution:** *If resistance is encountered, do not hammer the sprocket onto the crankshaft. It may eventually move onto the shaft, but it may be cracked in the process and fail later, causing extensive engine damage.*

10   Loop the new chain over the camshaft sprocket, then turn the sprocket until the timing mark is in the 12 o' clock position. Mesh

the chain with the crankshaft sprocket and position the camshaft sprocket on the end of the cam. If necessary, turn the camshaft so the dowel pin fits into the sprocket hole with the timing mark in the 12 o' clock position. When correctly installed, the marks on the sprockets will be aligned as shown in the accompanying **illustration**. **Note:** *The number one piston must be at TDC on the compression stroke as the chain and sprockets are installed (see Paragraph 3 above).*

11   Apply Loctite to the camshaft sprocket bolt threads, then install and tighten them to the specified torque **(see illustration)**. Lubricate the chain with clean engine oil.

12   Check the oil pan-to-block joints to make sure that all excess sealant is removed **(see illustration)**.

13   Apply RTV sealant to the U-shaped channel on the bottom of the cover, then position a new rubber oil pan seal in the channel (the sealant should hold it in place as

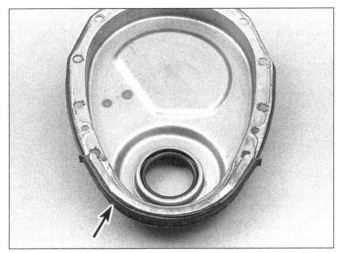

10.13 Use RTV sealant to retain the new rubber oil pan seal (arrow) in the timing chain cover

10.23a On spline type drives, paint a match mark to line up the spline shaft sticking out of the front cover for faster assembly, although it is keyed and will only go into the cover one way

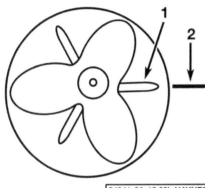

24041-2A-10.23b HAYNES

10.23b The later type distributor drive uses a long pin on the camshaft which fits into a slot (1) on the back of the distributor - align it with the mark (2) for TDC assembly

10.24 Remove the front cover bolts and remove the cover

the cover is installed) **(see illustration)**.
14   Apply a thin layer of RTV sealant to both sides of the new gasket, then position it on the engine block (the dowel pins and sealant will hold it in place).
15   Since the oil pan seal in the bottom of the cover must be compressed in order to position the cover over the dowel pins and thread the bolts into the block, the cover is nearly impossible to install unless the oil pan is removed first. It can be done, but it is very difficult, time consuming and frustrating.
16   Refer to Section 12 and remove the oil pan, then install the timing chain cover. After the cover bolts have been tightened securely, reinstall the oil pan.
17   Lubricate the oil seal contact surface of the vibration damper hub with moly-base grease or clean engine oil, then install the damper on the end of the crankshaft. The keyway in the damper must be aligned with the Woodruff key in the crankshaft nose. If the damper cannot be seated by hand, slip a large washer over the bolt, install the bolt and

tighten it to push the damper into place. Remove the large washer and tighten the bolt to the specified torque.
18   The remaining installation steps are the reverse of removal.

1992 and later models

Refer to illustrations 10.23a, 10.23b, 10.24, 10.26, 10.29 and 10.33

Removal

19   Refer to Chapter 3 and remove the water pump.
20   Refer to Section 3 and position the number one piston at TDC on the compression stroke. **Caution:** *Once this has been done, do not turn the crankshaft until the timing chain and sprockets have been reinstalled.*
21   Remove the crankshaft balancer and balancer hub (see Section 14).
22   The oil pan bolts will have to be loosened and the pan lowered slightly for the front cover to be removed. If the pan has been in place for an extended period of time

it' s likely the pan gasket will break when the pan is lowered. In this case the pan should be removed and a new gasket installed (see Section 12).
23   Remove the distributor (see Chapter 5). There are two types of distributor drives on the V8 engines: the spline-type or "type 1", and the pin-type or "type 2" (on 1995 and later models). On the spline-type, the distributor is driven by a splined shaft coming out of the front cover and driven by the camshaft gear. On these models, the short splined shaft may come out of the front cover when the distributor is removed. It will only go back in one way and the distributor will only go back onto it one way, due to key alignment splines. These distributors should not be turned while off the engine. On the later models, the camshaft has a long pin sticking out, which meshes with a slot in the backside of the distributor. Alignment of these types is simpler **(see illustrations)**.
24   Unbolt and remove the front cover **(see illustration)**.

10.26 Unbolt and remove the camshaft sprocket lowering it to remove the sprocket and chain as an assembly

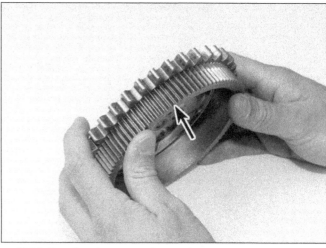

10.29 Examine the camshaft sprocket teeth and the integral water pump drive gear teeth (arrow) on the back for wear

25   On 1996 models, remove the crankshaft position sensor reluctor ring, which is a disc over the crankshaft snout, noting which side faces the engine.

26   Remove the three bolts from the end of the camshaft, then detach the camshaft sprocket and chain as an assembly **(see illustration)**.

27   Using a jaw-type puller, remove the crankshaft sprocket.

Inspection

28   The timing chain should be replaced with a new one if the engine has high mileage, the chain has visible damage, or total freeplay midway between the sprockets exceeds one-inch. Failure to replace a worn timing chain may result in erratic engine performance, loss of power and decreased fuel mileage. Loose chains can "jump" timing. In the worst case, chain "jumping" or breakage will result in severe engine damage.

29   Inspect the timing sprockets and water-pump drive gear (on the back of the cam gear) for worn, non-concentric teeth or wear in the valleys between the teeth **(see illustration)**.

Installation

30   Use a gasket scraper to remove all traces of old gasket material and sealant from the cover and engine block. Stuff a shop rag into the opening at the front of the oil pan to keep debris out of the engine. Clean the cover and block sealing surfaces with lacquer thinner or acetone. **Caution:** *Be careful when scraping the front cover; gouges in the soft aluminum could result in oil leaks.*

31   Replace the O-ring on the outer end of the water-pump driveshaft (see Chapter 3).

32   The timing chain must be replaced as a set with the camshaft and crankshaft sprockets. Never put a new chain on old sprockets. Align the sprocket with the Woodruff key and press the sprocket onto the crankshaft with a large socket or tap it gently into place until it is completely seated. **Caution:** *If resistance is encountered, do not hammer the sprocket*

10.33 Check the timing marks (arrows) before bolting the camshaft sprocket to the camshaft. The larger arrow indicates the splined distributor shaft in place on the cam gear

onto the crankshaft. It may eventually move onto the shaft, but it may be cracked in the process and fail later, causing extensive engine damage.

33   Loop the new chain over the camshaft sprocket, then turn the sprocket until the timing mark is in the 6 o' clock position **(see illustration)**. Mesh the chain with the crankshaft sprocket and position the camshaft sprocket on the end of the cam. If necessary, turn the camshaft so the dowel pin fits into the sprocket hole with the timing mark in the 6 o' clock position. **Note:** *If necessary, turn the water-pump driveshaft slightly to mesh its gear with the water-pump drive gear on the back of the camshaft sprocket.*

34   Apply non-hardening thread locking compound to the camshaft sprocket bolt threads, then install and tighten them to the torque listed in this Chapter' s Specifications. Lubricate the chain with clean engine oil.

35   The one-piece oil pan gasket should be checked for cracks and deformation before installing the front cover. If the gasket has deteriorated it must be replaced before reinstalling the cover.

36   Apply a thin layer of RTV sealant to both

sides of the new cover gasket, then position it on the engine block. The dowel pins and sealant will hold it in place. **Caution:** *Wrap some plastic tape around the splines on the water-pump driveshaft to protect the cover seal during installation.*

37   Install the front cover on the block, tightening the bolts finger-tight.

38   If the oil pan was removed, reinstall it (see Section 12). If it was only loosened, tighten the oil pan bolts, bringing the oil pan up against the front cover.

39   Tighten the front cover bolts to the torque listed in this Chapter' s Specifications.

40   Lubricate the oil seal contact surface of the crankshaft balancer hub with multi-purpose grease or clean engine oil, then install it on the end of the crankshaft. The keyway in the balancer hub must be aligned with the Woodruff key in the crankshaft nose. If the hub cannot be seated by hand, slip a large washer over the bolt, install the bolt and tighten it to push the hub into place. Remove the large washer and tighten the bolt to the torque listed in this Chapter' s Specifications.

41   If plastic tape was used to install the water-pump-driveshaft O-ring (see Chapter 3) remove the plastic tape from the water-

11.3 When checking the camshaft lobe lift, the dial indicator plunger must be positioned directly above the pushrod

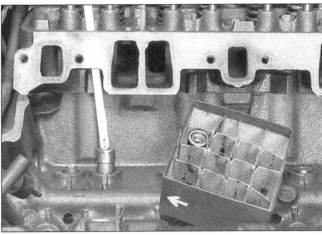

11.11 The lifters on an engine that has accumulated many miles may have to be removed with a special tool - be sure to store the lifters in an organized manner to make sure they are reinstalled in their original locations

pump driveshaft before installing the water pump. The remaining installation steps are the reverse of removal. **Caution:** *When reinstalling spline-type distributors, the distributor must go on smoothly and without much effort. It is possible to install the spline-type with improper alignment and force it on and tighten the bolts, but this will cause expensive damage to the parts. Wiggle the distributor slightly until you feel the spline actually engage fully before bolting down the distributor.*

11 Camshaft, bearings and lifters - removal, inspection and installation

Refer to illustrations 11.3, 11.11, 11.12, 11.15, 11.18a, 11.18b, 11.18c, 11.20, 11.23 and 11.25

Camshaft lobe lift check

1 In order to determine the extent of cam lobe wear, the lobe lift should be checked prior to camshaft removal. Refer to Section 3 and remove the rocker arm covers.
2 Position the number one piston at TDC on the compression stroke (see Section 9).
3 Beginning with the number one cylinder valves, mount a dial indicator on the engine and position the plunger against the top surface of the first rocker arm. The plunger should be directly above and in line with the pushrod **(see illustration)**.
4 Zero the dial indicator, then very slowly turn the crankshaft in the normal direction of rotation until the indicator needle stops and begins to move in the opposite direction. The point at which it stops indicates maximum cam lobe lift.
5 Record this figure for future reference, then reposition the piston at TDC on the compression stroke.
6 Move the dial indicator to the remaining number one cylinder rocker arm and repeat

the check. Be sure to record the results for each valve.
7 Repeat the check for the remaining valves. Since each piston must be at TDC on the compression stroke for this procedure, work from cylinder-to-cylinder following the firing order sequence.
8 After the check is complete, compare the results to the Specifications. If camshaft lobe lift is less than specified, cam lobe wear has occurred and a new camshaft should be installed.

Removal

9 Refer to the appropriate Sections and remove the intake manifold, the rocker arms, the pushrods and the timing chain and camshaft sprocket. The radiator should be removed as well (Chapter 3).
10 There are several ways to extract the lifters from the bores. A special tool designed to grip and remove lifters is manufactured by many tool companies and is widely available, but it may not be required in every case. On newer engines without a lot of varnish buildup, the lifters can often be removed with a small magnet or even with your fingers. A

machinist's scribe with a bent end can be used to pull the lifters out by positioning the point under the retainer ring in the top of each lifter. **Caution:** *Do not use pliers to remove the lifters unless you intend to replace them with new ones (along with the camshaft). The pliers will damage the precision machined and hardened lifters, rendering them useless.*
11 Before removing the lifters, arrange to store them in a clearly labeled box to ensure that they are reinstalled in their original locations. Remove the lifters and store them where they will not get dirty **(see illustration)**. Do not attempt to withdraw the camshaft with the lifters in place.
12 Thread a 6-inch long 5/16-18 bolt into one of the camshaft sprocket bolt holes to use as a handle when removing the camshaft from the block **(see illustration)**.
13 Carefully pull the camshaft out. Support the cam near the block so the lobes do not nick or gouge the bearings as it is withdrawn.

Inspection

Camshaft and bearings

14 After the camshaft has been removed from the engine, cleaned with solvent and

11.12 A long bolt can be threaded into one of the camshaft bolt holes to provide a handle for removal and installation of the camshaft

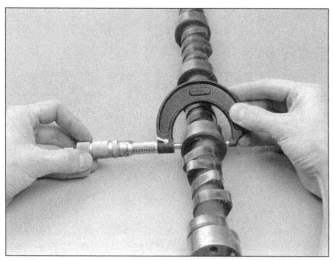

11.15 The camshaft bearing journal diameter is checked to pinpoint excessive wear and out-of-round conditions

11.18a The foot of each lifter should be slightly convex - the side of another lifter can be used as a straightedge to check it; if it appears flat, it is worn and must not be reused

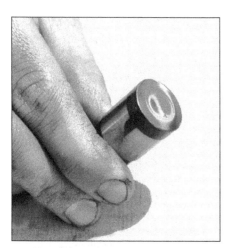

11.18b If the bottom of any lifter is worn concave, scratched or galled, replace the entire set with new lifters

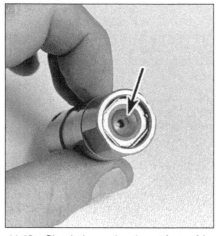

11.18c Check the pushrod seat (arrow) in the top of each lifter for wear

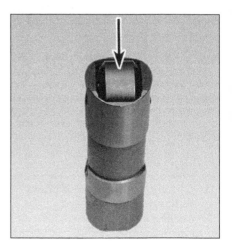

11.20 The roller on roller lifters must turn freely - check for wear and excessive play as well

dried, inspect the bearing journals for uneven wear, pitting and evidence of seizure. If the journals are damaged, the bearing inserts in the block are probably damaged as well. Both the camshaft and bearings will have to be replaced.

15   Measure the bearing journals with a micrometer (see illustration) to determine if they are excessively worn or out-of-round.

16   Check the camshaft lobes for heat discoloration, score marks, chipped areas, pitting and uneven wear. If the lobes are in good condition and if the lobe lift measurements are as specified, the camshaft can be reused.

Conventional lifters

17   Clean the lifters with solvent and dry them thoroughly without mixing them up.

18   Check each lifter wall, pushrod seat and foot for scuffing, score marks and uneven wear. Each lifter foot (the surface that rides on the cam lobe) must be slightly convex, although this can be difficult to determine by eye. If the base of the lifter is concave (see

illustrations), the lifters and camshaft must be replaced. If the lifter walls are damaged or worn (which is not very likely), inspect the lifter bores in the engine block as well. If the pushrod seats (see illustration) are worn, check the pushrod ends.

19   If new lifters are being installed, a new camshaft must also be installed. If a new camshaft is installed, then use new lifters as well. Never install used lifters unless the original camshaft is used and the lifters can be installed in their original locations.

Roller lifters

20   Check the rollers carefully for wear and damage and make sure they turn freely without excessive play (see illustration). The inspection procedure for conventional lifters also applies to roller lifters.

21   Unlike conventional lifters, used roller lifters can be reinstalled with a new camshaft and the original camshaft can be used if new lifters are installed.

Bearing replacement

22   Camshaft bearing replacement requires special tools and expertise that place it outside the scope of the home mechanic. Take the block to an automotive machine shop to ensure that the job is done correctly.

Installation

23   Lubricate the camshaft bearing journals and cam lobes with moly-base grease or engine assembly lube (see illustration).

24   Slide the camshaft into the engine. Support the cam near the block and be careful not to scrape or nick the bearings.

25   Turn the camshaft until the dowel pin is in the 9 o' clock position (see illustration).

26   Refer to Section 10 and install the timing chain and sprockets.

27   Lubricate the lifters with clean engine oil and install them in the block. If the original lifters are being reinstalled, be sure to return them to their original locations. If a new camshaft was installed, be sure to install new

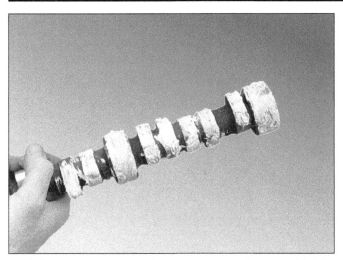

11.23 Be sure to apply moly-base grease or engine assembly lube to the cam lobes and bearing journals before installing the camshaft

11.25 After the camshaft is in place, turn it until the dowel pin (arrow) is in the 9 o' clock position as shown here

lifters as well (except for engines with roller lifters).
28   The remaining installation steps are the reverse of removal.
29   Before starting and running the engine, change the oil and install a new oil filter (see Chapter 1).

12   Oil pan - removal and installation

Note: *The following procedure is based on the assumption that the engine is in place in the vehicle. If it has been removed, merely unbolt the oil pan and detach it from the block.*

Removal

1   Disconnect the negative battery cable from the battery, then refer to Chapter 1 and drain the oil. On 1986 and later vehicles, remove the oil filter and the oil cooler adapter from the block.
2   Refer to Chapter 5 and remove the starter motor.
3   Remove the cover from the lower part of the bellhousing. On 1986 and later vehicles, disconnect the transmission oil cooler lines at the cover.
4   On later models, loosen the clamp and detach the catalytic converter AIR pipe from the manifold. Disconnect the oil cooler line at the oil pan, then remove the ESC sensor shield and the front crossmember braces.
5   Remove the oil pan mounting bolts/nuts. Later models are equipped with a reinforcement strip on each side of the oil pan, which may come loose after the bolts/nuts are removed.
6   Carefully separate the pan from the block. Do not pry between the block and pan or damage to the sealing surfaces may result and oil leaks may develop. Later models are equipped with a one-piece rubber gasket that can be reused if it isn't damaged during

removal. You may have to turn the crankshaft slightly to maneuver the front of the pan past the crank counterweights.

Installation

7   Use a gasket scraper to remove all traces of old gasket material and sealant from the pan and block (this doesn't apply to later models with rubber gaskets).
8   Wipe the sealing surfaces with a cloth saturated with lacquer thinner or acetone. Make sure the bolt holes in the block are clean.
9   Check the oil pan flange for distortion, particularly around the bolt holes. If necessary, place the pan on a block of wood and use a hammer to flatten and restore the gasket surface.
10   Remove the old rubber seals from the rear main bearing cap and the timing chain cover, then clean the grooves and install new seals (this doesn't apply to later models with rubber gaskets). Use RTV sealant or "Gasgacinch" contact cement to hold the new seals in place, then apply a bead of RTV sealant to the block-to-seal junctions. Use the same sealant to attach the new side gaskets to the oil pan.
11   The one-piece rubber gasket used on later models should be checked carefully and replaced with a new one if damage is noted. Apply a small amount of RTV sealant to the corners of the semi-circular cutouts at both ends of the pan, then attach the rubber gasket to the pan.
12   Carefully position the pan against the block and install the bolts/nuts finger tight (don' t forget the reinforcement strips, if used). Make sure the seals and gaskets haven' t shifted, then tighten the bolts/nuts in three steps to the specified torque. Start at the center of the pan and work out toward the ends in a spiral pattern.
13   The remaining steps are the reverse of removal. **Caution:** *Don' t forget to refill the*

engine with oil before starting it (see Chapter 1).
14   Start the engine and check carefully for oil leaks at the oil pan.

13   Oil pump - removal and installation

1   Remove the oil pan as described in Section 12.
2   While supporting the oil pump, remove the pump-to-rear main bearing cap bolt. On some models, the oil pan baffle must be removed first, since it is also held in place by the pump mounting bolt.
3   Lower the pump and remove it along with the pump driveshaft.
4   If a new oil pump is installed, make sure the pump driveshaft is mated with the shaft inside the pump.
5   Position the pump on the engine and make sure the slot in the upper end of the driveshaft is aligned with the tang on the lower end of the distributor shaft. The distributor drives the oil pump, so it is absolutely essential that the components mate properly.
6   Install the mounting bolt and tighten it to the specified torque.
7   Install the oil pan.

14   Crankshaft oil seals - replacement

Front seal

Refer to illustrations 14.2, 14.4, 14.9, 14.11a, 14.11b, 14.12 and 14.14

1991 and earlier

1   Remove the timing chain cover as described in Section 10.
2   Use a punch or screwdriver and hammer to drive the seal out of the cover from the

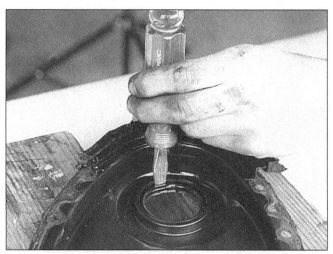

14.2 While supporting the cover near the seal bore, drive the old seal out from the inside with a hammer and punch or screwdriver

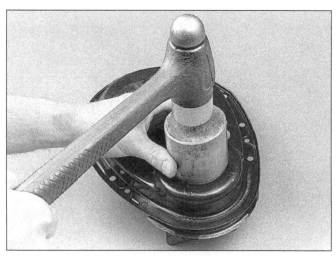

14.4 Clean the bore, then apply a small amount of oil to the outer edge of the seal and drive it squarely into the opening with a large socket and a hammer - do not damage the seal in the process!

14.9 Use a 2-pin spanner to hold the balancer while the three balancer bolts are removed (arrows); Note: The cast-in TDC arrow on the balancer (pointing at top TDC location)

14.11a The portion of the crank snout exposed inside the hub is too small for a standard puller' s beveled tip to fit into, so use a long Allen bolt that fits inside the crank threads (not threaded into them, but smaller) - The puller will then press against the socket-head of the Allen bolt - Arrows indicate the TDC mark on the hub and the cover rib it must align with, since the hub has no keyway

backside. Support the cover as close to the seal bore as possible **(see illustration)**. Be careful not to distort the cover or scratch the wall of the seal bore. If the engine has accumulated a lot of miles, apply penetrating oil to the seal-to-cover joint on each side and allow it to soak in before attempting to drive the seal out.

3   Clean the bore to remove any old seal material and corrosion. Support the cover on blocks of wood and position the new seal in the bore with the open end of the seal facing in. A small amount of oil applied to the outer edge of the new seal will make installation easier - don' t overdo it!

4   Drive the seal into the bore with a large socket and hammer until it is completely seated **(see illustration)**. Select a socket that is the same outside diameter as the seal (a section of pipe can be used if a socket is not available).

5   Reinstall the timing chain cover.

1992 and later models

6   Disconnect the negative cable from the battery. **Note:** *Unless you have a factory puller for the balancer hub, this procedure will be time consuming, as the distributor and water pump may have to be removed (see Step 11).*

7   Remove the serpentine drivebelt belt (see Chapter 1).

8   Raise the vehicle and support it securely on jackstands.

9   Remove the three balancer bolts and the balancer **(see illustration)**. A puller is not required to remove the balancer.

10   Apply matchmarks on the balancer hub and the front cover, and remove the balancer hub bolt. **Caution:** *Once the matchmarks have been applied, do not turn the crankshaft until the balancer has been reinstalled. When loosening this bolt, prevent the crankshaft from turning by wedging a large screwdriver*

into the ring gear teeth of the flywheel/drive-plate.

11   Use a bolt-type puller to remove the balancer hub **(see illustrations)**. Because the bolt holes in the hub are larger than most standard pullers will allow, you will have to use three, four-inch-long (grade 8) 5/16-inch bolts and nuts to use a standard puller, along with a four-inch-long (grade 8) 5/16-inch Allen bolt to fit into the crank snout. Because of the clearance around the distributor, it is safest to remove the water pump (see Chapter 3) and distributor (see Chapter 5) first, which also exposes the portion of the front cover with the "TDC" rib for aligning the key-less hub when reinstalling it.

12   Carefully pry the seal out of the cover with a seal removal tool or a large screwdriver **(see illustration)**. Be careful not to distort the cover or scratch the wall of the seal bore.

13   Clean the bore to remove any old seal

14.11b Most standard pullers will have to be adapted with four 5/16 inch bolts and nuts to attached to the balancer hub, with the puller's beveled tip fitting into the Allen bolt

14.12 Carefully pry the old seal out of the front cover - don't damage the crank surface

14.14 Use a special seal installing tool or drive the seal in carefully and evenly with a large socket and hammer

14.19 Drive one end of the seal in until the seal protrudes enough on the other side to be gripped with pliers and removed

material and corrosion. Position the new seal in the bore with the open end of the seal facing IN. A small amount of oil applied to the outer edge of the new seal will make installation easier - don't overdo it!

14 Drive the seal into the bore with a seal driving tool or a large socket and hammer until it's completely seated (see illustration). Select a socket that's the same outside diameter as the seal.

15 Lubricate the seal lips with engine oil and reinstall the balancer hub, lining up the matchmarks. The remaining installation steps are the reverse of removal. Note: *If the balancer is being replaced, be sure to transfer the bolt-on balancer weights from the old balancer to the same positions on the new balancer.*

Rear seal

Refer to illustrations 14.19, 14.21, 14.22a, 14.22b, 14.23, 14.24, and 14.26

1984 and 1985 models

16 The rear main seal on these models can

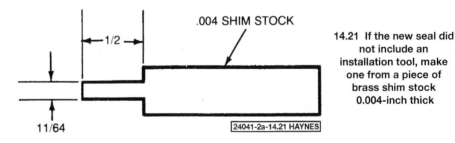

.004 SHIM STOCK

1/2

11/64

24041-2a-14.21 HAYNES

14.21 If the new seal did not include an installation tool, make one from a piece of brass shim stock 0.004-inch thick

be replaced with the engine in the vehicle. Refer to the appropriate Sections and remove the oil pan and oil pump.

17 Remove the bolts and detach the rear main bearing cap from the engine.

18 The seal section in the bearing cap can be pried out with a screwdriver.

19 To remove the seal section in the block, tap on one end with a hammer and brass punch or wood dowel until the other end pro-

trudes far enough to grip it with a pair of pliers and pull it out (see illustration). Be very careful not to nick or scratch the crankshaft journal or seal surface as this is done.

20 Inspect the bearing cap and engine block mating surfaces, as well as the cap seal grooves, for nicks, burrs and scratches. Remove any defects with a fine file or deburring tool.

21 A small seal installation tool is usually

14.22a Using the tool like a "shoehorn", attach the seal section to the bearing cap . . .

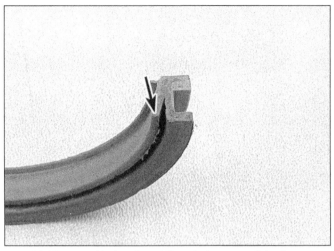

14.22b . . . with the oil seal lip pointing toward the front of the engine

14.23 Position the tool to protect the backside of the seal as it passes over the sharp edge of the ridge - note that the seal straddles the ridge

14.24 Make sure the seal lip faces the front of the engine and hold the tool in place to protect the seal as it is installed

included when a new seal is purchased. If you didn't receive one, they can also be purchased separately at most auto parts stores or you can make one from an old feeler gauge or a piece of brass shim stock **(see illustration)**.

22   Using the tool, install one seal section in the cap with the lip facing the front of the engine (if the seal has two lips, the one with the helix must face the front) **(see illustrations)**. The ends should be flush with the mating surface of the cap. Make sure it is completely seated.

23   Position the narrow end of the tool so that it will protect the backside of the seal as it passes over the sharp edge of the ridge in the block **(see illustration)**.

24   Lubricate the seal lips and the groove in the backside with moly-base grease or clean engine oil - do not get any lubricant on the seal ends. Insert the seal into the block, over the tool **(see illustration)**. **Caution:** *Make sure that the lip points toward the front of the engine when the seal is installed.*

25   Push the seal into place, using the tool like a "shoehorn". Turning the crankshaft may help to draw the seal into place. When both ends of the seal are flush with the block surface, remove the tool.

26   Apply a thin, even coat of anaerobic-type gasket sealant to the areas of the cap or block indicated in the accompanying **illustration**. Do not get any sealant on the bearing face, crankshaft journal, seal ends or seal lips. Also, lubricate the cap seal lips with moly-base grease or clean engine oil.

27   Carefully position the bearing cap on the block, install the bolts and tighten them to 10-to-12 ft-lbs only. Tap the crankshaft forward and backward with a lead or brass hammer to line up the main bearing and crankshaft thrust surfaces, then tighten the rear bearing cap bolts to the specified torque.

28   Install the oil pump and oil pan.

1986 and later models

29   Later models are equipped with a one-piece seal that requires an entirely different installation procedure. The transmission and oil pan must be removed to gain access to the seal housing. Once the housing is removed, the seal replacement procedure in Chapter 2, Part B, Section 22 can be followed.

**15   Engine mounts - check and r
     replacement**

---

1   Engine mounts seldom require attention, but broken or deteriorated mounts should be replaced immediately or the added strain placed on the driveline components may cause damage.

Check

2   During the check, the engine must be raised slightly to remove the weight from the mounts. On 1991 and earlier models, refer to Chapter 1 and remove the distributor cap before raising the engine.

3   Raise the vehicle and support it securely on jackstands, then position the jack under the engine oil pan. Place a large block of wood between the jack head and the oil pan, then carefully raise the engine just enough to take the weight off the mounts.

4   Check the mounts to see if the rubber is cracked, hardened or separated from the metal plates. Sometimes the rubber will split right down the center. Rubber preservative or WD-40 should be applied to the mounts to slow deterioration.

5   Check for relative movement between the mount plates and the engine or frame (use a large screwdriver or pry bar to attempt to move the mounts). If movement is noted, lower the engine and tighten the mount fasteners.

Replacement

6   Disconnect the negative battery cable from the battery, then raise the vehicle and support it securely on jackstands.

**APPLY SEALANT
TO THESE TWO AREAS**

14.26 Before installing the rear main bearing cap, apply the specified sealant to the shaded areas of the block (or the equivalent areas on the cap)

7   Remove the nut and withdraw the mount through bolt from the frame bracket.
8   If you are replacing the right-hand mount, disconnect the catalytic converter AIR pipe at the exhaust manifold, exhaust pipe and converter.

9   Raise the engine slightly, then remove the mount-to-block bolts and detach the mount.
10   Installation is the reverse of removal. Use Locktite on the mount bolts and be sure to tighten them securely.

Notes

Chapter 2 Part B
General engine overhaul procedures

Contents

Specifications

General

Bore and stroke	4.000 x 3.480 inches
Oil pressure	50 to 65 psi @ 2000 rpm
Compression pressure	Lowest reading cylinder must be at least 70% of highest reading
cylinder (100 psi minimum)	

Engine block

Cylinder bore

Diameter	3.9995 to 4.0025 inches
Taper limit	0.001 inch
Out-of-round limit	0.002 inch

Pistons and rings

Piston-to-cylinder bore clearance

Standard	0.0025 to 0.0035 inch
Service limit	0.0045 inch

Piston ring-to-groove side clearance

Standard

Compression rings	0.0012 to 0.0032 inch
Oil control ring	0.002 to 0.007 inch

Service limit

Compression rings	0.0042 inch
Oil control ring	0.008 inch

Pistons and rings (continued)

Piston ring end gap
    Standard
        Top compression ring.. 0.010 to 0.020 inch
        Second compression ring.. 0.010 to 0.025 inch
        Oil control ring .... 0.015 to 0.055 inch
    Service limit
        Top compression ring.. 0.030 inch
        Second compression ring.. 0.035 inch
        Oil control ring .... 0.065 inch
Piston pin
    Diameter.. 0.9270 to 0.9273 inch
    Pin-to-piston clearance limit .... 0.001 inch
    Pin-to-rod interference fit... 0.0008 to 0.0016 inch

Crankshaft and flywheel

Main journal
    Diameter
        No. 1 journal.. 2.4484 to 2.4493 inches
        No. 2, 3 and 4 journals... 2.4481 to 2.4490 inches
        No. 5 journal.. 2.4479 to 2.4488 inches
    Taper limit.. 0.001 inch
    Out-of-round limit... 0.001 inch
Main bearing oil clearance
    Standard
        No. 1 journal.. 0.0008 to 0.0020 inch
        No. 2, 3 and 4 journals... 0.0011 to 0.0023 inch
        No. 5 journal.. 0.0017 to 0.0032 inch
    Service limit
        No. 1 journal.. 0.001 to 0.0015 inch
        No. 2, 3 and 4 journals... 0.001 to 0.0025 inch
        No. 5 journal.. 0.0025 to 0.0035 inch
Connecting rod journal
    Diameter.. 2.0988 to 2.0998 inches
    Taper limit.. 0.001 inch
    Out-of-round limit... 0.001 inch
Connecting rod bearing oil clearance
    Standard.. 0.0013 to 0.0035 inch
    Service limit.. 0.0035 inch
Connecting rod end play .... 0.006 to 0.014 inch
Crankshaft end play... 0.002 to 0.006 inch
Cylinder head and valve train
    Head warpage limit .... 0.003 inch per 6 inch span/0.006 inch overall
    Valve seat angle 46-degrees
Valve seat width
    Intake .... 1/32 to 1/16 (0.0313 to 0.0625) inch
    Exhaust .... 1/16 to 3/32 (0.0625 to 0.0938) inch
Valve seat runout limit .... 0.002 inch
Valve face angle.. 45-degrees
Minimum valve margin width.. 1/32 inch
Valve stem-to-guide clearance
    Standard.. 0.0010 to 0.0027 inch
    Service limit
        Intake .... 0.0037 inch
        Exhaust .... 0.0047 inch
Valve spring free length .... 2.030 inches
Valve spring damper (inner spring) free length..................................... 1.860 in
Valve spring installed height
    Intake .... 1-23/32 (1.7188) inches
    Exhaust .... 1-19/32 (1.5938) inches

Torque specifications*

	Ft-lbs
Rocker arm studs	50
Main bearing cap bolts	80
Connecting rod cap nuts	45
Oil pump bolts	65

Additional torque specifications can be found in Chapter 2, Part A

1 General information

Included in this portion of Chapter 2 are the general overhaul procedures for the cylinder head and internal engine components. The information ranges from advice concerning preparation for an overhaul and the purchase of replacement parts to detailed, step-by-step procedures covering removal and installation of internal engine components and the inspection of parts.

The following Sections have been written based on the assumption that the engine has been removed from the vehicle. For information concerning in-vehicle engine repair, as well as removal and installation of the external components necessary for the overhaul, see Part A of this Chapter and Section 7 of this Part.

The specifications included here in Part B are only those necessary for the inspection and overhaul procedures which follow. Refer to Part A for additional specifications.

2 Engine removal - methods and precautions

Warning: *Later models are equipped with airbags. Always disable the airbag system before working in the vicinity of the steering column, instrument panel or impact sensors to avoid the possibility of accidental deployment of the airbag, which could cause personal injury (see Chapter 12).*

If you have decided that an engine must be removed for overhaul or major repair work, several preliminary steps should be taken.

Locating a suitable work area is extremely important. A shop is, of course, the most desirable place to work. Adequate work space, along with storage space for the vehicle, will be needed. If a shop or garage is not available, at the very least a flat, level, clean work surface made of concrete or asphalt is required.

Cleaning the engine compartment and engine before beginning the removal procedure will help keep tools clean and organized.

An engine hoist or A-frame will also be necessary. Make sure that the equipment is rated in excess of the combined weight of the engine and its accessories. Safety is of primary importance, considering the potential hazards involved in lifting the engine out of the vehicle.

If the engine is being removed by a novice, a helper should be available. Advice and aid from someone more experienced would also be helpful. There are many instances when one person cannot simultaneously perform all of the operations required when lifting the engine out of the vehicle.

Plan the operation ahead of time. Arrange for or obtain all of the tools and equipment you will need prior to beginning the job. Some of the equipment necessary to perform engine removal and installation safely and with relative ease are (in addition to an engine hoist) a heavy duty floor jack, complete sets of wrenches and sockets as described in the front of this manual, wooden blocks and plenty of rags and cleaning solvent for mopping up spilled oil, coolant and gasoline. If the hoist is to be rented, make sure that you arrange for it in advance and perform beforehand all of the operations possible without it. This will save you money and time.

Plan for the vehicle to be out of use for a considerable amount of time. A machine shop will be required to perform some of the work that the do-it-yourselfer cannot accomplish due to a lack of special equipment. These shops often have a busy schedule, so it would be wise to consult them before removing the engine in order to accurately estimate the amount of time required to rebuild or repair components that may need work.

Always use extreme caution when removing and installing the engine. Serious injury can result from careless actions. Plan ahead, take your time and a job of this nature, although major, can be accomplished successfully.

3 Engine overhaul - general information

Refer to illustration 3.4

It is not always easy to determine when, or if, an engine should be completely overhauled, as a number of factors must be considered.

High mileage is not necessarily an indication that an overhaul is needed, while low mileage does not preclude the need for an overhaul. Frequency of servicing is probably the most important consideration. An engine that has had regular and frequent oil and filter changes, as well as other required maintenance, will most likely give many thousands of miles of reliable service. Conversely, a neglected engine may require an overhaul very early in its life.

Excessive oil consumption is an indication that piston rings and/or valve guides are in need of attention. Make sure that oil leaks are not responsible before deciding that the rings and/or guides are bad. Have a cylinder compression or leakdown test performed by an experienced tune-up mechanic to determine the extent of the work required.

If the engine is making obvious knocking or rumbling noises, the connecting rod and/or main bearings are probably at fault. Check the oil pressure with a gauge installed in place of the oil temperature sending unit **(see illustration)** and compare it to the Specifications. If it is extremely low, the bearings and/or oil pump are probably worn out.

Loss of power, rough running, excessive valve train noise and high fuel consumption rates may also point to the need for an overhaul, especially if they are all present at the same time. If a complete tune-up does not remedy the situation, major mechanical work is the only solution.

An engine overhaul involves restoring the internal parts to the specifications of a new engine. During an overhaul, the piston rings are replaced and the cylinder walls are reconditioned (rebored and/or honed). If a rebore is done, new pistons are required. The main bearings, connecting rod bearings and camshaft bearings are generally replaced with new ones and, if necessary, the crankshaft may be reground to restore the journals. Generally, the valves are serviced as well, since they are usually in less-than-perfect condition at this point. While the engine is being overhauled, other components, such as the distributor, starter and alternator, can be rebuilt as well. The end result should be a like new engine that will give many trouble free miles.

Before beginning the engine overhaul, read through the entire procedure to familiarize yourself with the scope and requirements of the job. Overhauling an engine is not difficult, but it is time consuming. Plan on the vehicle being tied up for a minimum of two weeks, especially if parts must be taken to an automotive machine shop for repair or reconditioning. Check on availability of parts and make sure that any necessary special tools

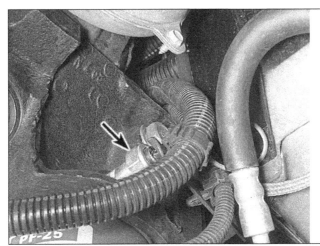

3.4 The oil pressure can be checked - after raising the vehicle and supporting it safely on jackstands - by removing the oil temperature sending unit (arrow) and installing a pressure gauge in the hole

4.3 Before cranking the engine when checking the compression, disconnect the wire from the BAT terminal on the distributor cap (the connector is out of sight and difficult to reach - the terminal is shown here with the distributor cap removed to clarify its location)

4.4 A compression gauge with a threaded fitting for the plug hole is preferred over the type that requires hand pressure to maintain the seal at the plug hole

and equipment are obtained in advance. Most work can be done with typical hand tools, although a number of precision measuring tools are required for inspecting parts to determine if they must be replaced. Often an automotive machine shop will handle the inspection of parts and offer advice concerning reconditioning and replacement. **Note:** *Always wait until the engine has been completely disassembled and all components, especially the engine block, have been inspected before deciding what service and repair operations must be performed by an automotive machine shop. Since the block's condition will be the major factor to consider when determining whether to overhaul the original engine or buy a rebuilt one, never purchase parts or have machine work done on other components until the block has been thoroughly inspected. As a general rule, time is the primary cost of an overhaul, so it does not pay to install worn or substandard parts.*

As a final note, to ensure maximum life and minimum trouble from a rebuilt engine, everything must be assembled with care in a spotlessly clean environment.

4   Cylinder compression check

Refer to illustrations 4.3 and 4.4

1   A compression check will tell you what mechanical condition the upper end (pistons, rings, valves, head gaskets) of your engine is in. Specifically, it can tell you if the compression is down due to leakage caused by worn piston rings, defective valves and seats or a blown head gasket. **Note:** *The engine must be at normal operating temperature for this check and the battery must be fully charged.*

2   Begin by cleaning the area around the spark plugs before you remove them (compressed air works best for this). This will prevent dirt from getting into the cylinders as the compression check is being done. Remove all of the spark plugs from the engine.

3   Block the throttle wide open and dis-

connect the wire from the BAT terminal on the distributor cap **(see illustration)**. Also, unplug the four wire distributor harness connector. These connectors are difficult to undo, since the BAT terminal on the cap is hidden by part of the firewall and the four wire connector is under the distributor - both are difficult to reach.

4   With the compression gauge in the number one spark plug hole **(see illustration)**, crank the engine over at least four compression strokes and watch the gauge. The compression should build up quickly in a healthy engine. Low compression on the first stroke, followed by gradually increasing pressure on successive strokes, indicates worn piston rings. A low compression reading on the first stroke, which does not build up during successive strokes, indicates leaking valves or a blown head gasket (a cracked head could also be the cause). Record the highest gauge reading obtained.

5   Repeat the procedure for the remaining cylinders and compare the results to the Specifications.

6   Add some engine oil (about three squirts from a plunger-type oil can) to each cylinder, through the spark plug hole, and repeat the test.

7   If the compression increases after the oil is added, the piston rings are definitely worn. If the compression does not increase significantly, the leakage is occurring at the valves or head gasket. Leakage past the valves may be caused by burned valve seats and/or faces or warped, cracked or bent valves.

8   If two adjacent cylinders have equally low compression, there is a strong possibility that the head gasket between them is blown. The appearance of coolant in the combustion chambers or the crankcase would verify this condition.

9   If the compression is unusually high, the combustion chambers are probably coated with carbon deposits. If that is the case, the cylinder heads should be removed and decarbonized.

10   If compression is way down or varies greatly between cylinders, it would be a good idea to have a leak-down test performed by an automotive repair shop. This test will pinpoint exactly where the leakage is occurring and how severe it is.

5   Engine rebuilding alternatives

The do-it-yourselfer is faced with a number of options when performing an engine overhaul. The decision to replace the engine block, piston/connecting rod assemblies and crankshaft depends on a number of factors, with the number one consideration being the condition of the block. Other considerations are cost, access to machine shop facilities, parts availability, time required to complete the project and the extent of prior mechanical experience on the part of the do-it-yourselfer.

Some of the rebuilding alternatives include:

Individual parts - If the inspection procedures reveal that the engine block and most engine components are in reusable condition, purchasing individual parts may be the most economical alternative. The block, crankshaft and piston/connecting rod assemblies should all be inspected carefully. Even if the block shows little wear, the cylinder bores should be surface honed.

Crankshaft kit - This rebuild package consists of a reground crankshaft and a matched set of pistons and connecting rods. The pistons will already be installed on the connecting rods. Piston rings and the necessary bearings will be included in the kit. These kits are commonly available for standard cylinder bores, as well as for engine blocks that have been bored to a regular oversize.

Short block - A short block consists of an engine block with a crankshaft and piston/connecting rod assemblies already installed. All new bearings are incorporated and all clearances will be correct. The exist-

6.36 When removing the engine, all hoses, cables and wires, as well as several components, must be detached - they include, but are not limited to, the items shown here

1 Throttle/cruise control cables	6 Battery cables (positive cable hidden from view)	10 Power steering reservoir
2 Fuel injector wire harness left side shown)	7 Alternator	11 Drivebelt pulley/vibration damper
3 Distributor	8 Drivebelt	12 Upper radiator hose
4 Windshield wiper motor	9 Power steering pump	13 Intake duct/MAF sensor
5 Brake booster vacuum hose		14 AIR switching valve

ing camshaft, valve train components, cylinder heads and external parts can be bolted to the short block with little or no machine shop work necessary.

Long block - A long block consists of a short block plus an oil pump, oil pan, cylinder heads, rocker arm covers, camshaft and valve train components, timing sprockets and chain and timing cover. All components are installed with new bearings, seals and gaskets incorporated throughout. The installation of manifolds and external parts is all that is necessary.

Give careful thought to which alternative is best for you and discuss the situation with local automotive machine shops, auto parts dealers or parts store countermen before ordering or purchasing replacement parts.

6   Engine - removal and installation

Refer to illustrations 6.36, 6.52 and 6.66
Warning 1: *Gasoline is extremely flammable, so take extra precautions when you work on any part of the fuel system. Don't smoke or allow open flames or bare light bulbs near the*

work area, and don't work in a garage where a natural gas-type appliance (such as a water heater or clothes dryer) with a pilot light is present. Since gasoline is carcinogenic, wear latex gloves when there's a possibility of being exposed to fuel, and, if you spill any fuel on your skin, rinse it off immediately with soap and water. Mop up any spills immediately and do not store fuel-soaked rags where they could ignite. The fuel system on fuel-injected models is under constant pressure, so, if any fuel lines are to be disconnected, the fuel pressure in the system must be relieved first (see Chapter 4 for more information). When you perform any kind of work on the fuel system, wear safety glasses and have a Class B type fire extinguisher on hand.*
Warning 2: *The air conditioning system is under high pressure! Have a dealer service department or service station discharge the system before disconnecting any air conditioning system hoses or fittings.*
Warning 3: *Some of the models covered by this manual are equipped with Supplemental Inflatable Restraints (SIR), more commonly known as airbags. Always disable the airbag system before working in the vicinity of any*

airbag system components to avoid the possibility of accidental deployment of the airbag, which could cause personal injury (see Chapter 12).

1984 only

1   Disconnect the negative battery cable from the battery, then refer to Chapter 1, drain the cooling system and remove the drivebelt from the front of the engine.
2   Remove the air cleaner assembly. Refer to Chapter 11 and remove the hood.
3   Detach the AIR system hoses at the exhaust manifold check valves.
4   Refer to Chapter 3 and remove the air conditioning compressor and the idler pulley.
5   Refer to Chapter 6 and remove the AIR pump.
6   Refer to Chapter 5 and remove the alternator.
7   Refer to Chapter 10 and remove the power steering pump. The upper radiator hose must be disconnected from the thermostat housing and the power steering reservoir brace. The AIR system line must be detached from the intake manifold and the power steering reservoir brace.

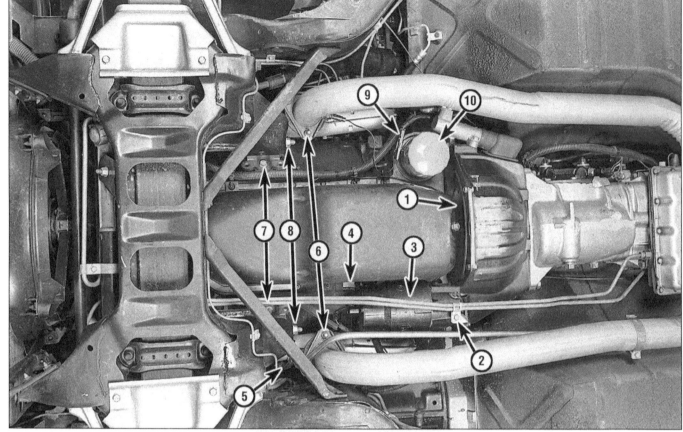

6.52 Several items on the underside of the vehicle must be disconnected/detached prior to lifting the engine out of the vehicle - they include, but are not limited to, the components shown here

1   Flywheel/driveplate cover
2   Transmission oil cooler line bracket
3   Starter motor
4   ESC (knock) sensor wire connector
5   Catalytic converter AIR pipe-to-manifold mount

6   Exhaust pipe-to-manifold junctions
7   Engine mount-to-block bolts (others hidden by mount)
8   Engine mount through bolts
9   Oxygen sensor wire connector
10 Engine oil filter

8   Refer to Chapter 4 and relieve the fuel pressure, then detach the fuel inlet and return lines from the TBI units. Separate the fuel line bracket from the front lower right side of the engine.
9   Separate the lower radiator hose and the heater hose from the water pump fittings.
10   Move the hoses and fuel lines aside, then disconnect the accelerator, TV and cruise control cables from the brackets and TBI unit (refer to Chapter 4 for more information).
11   Detach the brake vacuum hose and PCV system hose from the intake cover.
12   Disconnect the wires from the TBI units, the front ground stud, the coolant temperature sensor and the EGR solenoid. It would be a good idea to label all of the wires and terminals to ensure that the wires are reconnected correctly.
13   Label and detach all vacuum lines at the engine.
14   Detach the wire harness from the rocker arm cover, then move the air management valve, AIR pipe and wire harness to the right rear side of the engine.
15   Detach the tach filter and ground wires

from the intake cover stud.
16   Refer to Chapter 1 and remove the distributor cap and spark plug wires as an assembly.
17   Remove the heater hose from the intake manifold fitting.
18   Refer to Chapter 5 and remove the distributor.
19   Remove the oil pressure sending unit.
20   Remove the bolt and detach the wire bracket from the rear of the intake manifold.
21   Refer to Chapter 2A and remove the vibration damper from the front of the crankshaft.
22   Raise the vehicle and support it securely on jackstands, then detach the crossover pipe at the manifold and the converter. Separate the hanger from the crossover pipe.
23   Detach the converter AIR pipe from the right exhaust manifold.
24   Remove the starter wires and the knock sensor wire.
25   Disconnect the coolant temperature sensor wire at the head and the bracket on the block.
26   Detach the wire harness from the oil pan, the front of the block, the coolant tem-

perature sensor, the oil temperature sensor and the oxygen sensor.
27   Detach the battery and engine ground wires from the rear of the block, above the oil filter.
28   If your vehicle is equipped with an automatic transmission, remove the driveplate cover, then remove the torque converter mounting bolts. The converter will have to be turned to expose each of the bolts.
29   Support the engine with a hoist, then remove the bellhousing bolts from the right side (remove the lower and upper bolts to gain access to the center bolt).
30   Remove the bolts from the left side (including the ground wire).
31   Remove the engine mount-to-block bolts.
32   Position a jack under the transmission and raise it just enough to support it. A block of wood should be used to prevent the jack head from bearing against the transmission case or oil pan.
33   Pull the engine forward slightly, then detach the wires from the rear of the left-hand cylinder head.
34   Move the engine forward to detach it

6.66 Depending on the type of hoist you are using, it may be a good idea to remove the rocker arm covers and fuel injection components from the engine so they don't get damaged by the chain used to lift the engine out of the vehicle

from the transmission, then make one last check to be sure all wires, cables and hoses have been disconnected before carefully lifting it out of the engine compartment.

35   Installation can be accomplished by reversing the removal procedure.

1985 on

36   Disconnect the battery cables from the battery, then refer to Chapter 4 and detach the throttle and cruise control cables from the engine **(see illustration)**.

37   Refer to Chapter 5 and remove the distributor.

38   Refer to Chapter 12 and remove the wiper motor.

39   Remove the air intake duct and MAF sensor (if equipped).

40   Disconnect the brake booster vacuum hose and the PCV line-to-canister hose.

41   Label and disconnect the vacuum hoses and wires attached to the engine.

42   Disconnect the fuel injector wires and detach the harness from the engine.

43   Refer to Chapter 1 and drain the cooling system, then detach the radiator hoses and heater hoses from the engine fittings. Also, remove the drivebelt from the pulleys.

44   Refer to Chapter 4 and relieve the fuel pressure, then detach the fuel lines at the fuel rail fittings.

45   Disconnect the AIR switching valve at the air conditioning compressor.

46   Remove the catalytic converter AIR pipe.

47   Refer to Chapter 3 and remove the air conditioning compressor. Detach the accumulator at the fan shroud and brace.

48   Disconnect the fuel lines at the engine block fittings.

49   Refer to Chapter 5 and remove the alternator.

50   Refer to Chapter 10 and remove the power steering pump. Detach the power steering reservoir at the fan shroud and brace.

51   Remove the water pump pulley (if equipped), then refer to Chapter 2A and remove the vibration damper from the end of the crankshaft.

52   Raise the vehicle and support it securely on jackstands, then disconnect the wires from the oxygen sensor, the ESC sensor and the coolant temperature sensor **(see illustration)**.

53   Detach the temperature sensor and ground wires from the engine block.

54   Detach the catalytic converter AIR pipe from the right-hand exhaust manifold (if equipped).

55   Detach the transmission oil cooler lines from the flywheel cover bracket.

56   Refer to Chapter 5 and remove the

starter motor.

57   Refer to Chapter 1, drain the engine oil and remove the oil filter.

58   Detach the exhaust pipes from the exhaust manifolds and the converter hanger (refer to Chapter 4).

59   Detach the oil cooler adapter from the block and the line from the oil pan.

60   Remove the flywheel/driveplate cover.

61   If your vehicle is equipped with an automatic transmission, remove the torque converter-to-driveplate bolts. The converter will have to be turned to expose the upper bolts.

62   Support the engine with a hoist, then remove the bellhousing bolts.

63   Remove the engine mount through bolts.

64   Remove the engine mount-to-block bolts.

65   Position a jack under the transmission and raise it just enough to support it. A block of wood should be used to prevent the jack head from bearing against the transmission case or oil pan.

66   Move the engine forward to detach it from the transmission, then make one last check to be sure all wires, cables and hoses have been disconnected before carefully lifting it out of the engine compartment **(see illustration)**.

67   Installation can be accomplished by reversing the removal procedure.

8.2 A small plastic bag, with an appropriate label, can be used to store the valve train components so they can be kept together and reinstalled in the correct guide

8.3a Use a valve spring compressor to compress the spring, then remove the keepers from the valve stem

7   Engine overhaul - disassembly sequence

1   It is much easier to disassemble and work on the engine if it is mounted on a portable engine stand. These stands can often be rented quite cheaply from an equipment rental yard. Before the engine is mounted on a stand, the flywheel/driveplate should be removed from the engine (refer to Chapter 8).

2   If a stand is not available, it is possible to disassemble the engine with it blocked up on a sturdy workbench or on the floor. Be extra careful not to tip or drop the engine when working without a stand.

3   If you are going to obtain a rebuilt engine, all external components must come off first, to be transferred to the replacement engine, just as they will if you are doing a complete engine overhaul yourself. These include:

Alternator and brackets
Emissions control components
Distributor, spark plug wires and spark plugs
Thermostat and housing cover
Water pump
EFI components
Intake/exhaust manifolds
Oil filter
Engine mounts
Clutch and flywheel/driveplate

   Note: *When removing the external components from the engine, pay close attention to details that may be helpful or important during installation. Note the installed position of gaskets, seals, spacers, pins, washers, bolts and other small items.*

4   If you are obtaining a short block, which consists of the engine block, crankshaft, pistons and connecting rods all assembled, then the cylinder heads, oil pan and oil pump will have to be removed as well. See Engine rebuilding alternatives for additional informa-

tion regarding the different possibilities to be considered.

5   If you are planning a complete overhaul, the engine must be disassembled and the internal components removed in the following order:

Rocker arm covers
Exhaust and intake manifolds
Rocker arms and pushrods
Valve lifters
Cylinder heads
Timing chain cover
Timing chain and sprockets
Camshaft
Oil pan
Oil pump
Piston/connecting rod assemblies
Crankshaft and main bearings

6   Before beginning the disassembly and overhaul procedures, make sure the following items are available:

Common hand tools
Small cardboard boxes or plastic bags for storing parts
Gasket scraper
Ridge reamer
Vibration damper puller
Micrometers
Telescoping gauges
Dial indicator set
Valve spring compressor
Cylinder surfacing hone
Piston ring groove cleaning tool
Electric drill motor
Tap and die set
Wire brushes
Oil gallery brushes
Cleaning solvent

8   Cylinder head - disassembly

Refer to illustrations 8.2, 8.3a and 8.3b
Note: *New and rebuilt cylinder heads are commonly available for most engines at dealerships and auto parts stores. Due to the fact*

that some specialized tools are necessary for the disassembly and inspection procedures, and replacement parts may not be readily available, it may be more practical and economical for the home mechanic to purchase replacement heads rather than taking the time to disassemble, inspect and recondition the originals.

1   Cylinder head disassembly involves removal of the intake and exhaust valves and related components. If they are still in place, remove the rocker arm nuts, pivot balls and rocker arms from the cylinder head studs. Label the parts or store them separately so they can be reinstalled in their original locations.

2   Before the valves are removed, arrange to label and store them, along with their related components, so they can be kept separate and reinstalled in the same valve guides they are removed from **(see illustration)**.

3   Compress the springs on the first valve with a spring compressor and remove the keepers **(see illustration)**. Carefully release the valve spring compressor and remove the retainer (exhaust valves have rotators), the shield, the springs and the spring seat. Next, remove the O-ring seal from the upper end of the valve stem (just under the keeper groove) and the umbrella-type seal from the guide, then pull the valve from the head. If the valve binds in the guide (won't pull through), push it back into the head and deburr the area around the keeper groove with a fine file or whetstone **(see illustration)**.

4   Repeat the procedure for the remaining valves. Remember to keep all the parts for each valve together so they can be reinstalled in the same locations.

5   Once the valves and related components have been removed and stored in an organized manner, the head should be thoroughly cleaned and inspected. If a complete engine overhaul is being done, finish the engine disassembly procedures before beginning the cylinder head cleaning and inspection process.

8.3b If the valve won't pull through the guide, deburr the edge of the stem end and the area around the top of the keeper groove with a file

9.12 Check the cylinder head gasket surface for warpage by trying to slip a feeler gauge under the straightedge (see the Specifications for the maximum warpage allowed and use a feeler gauge of that thickness)

9 Cylinder head - cleaning and inspection

Refer to illustrations 9.12, 9.14, 9.16, 9.17, 9.18 and 9.19

1   Thorough cleaning of the cylinder heads and related valve train components, followed by a detailed inspection, will enable you to decide how much valve service work must be done during the engine overhaul.

Cleaning

2   Scrape away all traces of old gasket material and sealing compound from the head gasket, intake manifold and exhaust manifold sealing surfaces. Be very careful not to gouge the cylinder head, particularly if it is made of aluminum. Special gasket removal solvents, which soften gaskets and make removal much easier, are available at auto parts stores.
3   Remove any built up scale from the coolant passages.
4   Run a stiff wire brush through the various holes to remove any deposits that may have formed in them.
5   Run an appropriate size tap into each of the threaded holes to remove any corrosion and thread sealant that may be present. If compressed air is available, use it to clear the holes of debris produced by this operation.
6   Clean the rocker arm pivot stud threads with a wire brush.
7   Clean the cylinder head with solvent and dry it thoroughly. Compressed air will speed the drying process and ensure that all holes and recessed areas are clean. **Note:** *Decarbonizing chemicals are available and may prove very useful when cleaning cylinder heads and valve train components. They are very caustic and should be used with caution. Be sure to follow the instructions on the container.*
8   Clean the rocker arms, pivot balls, nuts and pushrods with solvent and dry them thor-

9.14 A dial indicator can be used to determine the valve stem-to-guide clearance (move the valve stem as indicated by the arrows)

oughly (don't mix them up during the cleaning process). Compressed air will speed the drying process and can be used to clean out the oil passages.
9   Clean all the valve springs, shields, keepers and retainers (or rotators) with solvent and dry them thoroughly. Do the components from one valve at a time to avoid mixing up the parts.
10   Scrape off any heavy deposits that may have formed on the valves, then use a motorized wire brush to remove deposits from the valve heads and stems. Again, make sure the valves do not get mixed up.

Inspection

Cylinder head

11   Inspect the head very carefully for cracks, evidence of coolant leakage and other damage. If cracks are found, a new cylinder head should be obtained.
12   Using a straightedge and feeler gauge, check the head gasket mating surface for warpage **(see illustration)**. If the warpage exceeds the specified limit, it can be resurfaced at an automotive machine shop. **Note:**

If the heads are resurfaced, the intake manifold flanges will also require machining.
13   Examine the valve seats in each of the combustion chambers. If they are pitted, cracked or burned, the head will require valve service that is beyond the scope of the home mechanic.
14   Check the valve stem-to-guide clearance by measuring the lateral movement of the valve stem with a dial indicator attached securely to the head **(see illustration)**. The valve must be in the guide and approximately 1/16-inch off the seat. The total valve movement indicated by the gauge needle must be divided by two to obtain the actual clearance. After this is done, if there is still some doubt regarding the condition of the valve guides they should be checked by an automotive machine shop (the cost should be minimal).

Valves

15   Carefully inspect each valve face for uneven wear, deformation, cracks, pits and burned spots. Check the valve stem for scuffing and galling and the neck for cracks. Rotate the valve and check for any obvious

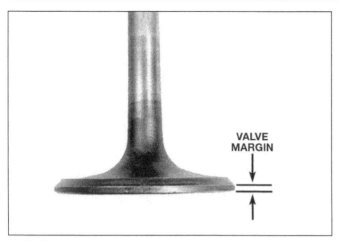

9.16 The margin width on each valve must be as specified (if no margin exists, the valve cannot be reused)

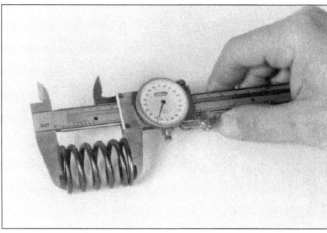

9.17 Measure the free length of each valve spring with a dial or vernier caliper

indication that it is bent. Look for pits and excessive wear on the end of the stem. The presence of any of these conditions indicates the need for valve service by an automotive machine shop.

16   Measure the margin width on each valve **(see illustration)**. Any valve with a margin narrower than 1/32-inch will have to be replaced with a new one.

Valve components

17   Check each valve spring for wear (on the ends) and pits. Measure the free length and compare it to the Specifications **(see illustration)**. Any springs that are shorter than specified have sagged and should not be reused. The tension of all springs should be checked with a special fixture before deciding that they are suitable for use in a rebuilt engine (take the springs to an automotive machine shop for this check).

18   Stand each spring on a flat surface and check it for squareness **(see illustration)**. If any of the springs are distorted or sagged, replace all of them with new parts.

19   Check the spring retainers (or rotators) and keepers for obvious wear and cracks. Any questionable parts should be replaced

with new ones, as extensive damage will occur if they fail during engine operation. Make sure the rotators operate smoothly with no binding or excessive play **(see illustration)**.

Rocker arm components

20   Check the rocker arm faces (the areas that contact the pushrod ends and valve stems) for pits, wear, galling, score marks and rough spots. Check the rocker arm pivot contact areas and pivot balls as well. Look for cracks in each rocker arm and nut.

21   Inspect the pushrod ends for scuffing and excessive wear. Roll each pushrod on a flat surface, such as a piece of plate glass, to determine if it is bent.

22   Check the rocker arm studs in the cylinder heads for damaged threads and secure installation.

23   Any damaged or excessively worn parts must be replaced with new ones.

24   If the inspection process indicates that the valve components are in generally poor condition and worn beyond the limits specified, which is usually the case in an engine that is being overhauled, reassemble the valves in the cylinder head and refer to Sec-

tion 10 for valve servicing recommendations.

25   If the inspection turns up no excessively worn parts, and if the valve faces and seats are in good condition, the valve train components can be reinstalled in the cylinder head without major servicing. Refer to the appropriate Section for the cylinder head reassembly procedure.

10 Valves - servicing

1   Because of the complex nature of the job and the special tools and equipment needed, servicing of the valves, the valve seats and the valve guides, commonly known as a valve job, is best left to a professional.

2   The home mechanic can remove and disassemble each head, do the initial cleaning and inspection, then reassemble and deliver the heads to a dealer service department or an automotive machine shop for the actual valve servicing.

3   The dealer service department, or automotive machine shop, will remove the valves and springs, recondition or replace the valves and valve seats, recondition the valve guides, check and replace the valve springs, spring

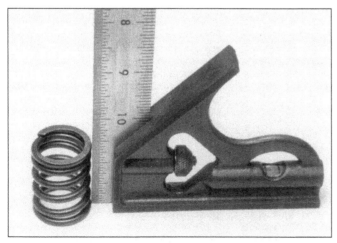

9.18 Check each valve spring for squareness

9.19 The exhaust valve rotators can be checked by turning the inner and outer sections in opposite directions to check for smooth movement and excessive play

11.6a Make sure the O-ring seal under the retainer is seated in the groove and not twisted before installing the keepers

11.6b Apply a small dab of grease to each keeper as shown here before installation - it will hold them in place on the valve stem as the spring is released

retainers or rotators and keepers (as necessary), replace the valve seals with new ones, reassemble the valve components and make sure the installed spring height is correct. The cylinder head gasket surface will also be resurfaced if it is warped.

4   After the valve job has been performed by a professional, the head will be in like new condition. When the head is returned, be sure to clean it again before installation on the engine to remove any metal particles and abrasive grit that may still be present from the valve service or head resurfacing operations. Use compressed air, if available, to blow out all the oil holes and passages.

11   Cylinder head - reassembly

Refer to illustrations 11.6a, 11.6b, 11.8 and 11.9

1   Regardless of whether or not the heads were sent to an automotive repair shop for valve servicing, make sure they are clean before beginning reassembly.

2   If the heads were sent out for valve servicing, the valves and related components will already be in place. Begin the reassembly procedure with Step 8.

3   Install new seals on each of the intake valve guides. Using a hammer and a deep socket, gently tap each seal into place until it is completely seated on the guide. Do not twist or cock the seals during installation or they will not seal properly on the valve stems. The umbrella-type seals are installed over the exhaust valves after the valves are in place.

4   Beginning at one end of the head, lubricate and install the first valve. Apply moly-base grease or clean engine oil to the valve stem.

5   Drop the spring seat or shim(s) over the valve guide and set the valve springs, shield and retainer (or rotator) in place.

6   Compress the springs with a valve spring compressor and carefully install the O-ring oil seal in the lower groove of the valve stem. Make sure the seal is not twisted - it must lie perfectly flat in the groove **(see illustration)**. Position the keepers in the upper groove, then slowly release the compressor and make sure the keepers seat properly. Apply a small dab of grease to each keeper to hold it in place if necessary **(see illustration)**.

7   Repeat the procedure for the remaining valves. Be sure to return the components to their original locations - do not mix them up!

8   Once all the valves are in place in both

heads, the valve stem O-ring seals must be checked to make sure they do not leak. This procedure requires a vacuum pump and special adapter, so it may be a good idea to have it done by a dealer service department, repair shop or automotive machine shop. The adapter is positioned on each valve retainer or rotator and vacuum is applied with the hand pump **(see illustration)**. If the vacuum cannot be maintained, the seal is leaking and must be checked/replaced before the head is installed on the engine.

9   Check the installed valve spring height with a ruler graduated in 1/32-inch increments or a dial caliper. If the heads were sent out for service work, the installed height should be correct (but don' t automatically assume that it is). The measurement is taken from the top of each spring seat or shim(s) to the top of the oil shield (or the bottom of the retainer/rotator, the two points are the same) **(see illustration)**. If the height is greater than specified, shims can be added under the springs to correct it. **Caution:** *Do not, under any circumstances, shim the springs to the point where the installed height is less than specified.*

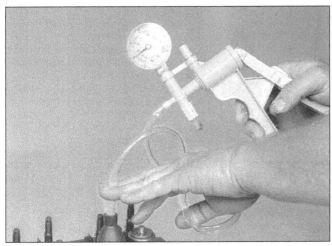

11.8 A special adapter and a vacuum pump are required to check the O-ring valve stem seals for leaks

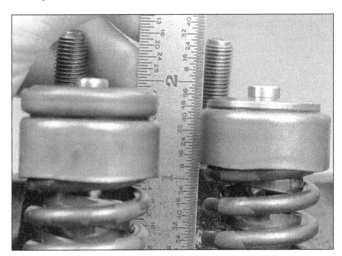

11.9 Be sure to check the valve spring installed height (the distance from the top of the seat/shims to the top of the shield)

12.1 A ridge reamer is required to remove the ridge from the top of the cylinder - do this before removing the pistons!

12.3 Check the connecting rod side clearance with a feeler gauge as shown

10   Apply moly-base grease to the rocker arm faces and the pivot balls, then install the rocker arms and pivots on the cylinder head studs. Thread the nuts on three or four turns only (when the heads are installed, the nuts will be tightened following a specific procedure).

12 Piston/connecting rod assembly - removal

Refer to illustrations 12.1, 12.3 and 12.5
Note: *Prior to removing the piston/connecting rod assemblies, remove the cylinder heads, the oil pan and the oil pump by referring to the appropriate Sections in Chapter 2, Part A.*
1   Completely remove the ridge at the top of each cylinder with a ridge reaming tool **(see illustration)**. Follow the manufacturer's instructions provided with the tool. Failure to remove the ridge before attempting to remove the piston/connecting rod assemblies will result in piston breakage.
2   After the cylinder ridges have been removed, turn the engine upside-down so the crankshaft is facing up.
3   Before the connecting rods are removed, check the end play with feeler gauges. Slide them between the first connecting rod and the crankshaft throw until the play is removed **(see illustration)**. The end play is equal to the thickness of the feeler gauge(s). If the end play exceeds the service limit, new connecting rods will be required. If new rods (or a new crankshaft) are installed, the end play may fall under the specified minimum (if it does, the rods will have to be machined to restore it - consult an automotive machine shop for advice if necessary). Repeat the procedure for the remaining connecting rods.
4   Check the connecting rods and caps for identification marks. If they are not plainly marked, use a small center punch to make

12.5 To prevent damage to the crankshaft journals and cylinder walls, slip sections of hose over the rod bolts before removing the pistons

the appropriate number of indentations on each rod and cap (1 - 8, depending on the cylinder they are associated with).
5   Loosen each of the connecting rod cap nuts 1/2-turn at a time until they can be removed by hand. Remove the number one connecting rod cap and bearing insert. Do not drop the bearing insert out of the cap. Slip a short length of plastic or rubber hose over each connecting rod cap bolt to protect the crankshaft journal and cylinder wall when the piston is removed **(see illustration)**. Push the connecting rod/piston assembly out through the top of the engine. Use a wooden hammer handle to push on the upper bearing insert in the connecting rod. If resistance is felt, double-check to make sure that all of the ridge was removed from the cylinder.
6   Repeat the procedure for the remaining cylinders. After removal, reassemble the connecting rod caps and bearing inserts in their respective connecting rods and install the cap nuts finger tight. Leaving the old bearing inserts in place until reassembly will help prevent the connecting rod bearing surfaces from being accidentally nicked or gouged.

13 Crankshaft - removal

Refer to illustrations 13.1, 13.3, 13.4a, 13.4b and 13.4c
Note: *The crankshaft can be removed only after the engine has been removed from the vehicle. It is assumed that the flywheel or driveplate, vibration damper, timing chain, oil pan, oil pump and piston/connecting rod assemblies have already been removed. If your engine is equipped with a one-piece rear main oil seal (first used on 1986 models), the seal housing must be unbolted and separated from the block before proceeding with crankshaft removal.*
1   Before the crankshaft is removed, check the end play. Mount a dial indicator with the stem in line with the crankshaft and just touching one of the crank throws **(see illustration)**.
2   Push the crankshaft all the way to the rear and zero the dial indicator. Next, pry the crankshaft to the front as far as possible and check the reading on the dial indicator. The distance that it moves is the end play. If it is

13.1 Checking crankshaft end play with a dial indicator

13.3 Checking crankshaft end play with a feeler gauge

13.4a Use a center punch or number stamping dies to mark the main bearing caps to ensure that they are reinstalled in their original locations on the block (make the punch marks near one of the bolt heads)

13.4b Mark the caps in order from the front of the engine to the rear (one mark for the front cap, two for the second one and so on) - the rear cap doesn't have to be marked since it cannot be installed in any other location

greater than specified, check the crankshaft thrust surfaces for wear. If no wear is evident, new main bearings should correct the end play.

3   If a dial indicator is not available, feeler gauges can be used. Gently pry or push the crankshaft all the way to the front of the engine. Slip feeler gauges between the crankshaft and the front face of the thrust main bearing to determine the clearance **(see illustration)**.

4   Check the main bearing caps to see if they are marked to indicate their locations. They should be numbered consecutively from the front of the engine to the rear. If they aren't, mark them with number stamping dies or a center punch **(see illustrations)**. Main bearing caps generally have a cast-in arrow, which points to the front of the engine **(see illustration)**. Loosen each of the main bearing cap bolts 1/4-turn at a time each, until they can be removed by hand.

5   Gently tap the caps with a soft-face hammer, then separate them from the engine

13.4c The arrow on the main bearing cap indicates the front of the engine

block. If necessary, use the bolts as levers to remove the caps. Try not to drop the bearing inserts if they come out with the caps.

6   Carefully lift the crankshaft out of the engine. It is a good idea to have an assistant available, since the crankshaft is quite heavy. With the bearing inserts in place in the engine block and main bearing caps, return the caps to their respective locations on the engine block and tighten the bolts finger tight.

14.1a A hammer and large punch can be used to drive the soft plugs into the block

14.1b Use Channelock pliers to lever each soft plug through the hole in the block

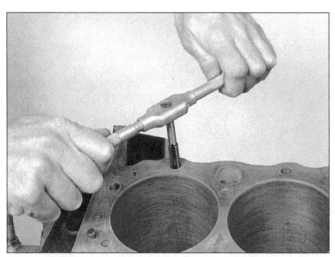

14.8 All bolt holes in the block - particularly the main bearing cap and head bolt holes - should be cleaned and restored with a tap (be sure to remove debris from the holes after this is done)

14.10 A large socket on an extension can be used to drive the new soft plugs into the bores

14   Engine block - cleaning

Refer to illustrations 14.1a, 14.1b, 14.8 and 14.10

1   Remove the soft plugs from the engine block. To do this, knock the plugs into the block, using a hammer and punch, then grasp them with large pliers and pull them back through the holes **(see illustrations)**.

2   Using a gasket scraper, remove all traces of gasket material from the engine block. Be very careful not to nick or gouge the gasket sealing surfaces.

3   Remove the main bearing caps and separate the bearing inserts from the caps and the engine block. Tag the bearings, indicating which cylinder they were removed from and whether they were in the cap or the block, then set them aside.

4   Using a 1/4-inch drive breaker bar or ratchet, remove all of the threaded oil gallery plugs from the rear of the block. The plugs are very tight - they may have to be drilled out and the holes retapped. Discard the plugs and use new ones when the engine is reassembled.

5   If the engine is extremely dirty it should be taken to an automotive machine shop to be steam cleaned or hot tanked.

6   After the block is returned, clean all oil holes and oil galleries one more time. Brushes specifically designed for this purpose are available at most auto parts stores. Flush the passages with warm water until the water runs clear, dry the block thoroughly and wipe all machined surfaces with a light, rust preventative oil. If you have access to compressed air, use it to speed the drying process and to blow out all the oil holes and galleries.

7   If the block is not extremely dirty or sludged up, you can do an adequate cleaning job with warm soapy water and a stiff brush. Take plenty of time and do a thorough job. Regardless of the cleaning method used, be sure to clean all oil holes and galleries very thoroughly, dry the block completely and

coat all machined surfaces with light oil.

8   The threaded holes in the block must be clean to ensure accurate torque readings during reassembly. Run the proper size tap into each of the holes to remove any rust, corrosion, thread sealant or sludge and to restore any damaged threads **(see illustration)**. If possible, use compressed air to clear the holes of debris produced by this operation. Now is a good time to clean the threads on the head bolts and the main bearing cap bolts as well.

9   Reinstall the main bearing caps and tighten the bolts finger tight.

10   After coating the sealing surfaces of the new soft plugs with RTV sealant, install them in the engine block **(see illustration)**. Make sure they are driven in straight and seated properly or leakage could result. Special tools are available for this purpose, but equally good results can be obtained using a large socket, with an outside diameter that will just slip into the soft plug, and a hammer.

11   Apply RTV sealant or Teflon sealing tape

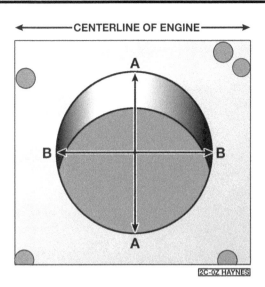

15.4a Measure the diameter of each cylinder at a right angle to the engine centerline (A), and parallel to the engine centerline (B) - out-of-round is the difference between A and B; taper is the difference between A and B at the top of the cylinder and A and B at the bottom of the cylinder

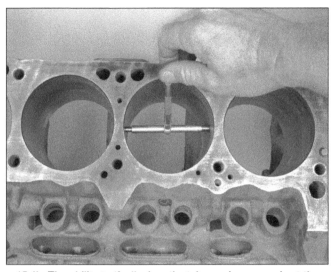

15.4b The ability to 'feel' when the telescoping gauge is at the correct point will be developed over time, so work slowly and repeat the check until you are satisfied that the bore measurement is accurate

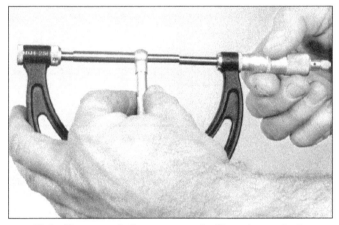

15.4c The gauge is then measured with a micrometer to determine the bore size

16.3a A 'bottle brush' hone will produce better results if you have never done cylinder honing before

to the new oil gallery plugs and thread them into the holes at the rear of the block. Make sure they are tightened securely.
12   If the engine is not going to be reassembled right away, cover it with a large plastic trash bag to keep it clean.

15   Engine block - inspection

Refer to illustrations 15.4a, 15.4b and 15.4c
1   Before the block is inspected, it should be cleaned as described in Section 14. Double-check to make sure that the ridge at the top of each cylinder has been completely removed.
2   Visually check the block for cracks, rust and corrosion. Look for stripped threads in the threaded holes. It is also a good idea to have the block checked for hidden cracks by an automotive machine shop that has the

special equipment to do this type of work. If defects are found, have the block repaired, if possible, or replaced.
3   Check the cylinder bores for scuffing and scoring.
4   Measure the diameter of each cylinder at the top (just under the ridge area), center and bottom of the cylinder bore, parallel to the crankshaft axis **(see illustrations)**. Next, measure each cylinder's diameter at the same three locations across the crankshaft axis. Compare the results to the Specifications. If the cylinder walls are badly scuffed or scored, or if they are out-of-round or tapered beyond the limits given in the Specifications, have the engine block rebored and honed at an automotive machine shop. If a rebore is done, oversize pistons and rings will be required.
5   If the cylinders are in reasonably good condition and not worn to the outside of the limits, and if the piston-to-cylinder clearances

can be maintained properly, then they do not have to be rebored. Honing is all that is necessary (Section 16).

16   Cylinder honing

Refer to illustrations 16.3a and 16.3b
1   Prior to engine reassembly, the cylinder bores must be honed so the new piston rings will seat correctly and provide the best possible combustion chamber seal. **Note:** *If you do not have the tools or do not want to tackle the honing operation, most automotive machine shops will do it for a reasonable fee.*
2   Before honing the cylinders, install the main bearing caps and tighten the bolts to the specified torque.
3   Two types of cylinder hones are commonly available - the flex hone or 'bottle brush' type and the more traditional surfacing

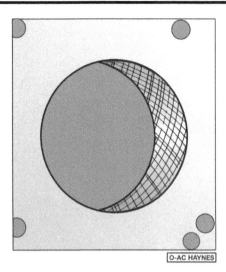

16.3b The cylinder hone should leave a smooth, crosshatch pattern with the lines intersecting at approximately a 60-degree angle

17.4a The piston ring grooves can be cleaned with a special tool, as shown here . . .

hone with spring-loaded stones. Both will do the job, but for the less experienced mechanic the 'bottle brush' hone will probably be easier to use. You will also need plenty of light oil or honing oil, some rags and an electric drill motor. Proceed as follows:

a) *Mount the hone in the drill motor, compress the stones and slip it into the first cylinder* **(see illustration)**.

b) *Lubricate the cylinder with plenty of oil, turn on the drill and move the hone up-and-down in the cylinder at a pace that will produce a fine crosshatch pattern on the cylinder walls. Ideally, the crosshatch lines should intersect at approximately a 60-degrees angle* **(see illustration)**. *Be sure to use plenty of lubricant and do not take off any more material than is absolutely necessary to produce the desired finish.*

c) *Do not withdraw the hone from the cylinder while it is running. Instead, shut off the drill and continue moving the hone up-and-down in the cylinder until it comes to a complete stop, then compress the stones and withdraw the hone.*

d) *Wipe the oil out of the cylinder and repeat the procedure for the remaining cylinders.*

4   After the honing job is complete, chamfer the top edges of the cylinder bores with a small file so the rings will not catch when the pistons are installed. Be very careful not to nick the cylinder walls with the end of the file.

5   The entire engine block must be washed again very thoroughly with warm, soapy water to remove all traces of the abrasive grit produced during the honing operation. Be sure to run a brush through all oil holes and galleries and flush them with running water.

6   After rinsing, dry the block and apply a coat of light rust preventive oil to all machined surfaces. Wrap the block in a plastic trash bag to keep it clean and set it aside until reassembly.

17.4b . . . or a section of a broken ring

17   Piston/connecting rod assembly - inspection

Refer to illustrations 17.4a, 17.4b, 17.10 and 17.11

1   Before the inspection process can be carried out, the piston/connecting rod assemblies must be cleaned and the original piston rings removed from the pistons. **Note:** *Always use new piston rings when the engine is reassembled.*

2   Using a piston ring installation tool, carefully remove the rings from the pistons. Be careful not to nick or gouge the pistons in the process.

3   Scrape all traces of carbon from the top (known as the crown) of the piston. A handheld wire brush or a piece of fine emery cloth can be used once the majority of the deposits have been scraped away. Do not, under any circumstances, use a wire brush mounted in a drill motor to remove deposits from the pistons. The piston material is soft and will be eroded away by the wire brush.

4   Use a piston ring groove cleaning tool to remove carbon deposits from the ring grooves. If a tool is not available, a piece broken off the old ring will do the job. Be very careful to remove only the carbon deposits - don't remove any metal and do not nick or scratch the sides of the ring grooves **(see illustrations)**.

5   Once the deposits have been removed, clean the piston/rod assemblies with solvent and dry them with compressed air (if available). Make sure that the oil return holes in the back sides of the ring grooves are clear.

6   If the pistons are not damaged or worn excessively, and if the engine block is not rebored, new pistons will not be necessary. Normal piston wear appears as even vertical wear on the piston thrust surfaces and slight looseness of the top ring in its groove. New piston rings, on the other hand, should always be used when an engine is rebuilt.

7   Carefully inspect each piston for cracks around the skirt, at the pin bosses and at the ring lands.

8   Look for scoring and scuffing on the thrust faces of the skirt, holes in the piston crown and burned areas at the edge of the

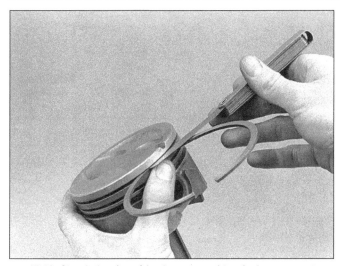

17.10 Check the ring side clearance with a feeler gauge at several points around the groove

17.11 Measure the piston diameter at a 90-degrees angle to the piston pin and in line with it

crown. If the skirt is scored or scuffed, the engine may have been suffering from overheating and/or abnormal combustion, which caused excessively high operating temperatures. The cooling and lubrication systems should be checked thoroughly. A hole in the piston crown is an indication that abnormal combustion (preignition) was occurring. Burned areas at the edge of the piston crown are usually evidence of spark knock (detonation). If any of the above problems exist, the causes must be corrected or the damage will occur again.

9   Corrosion of the piston, in the form of small pits, indicates that coolant is leaking into the combustion chamber and/or the crankcase. Again, the cause must be corrected or the problem may persist in the rebuilt engine.

10   Measure the piston ring side clearance by laying a new piston ring in each ring groove and slipping a feeler gauge in beside it **(see illustration)**. Check the clearance at three or four locations around each groove. Be sure to use the correct ring for each groove; they are different. If the side clearance is greater than specified, new pistons will have to be used.

11   Check the piston-to-bore clearance by measuring the bore (see Section 15) and the piston diameter. Make sure that the pistons and bores are correctly matched. Measure the piston across the skirt, at a 90-degrees angle to and in line with the piston pin **(see illustration)**. Subtract the piston diameter from the bore diameter to obtain the clearance. If it is greater than specified, the block will have to be rebored and new pistons and rings installed.

12   Check the piston-to-rod clearance by twisting the piston and rod in opposite directions. Any noticeable play indicates that there is excessive wear, which must be corrected. The piston/connecting rod assemblies should be taken to an automotive machine shop to have the pistons and rods rebored and new

pins installed.

13   If the pistons must be removed from the connecting rods for any reason, they should be taken to an automotive machine shop. While they are there have the connecting rods checked for bend and twist, since automotive machine shops have special equipment for this purpose. **Note:** *Unless new pistons and/or connecting rods must be installed, do not disassemble the pistons and connecting rods.*

14   Check the connecting rods for cracks and other damage. Temporarily remove the rod caps, lift out the old bearing inserts, wipe the rod and cap bearing surfaces clean and inspect them for nicks, gouges and scratches. After checking the rods, replace the old bearings, slip the caps into place and tighten the nuts finger tight.

---

18   Crankshaft - inspection

Refer to illustration 18.2

1   Clean the crankshaft with solvent and

dry it with compressed air (if available). Be sure to clean the oil holes with a stiff brush and flush them with solvent. Check the main and connecting rod bearing journals for uneven wear, scoring, pits and cracks. Check the rest of the crankshaft for cracks and other damage.

2   Using a micrometer, measure the diameter of the main and connecting rod journals and compare the results to the Specifications **(see illustration)**. By measuring the diameter at a number of points around each journal's circumference, you will be able to determine whether or not the journal is out-of-round. Take the measurement at each end of the journal, near the crank throws, to determine if the journal is tapered.

3   If the crankshaft journals are damaged, tapered, out-of-round or worn beyond the limits given in the Specifications, have the crankshaft reground by an automotive machine shop. Be sure to use the correct size bearing inserts if the crankshaft is reconditioned.

4   Refer to Section 19 and examine the main and rod bearing inserts.

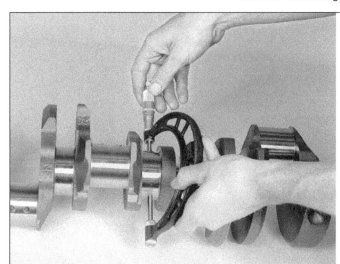

18.2 Measure the diameter of each crankshaft journal at several points to detect taper and out-of-round conditions

19   Main and connecting rod bearings - inspection

1   Even though the main and connecting rod bearings should be replaced with new ones during the engine overhaul, the old bearings should be retained for close examination, as they may reveal valuable information about the condition of the engine.

2   Bearing failure occurs because of lack of lubrication, the presence of dirt or other foreign particles, overloading the engine and corrosion. Regardless of the cause of bearing failure, it must be corrected before the engine is reassembled to prevent it from happening again.

3   When examining the bearings, remove them from the engine block, the main bearing caps, the connecting rods and the rod caps and lay them out on a clean surface in the same general position as their location in the engine. This will enable you to match any bearing problems with the corresponding crankshaft journal.

4   Dirt and other foreign particles get into the engine in a variety of ways. It may be left in the engine during assembly, or it may pass through filters or the PCV system. It may get into the oil, and from there into the bearings. Metal chips from machining operations and normal engine wear are often present. Abrasives are sometimes left in engine components after reconditioning, especially when parts are not thoroughly cleaned using the proper cleaning methods. Whatever the source, these foreign objects often end up embedded in the soft bearing material and are easily recognized. Large particles will not embed in the bearing and will score or gouge the bearing and journal. The best prevention for this cause of bearing failure is to clean all parts thoroughly and keep everything spotlessly clean during engine assembly. Frequent and regular engine oil and filter changes are also recommended.

5   Lack of lubrication (or lubrication breakdown) has a number of interrelated causes. Excessive heat (which thins the oil), overloading (which squeezes the oil from the bearing face) and oil leakage or throw off (from excessive bearing clearances, worn oil pump or high engine speeds) all contribute to lubrication breakdown. Blocked oil passages, which usually are the result of misaligned oil holes in a bearing shell, will also oil starve a bearing and destroy it. When lack of lubrication is the cause of bearing failure, the bearing material is wiped or extruded from the steel backing of the bearing. Temperatures may increase to the point where the steel backing turns blue from overheating.

6   Driving habits can have a definite effect on bearing life. Full throttle, low speed operation (lugging the engine) puts very high loads on bearings, which tends to squeeze out the oil film. These loads cause the bearings to flex, which produces fine cracks in the bearing face (fatigue failure). Eventually the bearing material will loosen in pieces and tear

away from the steel backing. Short trip driving leads to corrosion of bearings because insufficient engine heat is produced to drive off the condensed water and corrosive gases. These products collect in the engine oil, forming acid and sludge. As the oil is carried to the engine bearings, the acid attacks and corrodes the bearing material.

7   Incorrect bearing installation during engine assembly will lead to bearing failure as well. Tight fitting bearings leave insufficient bearing oil clearance and will result in oil starvation. Dirt or foreign particles trapped behind a bearing insert result in high spots on the bearing which lead to failure.

20   Engine overhaul - reassembly sequence

1   Before beginning engine reassembly, make sure you have all the necessary new parts, gaskets and seals as well as the following items on hand:

Common hand tools
A 1/2-inch drive torque wrench
Piston ring installation tool
Piston ring compressor
Short lengths of rubber or plastic hose to fit over connecting rod bolts
Plastigage
Feeler gauges
A fine-tooth file
New engine oil
Engine assembly lube or moly-base grease
RTV-type gasket sealant
Anaerobic-type gasket sealant
Thread locking compound

2   In order to save time and avoid problems, engine reassembly must be done in the following general order:

New camshaft bearings (must be done by automotive machine shop)
Piston rings
Crankshaft and main bearings
Piston/connecting rod assemblies
Oil pump and oil strainer
Camshaft and lifters

Cylinder heads pushrods and rocker arms
Timing chain and sprockets
Oil pan
Timing chain cover
Intake and exhaust manifolds
Rocker arm covers
Flywheel/driveplate

21   Piston rings - installation

Refer to illustrations 21.3, 21.4, 21.5, 21.9a, 21.9b and 21.12

1   Before installing the new piston rings, the ring end gaps must be checked. It is assumed that the piston ring side clearance has been checked and verified correct (Section 17).

2   Lay out the piston/connecting rod assemblies and the new ring sets so the ring sets will be matched with the same piston and cylinder during the end gap measurement and engine assembly.

3   Insert the top (number one) ring into the first cylinder and square it up with the cylinder walls by pushing it in with the top of the piston **(see illustration)**. The ring should be near the bottom of the cylinder, at the lower limit of ring travel.

4   To measure the end gap, slip feeler gauges between the ends of the ring until a gauge equal to the gap width is found **(see illustration)**. The feeler gauge should slide between the ring ends with a slight amount of drag. Compare the measurement to the Specifications. If the gap is larger or smaller than specified, double-check to make sure that you have the correct rings before proceeding.

5   If the gap is too small, it must be enlarged or the ring ends may come in contact with each other during engine operation, which can cause serious damage to the engine. The end gap can be increased by filing the ring ends very carefully with a fine file. Mount the file in a vise equipped with soft jaws, slip the ring over the file with the ends contacting the file face and slowly move the ring to remove material from the ends. When performing this operation, file only from the

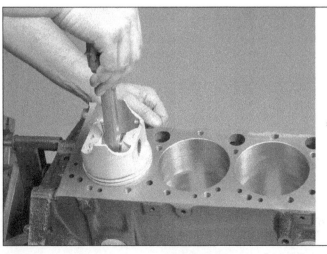

21.3 When checking piston ring end gap, the ring must be square in the cylinder bore (this is done by pushing the ring down with the top of a piston as shown)

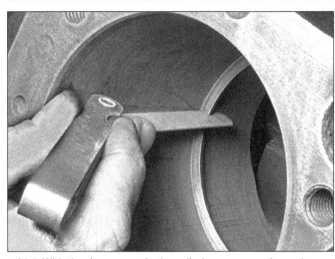

21.4 With the ring square in the cylinder, measure the end gap with a feeler gauge

21.5 If the end gap is too small, clamp a file in a vise and file the ring ends (from the outside in only) to enlarge the gap slightly

outside in **(see illustration)**.

6 Excess end gap is not critical unless it is greater than 0.040-inch. Again, double-check to make sure you have the correct rings for your engine.

7 Repeat the procedure for each ring that will be installed in the first cylinder and for each ring in the remaining cylinders. Remember to keep rings, pistons and cylinders matched up.

8 Once the ring end gaps have been checked/corrected, the rings can be installed on the pistons.

9 The oil control ring (lowest one on the piston) is installed first. It is composed of three separate components. Slip the spacer/expander into the groove **(see illustration)**. If an anti-rotation tang is used, make sure it is inserted into the drilled hole in the ring groove. Next, install the lower side rail. Do not use a piston ring installation tool on the oil ring side rails, as they may be damaged. Instead, place one end of the side rail into the groove between the spacer/expander and the ring land, hold it firmly in place and

slide a finger around the piston while pushing the rail into the groove **(see illustration)**. Next, install the upper side rail in the same manner.

10 After the three oil ring components have been installed, check to make sure that both the upper and lower side rails can be turned smoothly in the ring groove.

11 The number two (middle) ring is installed next. It is stamped with a mark that must face up, toward the top of the piston. **Note:** *Always follow the instructions printed on the ring package or box - different manufacturers may require different approaches. Do not mix up the top and middle rings, as they have different cross sections.*

12 Use a piston ring installation tool and make sure that the identification mark is facing the top of the piston, then slip the ring into the middle groove on the piston **(see illustration)**. Do not expand the ring any more than is necessary to slide it over the piston.

13 Install the number one (top) ring in the same manner. Make sure the mark is facing

21.9a Installing the spacer/expander in the oil control ring groove

up. Be careful not to confuse the number one and number two rings.

14 Repeat the procedure for the remaining pistons and rings.

21.9b DO NOT use a piston ring installation tool when installing the oil ring side rails

21.12 Installing the compression rings with a ring expander - the mark (arrow) must face up

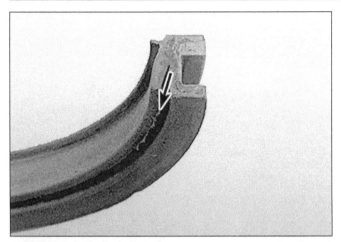

22.2 The rear main oil seal may have two lips - the oil seal (arrow) must point toward the front of the engine, which means that the dust seal will face out, toward the rear of the engine

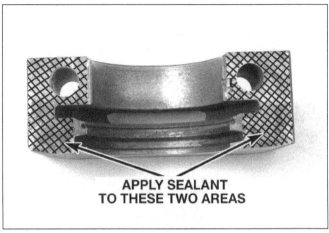

APPLY SEALANT TO THESE TWO AREAS

22.4 Before installing the rear main bearing cap, apply the specified sealant to the shaded areas of the block (or the equivalent areas on the cap)

22 Rear main oil seal installation

Refer to illustrations 22.2, 22.4 and 22.7 (additional illustrations can be found in Chapter 2, Part A)

1984 and 1985

1   Inspect the rear main bearing cap and engine block mating surfaces, as well as the seal grooves, for nicks, burrs and scratches. Remove any defects with a fine file or deburring tool.

2   Install one seal section in the block with the lip facing the front of the engine (if the seal has two lips, the one with the helix must face the front) **(see illustration)**. Leave one end protruding from the block approximately 1/4 to 3/8-inch and make sure it is completely seated.

3   Repeat the procedure to install the remaining seal half in the rear main bearing cap. In this case, leave the opposite end of the seal protruding from the cap the same distance the block seal is protruding from the block.

4   During final installation of the crankshaft (after the main bearing oil clearances have been checked with Plastigage) as described in Section 23, apply a thin, even coat of anaerobic-type gasket sealant to the areas of the cap or block indicated in the accompanying **illustration**. Do not get any sealant on the bearing face, crankshaft journal, seal ends or seal lips. Also, lubricate the seal lips with moly-base grease or clean engine oil.

1986 on

5   Later models are equipped with a one-piece seal that requires an entirely different installation procedure. The crankshaft must be installed first and the main bearing caps bolted in place, then the new seal should be installed in the housing and the housing bolted to the block.

6   Before installing the crankshaft, check the seal contact surface very carefully for

22.7 To remove the seal from the housing on 1986 and later models, insert the tip of the screwdriver into each of the three notches and lever the seal out

scratches and nicks that could damage the new seal lip and cause oil leaks. If the crankshaft is damaged, the only alternative is a new or different crankshaft.

7   The old seal can be removed from the housing by inserting a large screwdriver into the notches provided and prying it out **(see illustration)**. Be sure to note how far it is recessed into the housing bore before removing it; the new seal will have to be recessed an equal amount. Be very careful not to scratch or otherwise damage the bore in the housing or oil leaks could develop.

8   Make sure the housing is clean, then apply a thin coat of engine oil to the outer edge of the new seal. The seal must be pressed squarely into the housing bore, so hammering it into place is not recommended. If you do not have access to a press, sandwich the housing and seal between two smooth pieces of wood and press the seal into place with the jaws of a large vise. The pieces of wood must be thick enough to distribute the force evenly around the entire circumference of the seal. Work slowly and make sure the seal enters the bore squarely.

9   The seal lips must be lubricated with clean engine oil or moly-based grease before the seal/housing is slipped over the

crankshaft and bolted to the block. Use a new gasket - no sealant is required - and make sure the dowel pins are in place before installing the housing.

10   Tighten the nuts/screws a little at a time until they are all snug.

23 Crankshaft - installation and main bearing oil clearance check

Refer to illustrations 23.10 and 23.14

1   Crankshaft installation is the first step in engine reassembly. It is assumed at this point that the engine block and crankshaft have been cleaned, inspected and repaired or reconditioned. **Note:** *If your engine has a two-piece rear main oil seal (1984 and 1985 models), refer to Section 22 and install the oil seal sections in the cap and block before proceeding with crankshaft installation.*

2   Position the engine with the bottom facing up.

3   Remove the main bearing cap bolts and lift out the caps. Lay them out in the proper order to ensure that they are installed correctly.

4   If they are still in place, remove the old

23.10 Lay the Plastigage strips (arrow) on the main bearing journals, parallel to the crankshaft centerline

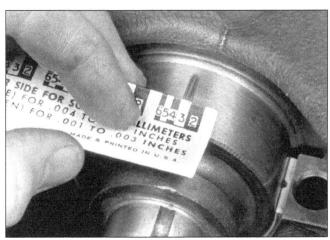

23.14 Compare the width of the crushed Plastigage to the scale on the container to determine the main bearing oil clearance (always take the measurement at the widest point of the Plastigage); be sure to use the correct scale - standard and metric scales are included

bearing inserts from the block and the main bearing caps. Wipe the main bearing surfaces of the block and caps with a clean, lint free cloth. They must be kept spotlessly clean.

5   Clean the backsides of the new main bearing inserts and lay one bearing half in each main bearing saddle in the block. Lay the other bearing half from each bearing set in the corresponding main bearing cap. Make sure the tab on the bearing insert fits into the recess in the block or cap. Also, the oil holes in the block must line up with the oil holes in the bearing insert. Do not hammer the bearing into place and do not nick or gouge the bearing faces. No lubrication should be used at this time.

6   The flanged thrust bearing must be installed in the rear cap and saddle.

7   Clean the faces of the bearings in the block and the crankshaft main bearing journals with a clean, lint free cloth. Check or clean the oil holes in the crankshaft, as any dirt here can go only one way - straight through the new bearings.

8   Once you are certain that the crankshaft is clean, carefully lay it in position (an assistant would be very helpful here) in the main bearings.

9   Before the crankshaft can be permanently installed, the main bearing oil clearance must be checked.

10   Trim several pieces of the appropriate size of Plastigage (they must be slightly shorter than the width of the main bearings) and place one piece on each crankshaft main bearing journal, parallel with the journal axis **(see illustration)**.

11   Clean the faces of the bearings in the caps and install the caps in their respective positions (do not mix them up) with the arrows pointing toward the front of the engine. Do not disturb the Plastigage.

12   Starting with the center main and working out toward the ends, tighten the main bearing cap bolts, in three steps, to the spec-

ified torque. Do not rotate the crankshaft at any time during this operation.

13   Remove the bolts and carefully lift off the main bearing caps. Keep them in order. Do not disturb the Plastigage or rotate the crankshaft. If any of the main bearing caps are difficult to remove, tap them gently from side-to-side with a soft-face hammer to loosen them.

14   Compare the width of the crushed Plastigage on each journal to the scale printed on the Plastigage container to obtain the main bearing oil clearance **(see illustration)**. Check the Specifications to make sure it is correct.

15   If the clearance is not as specified, the bearing inserts may be the wrong size (which means different ones will be required). Before deciding that different inserts are needed, make sure that no dirt or oil was between the bearing inserts and the caps or block when the clearance was measured. If the Plastigage was wider at one end than the other, the journal may be tapered (refer to Section 18).

16   Carefully scrape all traces of the Plastigage material off the main bearing journals and/or the bearing faces. Do not nick or scratch the bearing faces.

17   Carefully lift the crankshaft out of the engine. Clean the bearing faces in the block, then apply a thin, uniform layer of clean moly-base grease or engine assembly lube to each of the bearing surfaces. Be sure to coat the thrust faces as well as the journal face of the rear bearing.

18   If not already done, refer to Section 22 and install the rear main oil seal sections in the block and bearing cap. Lubricate the seal faces with moly-base grease, engine assembly lube or clean engine oil.

19   Make sure the crankshaft journals are clean, then lay the crankshaft back in place in the block. Clean the faces of the bearings in the caps, then apply lubricant to them. Install the caps in their respective positions with the

arrows pointing toward the front of the engine. Note that the rear cap must have a special sealant applied (see Section 22). Install the bolts.

20   Tighten all except the rear cap bolts (the one with the thrust bearing) to the specified torque (work from the center out and approach the final torque in three steps). Tighten the rear cap bolts to 10-to-12 ft-lbs. Tap the ends of the crankshaft forward and backward with a lead or brass hammer to line up the main bearing and crankshaft thrust surfaces. Retighten all main bearing cap bolts to the specified torque, starting with the center main and working out toward the ends.

21   On manual transmission equipped models, install a new pilot bearing in the end of the crankshaft (see Chapter 8).

22   Rotate the crankshaft a number of times by hand to check for any obvious binding.

23   The final step is to check the crankshaft end play with a feeler gauge or a dial indicator as described in Section 13. The end play should be correct if the crankshaft thrust faces are not worn or damaged and new bearings have been installed.

24   If your engine has a one-piece rear main oil seal, refer to Section 22 and install the new seal, then bolt the housing to the block.

---

24   Piston/connecting rod assembly - installation and rod bearing clearance check

---

Refer to illustrations 24.8, 24.9, 24.11 and 24.13

1   Before installing the piston/connecting rod assemblies the cylinder walls must be perfectly clean, the top edge of each cylinder must be chamfered, and the crankshaft must be in place.

2   Remove the connecting rod cap from the end of the number one connecting rod. Remove the old bearing inserts and wipe the

24.8 The notch in each piston must face the FRONT of the engine as the pistons are installed

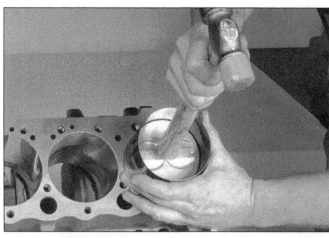

24.9 The piston can be driven (gently) into the cylinder bore with the end of a wooden hammer handle

24.11 Lay the Plastigage strips on each rod bearing journal, parallel to the crankshaft centerline

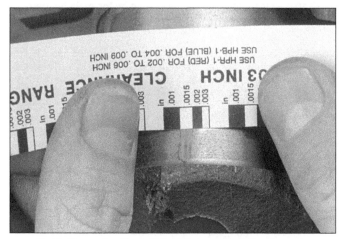

24.13 Measuring the width of the crushed Plastigage to determine the rod bearing oil clearance (be sure to use the correct scale - standard and metric scales are included)

bearing surfaces of the connecting rod and cap with a clean, lint free cloth. They must be kept spotlessly clean.

3   Clean the back side of the new upper bearing half, then lay it in place in the connecting rod. Make sure that the tab on the bearing fits into the recess in the rod. Do not hammer the bearing insert into place and be very careful not to nick or gouge the bearing face. Do not lubricate the bearing at this time.

4   Clean the backside of the other bearing insert and install it in the rod cap. Again, make sure the tab on the bearing fits into the recess in the cap, and do not apply any lubricant. It is critically important that the mating surfaces of the bearing and connecting rod are perfectly clean and oil free when they are assembled.

5   Position the piston ring gaps at 120-degrees intervals around the piston, then slip a section of plastic or rubber hose over each connecting rod cap bolt.

6   Lubricate the piston and rings with clean engine oil and attach a piston ring compressor to the piston. Leave the skirt protruding about 1/4-inch to guide the piston into the cylinder. The rings must be compressed until they are flush with the piston.

7   Rotate the crankshaft until the number one rod journal is at BDC (bottom dead center) and apply a coat of engine oil to the cylinder walls.

8   With the notch on top of the piston **(see illustration)** facing the front of the engine, gently insert the piston/connecting rod assembly into the number one cylinder bore and rest the bottom edge of the ring compressor on the engine block. Tap the top edge of the ring compressor to make sure it is contacting the block around its entire circumference.

9   Carefully tap on the top of the piston with the end of a wooden hammer handle **(see illustration)** while guiding the end of the connecting rod into place on the crankshaft journal. The piston rings may try to pop out of the ring compressor just before entering the cylinder bore, so keep some downward pressure on the ring compressor. Work slowly, and if any resistance is felt as the piston enters the cylinder, stop immediately. Find out what is hanging up and fix it before proceeding. Do not, for any reason, force the piston into the cylinder, as you will break a

ring and/or the piston.

10   Once the piston/connecting rod assembly is installed, the connecting rod bearing oil clearance must be checked before the rod cap is permanently bolted in place.

11   Cut a piece of the appropriate size Plastigage slightly shorter than the width of the connecting rod bearing and lay it in place on the number one connecting rod journal, parallel with the journal axis **(see illustration)**.

12   Clean the connecting rod cap bearing face, remove the protective hoses from the connecting rod bolts and install the rod cap. Make sure the mating mark on the cap is on the same side as the mark on the connecting rod. Install the nuts and tighten them to the specified torque, working up to it in three steps. **Note:** *Use a thin-wall socket to avoid erroneous torque readings that can result if the socket is wedged between the rod cap and nut. Do not rotate the crankshaft at any time during this operation.*

13   Remove the rod cap, being very careful not to disturb the Plastigage. Compare the width of the crushed Plastigage to the scale printed on the Plastigage container to obtain the oil clearance **(see illustration)**. Compare

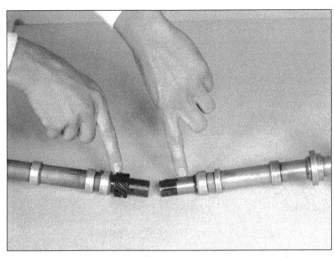

25.3 The pre-oil distributor (right) has the gear ground off and the advance weights (if equipped) removed

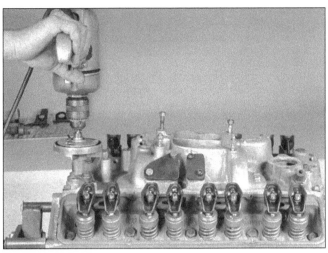

25.5 A drill motor connected to the modified distributor shaft drives the oil pump - make sure it turns clockwise as viewed from above

it to the Specifications to make sure the clearance is correct. If the clearance is not as specified, the bearing inserts may be the wrong size (which means different ones will be required). Before deciding that different inserts are needed, make sure that no dirt or oil was between the bearing inserts and the connecting rod or cap when the clearance was measured. Also, recheck the journal diameter. If the Plastigage was wider at one end than the other, the journal may be tapered (refer to Section 18).

14   Carefully scrape all traces of the Plastigage material off the rod journal and/or bearing face. Be very careful not to scratch the bearing face. Make sure the bearing faces are perfectly clean, then apply a uniform layer of clean moly-base grease or engine assembly lube to both of them. You will have to push the piston into the cylinder to expose the face of the bearing insert in the connecting rod - be sure to slip the protective hoses over the rod bolts first.

15   Slide the connecting rod back into place on the journal, remove the protective hoses from the rod cap bolts, install the rod cap and tighten the nuts to the specified torque. Again, work up to the torque in three steps.

16   Repeat the entire procedure for the remaining piston/connecting rod assemblies. Keep the back sides of the bearing inserts and the inside of the connecting rod and cap perfectly clean when assembling them. Make sure you have the correct piston for the cylinder and that the notch on the piston faces to the front of the engine when the piston is installed. Remember, use plenty of oil to lubricate the piston before installing the ring compressor. Also, when installing the rod caps for the final time, be sure to lubricate the bearing faces adequately.

17   After all the piston/connecting rod assemblies have been properly installed, rotate the crankshaft a number of times by hand to check for any obvious binding.

18   As a final step, the connecting rod end play must be checked. Refer to Section 12 for this procedure. Compare the measured end play to the Specifications to make sure it is correct. If it was correct before disassembly and the original crankshaft and rods were reinstalled, it should still be right. If new rods or a new crankshaft were installed, the end play may be too small. If so, the rods will have to be removed and taken to an automotive machine shop for resizing.

25   Pre-oiling engine after overhaul

Refer to illustrations 25.3, 25.5 and 25.6

1   After an overhaul it is a good idea to pre-oil the engine before it is installed in the vehicle and started for the first time. Pre-oiling will reveal any problems with the lubrication system at a time when corrections can be made easily and will prevent major engine damage. It will also allow the internal engine parts to be lubricated thoroughly in the normal fashion without the heavy loads associated with combustion placed on them.

2   The engine should be completely assembled with the exception of the distributor and rocker arm covers. The oil filter and oil pressure sending unit must be in place and the specified amount of oil must be in the crankcase (see Chapter 1).

3   A modified small block Chevrolet distributor will be needed for this procedure - a junkyard should be able to supply one for a reasonable price. In order to function as a pre-oil tool, the distributor must have the gear on the lower end of the shaft ground off **(see illustration)** and, if equipped, the advance weights on the upper end of the shaft removed.

4   Install the pre-oil distributor in place of the original distributor and make sure the lower end of the shaft mates with the upper end of the oil pump driveshaft. Turn the distributor shaft until they are aligned and the distributor body seats on the block. Install the distributor hold-down clamp and bolt.

5   Mount the upper end of the shaft in the chuck of an electric drill and use the drill to turn the pre-oil distributor shaft, which will drive the oil pump and circulate the oil throughout the engine **(see illustration)**. **Note:** *The drill must turn in a clockwise direction.*

6   It may take two or three minutes, but oil should soon start to flow out of all of the

25.6 Oil, assembly lube or grease will begin to flow from all of the rocker arm holes if the oil pump and lubrication system is functioning properly

rocker arm holes, indicating that the oil pump is working properly **(see illustration)**. Let the oil circulate for several seconds, then shut off the drill motor.

7   Remove the pre-oil distributor, then install the rocker arm covers. The distributor should be installed after the engine is installed in the vehicle, so plug the hole with a clean cloth.

26   Initial start-up and break-in after overhaul

1   Once the engine has been installed in the vehicle, double-check the engine oil and coolant levels.

2   With the spark plugs out of the engine and the pink ignition switch feed wire (connected to the BAT terminal on the distributor cap - see Section 4) disconnected, crank the engine until oil pressure registers on the gauge.

3   Install the spark plugs, hook up the plug wires and reconnect the ignition switch feed wire.

4   Start the engine. It may take a few moments for the gasoline to reach the injectors, but the engine should start without a great deal of effort.

5   After the engine starts, it should be allowed to warm up to normal operating temperature. While the engine is warming up, make a thorough check for fuel, oil and coolant leaks.

6   Shut the engine off and recheck the engine oil and coolant levels.

7   Drive the vehicle to an area with mini-mum traffic, accelerate from 30 to 50 mph, then allow the vehicle to slow to 30 mph with the throttle closed. Repeat the procedure 10 or 12 times. This will load the piston rings and cause them to seat properly against the cylinder walls. Check again for oil and coolant leaks.

8   Drive the vehicle gently for the first 500 miles (no sustained high speeds) and keep a constant check on the oil level. It is not unusual for an engine to use oil during the break-in period.

9   At approximately 500 to 600 miles, change the oil and filter.

10   For the next few hundred miles, drive the vehicle normally. Do not pamper it or abuse it.

11   After 2000 miles, change the oil and filter again and consider the engine fully broken in.

Chapter 3
Cooling, heating and air conditioning systems

Contents

Specifications

General

Coolant capacity	See Chapter 1
Radiator cap pressure rating	15 psi
Thermostat opening temperature	195-degrees F
Refrigerant type	
1993 and earlier	R-12
1994 and later	R-134a
Refrigerant capacity	
1993 and earlier	2.75 pounds
1994 and later	2.0 pounds

Torque specifications
	Ft-lbs
Water pump-to-block bolts and studs	25 to 35
Thermostat housing bolts	18 to 23

1 General information

Engine cooling system

The Corvette employs a pressurized engine cooling system with thermostatically controlled coolant circulation. An impeller type water pump mounted on the front of the block pumps coolant through the engine. The coolant flows around each cylinder and toward the rear of the engine. Cast-in coolant passages direct coolant around the intake and exhaust ports, near the spark plug areas and in close proximity to the exhaust valve guide inserts.

A wax pellet type thermostat is located in the front of the intake manifold, just beneath the throttle body on earlier models and located in the water pump housing on later models. During warm up, the closed thermostat prevents coolant from circulating through the radiator. When the engine reaches normal operating temperature, the thermostat opens and allows hot coolant to travel through the radiator, where it is cooled before returning to the engine.

The aluminum radiator is the crossflow type, with tanks on either side of the core.

The cooling system is sealed by a pressure type radiator cap. This raises the boiling point of the coolant and the higher boiling point of the coolant increases the cooling

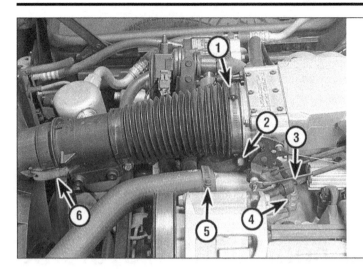

1   *Air intake duct hose clamp*
2   *Hose clamp on the smaller hose that is attached to the underside of the throttle body*
3   *Vacuum T-fitting from the vacuum hose under the throttle body*
4   *Ground wire connector from the thermostat housing stud*
5   *Radiator hose clamp*
6   *Mass air flow sensor retaining clips (one on each side of the sensor)*

efficiency of the radiator.

If the system pressure exceeds the cap pressure relief value, the excess pressure in the system forces the spring-loaded valve inside the cap off its seat and allows the coolant to escape through the overflow tube into a coolant reservoir. When the system cools the excess coolant is automatically drawn from the reservoir back into the radiator.

The coolant reservoir does double duty as both the point at which fresh coolant is added to the cooling system to maintain the proper fluid level and as a holding tank for overheated coolant. This type of cooling system is known as a closed design because coolant that escapes past the pressure cap is saved and reused.

Heating system

The heating system consists of a blower fan and heater core located under the dashboard, the inlet and outlet hoses connecting the heater core to the engine cooling system and the heater/air conditioning control head on the dashboard. Hot engine coolant is circulated through the heater core at all times. When the heater mode is activated, a flap door opens to expose the heater box to the passenger compartment. A fan switch on the control head activates the blower motor, which forces air through the core, heating the air.

Air conditioning system

The air conditioning system consists of a condenser mounted in front of the radiator, an evaporator mounted under the dash, a compressor mounted on the engine, a filter-drier (accumulator) which contains a high pressure relief valve and the plumbing connecting all of the above.

A blower fan forces the warmer air of the passenger compartment through the evaporator core (sort of a radiator-in-reverse), transferring the heat from the air to the refrigerant. The liquid refrigerant boils off into low pressure vapor, taking the heat with it when it leaves the evaporator.

Heat Vent Air Conditioning (HVAC) system

Both heating and air conditioning are integrated into one system. Air entering the vehicle passes through the air conditioning evaporator, then through or around the heater core. This design is known as a reheat or Heat Vent Air Conditioning (HVAC) system.

The HVAC system control components include an air conditioning control head assembly (controller), the programmer assembly, the blower power control module, the in-car temperature sensor, the outside air sensor and the air distribution assembly (the ducting) housing the air flow control valves, the evaporator and the heater core.

Electronic climate control

In the middle of 1986, an optional Electronic Climate Control (ECC) system became available. Although similar to the HVAC system, the ECC system also includes such features as a digital microprocessor (MPU) with a programmable memory, a diagnostic mode for self-diagnosis, both in-car and outside air temperature sensors signal for improved temperature compensation, a digital display, an English-metric conversion capability and a rear window defogger key pad switch.

While earlier HVAC systems use an electrically-operated, rheostat controlled motor to position the temperature door, the ECC HVAC system activates the Auto, Bi-level, Econ, Heat and Def modes through the use of vacuum routed through the programmer.

---

2   Antifreeze - general information

Warning: *Do not allow antifreeze to come in contact with your skin or painted surfaces of the vehicle. Flush contacted areas immediately with plenty of water. Do not store new coolant or leave old coolant lying around where it is easily accessible to children and pets, because they are attracted by its sweet taste. Ingestion of even a small amount can*

be fatal. Wipe up the garage floor and drip pan coolant spills immediately. Keep antifreeze containers covered and repair leaks in your cooling system immediately. Antifreeze is flammable - be sure to read the precautions on the container.
Note: *Non-Toxic coolant is available at local auto parts stores. Although the coolant is non-toxic when fresh, proper disposal is still required.*

The cooling system should be filled with the recommended antifreeze solution that will prevent freezing down to at least -20-degrees F (see Chapter 1). It also provides protection against corrosion and increases the coolant boiling point.

The cooling system should be drained, flushed and refilled according to the maintenance schedule (see Chapter 1). The use of antifreeze solutions for periods of longer than recommended is likely to cause damage and encourage the formation of rust and scale in the system.

Before adding antifreeze to the system, check all hose connections. Antifreeze can leak through very minute openings.

The exact mixture of antifreeze to water that you should use depends on the relative weather conditions. The mixture should contain at least 50 percent antifreeze, but should never contain more than 70 percent antifreeze.

---

3   Thermostat - replacement

1991 and earlier models

Refer to illustrations 3.3, 3.7 and 3.12

1   Disconnect the cable from the negative terminal of the battery. **Caution:** *On models equipped with a Delco Loc II audio system, be sure the lockout feature is turned off before performing any procedure which requires disconnecting the battery.*
2   Drain the coolant (Chapter 1).
3   Loosen the hose clamp at the air intake duct and pop off the retaining clips on the side of the mass air flow sensor meter body

3.7 If the entire stud starts to back out when you start to loosen the locknut, hold the jam nut (the lower nut) with a pair of needle nose-type vise grip pliers

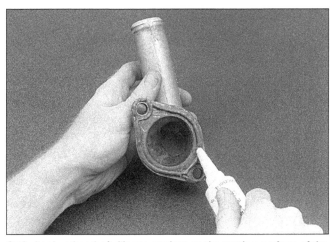

3.12 Apply a bead of silicone sealant on the mating surface of the thermostat housing

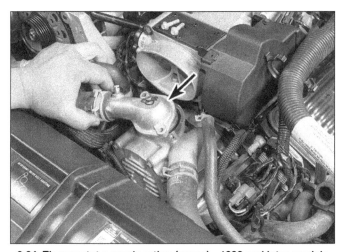

3.24 Thermostat cover location (arrow) - 1992 and later models

3.25 Note how the thermostat is installed, then remove the thermostat (the spring end points toward the engine)

(see illustration). Pull the air flow sensor away from the air cleaner housing enough to disconnect the electrical connection on the bottom of the sensor, then remove the air intake duct and mass air flow sensor and set them aside. **Caution:** *The mass air flow sensor is delicate - handle it carefully.*

4   Disconnect the radiator hose from the thermostat housing.

5   Disconnect the smaller coolant hose from the underside of the throttle body.

6   Disconnect the vacuum T-fitting from the vacuum hose under the throttle body.

7   Remove the thermostat housing bolt and stud. **Note:** *Should the stud start to back out when the lock (upper) nut is loosened, put a pair of needle-nose locking-type pliers on the jam (lower) nut to prevent the stud from turning* **(see illustration).**

8   Remove the thermostat housing.

9   Remove the thermostat. Note the shape of the top of the old thermostat to ensure that the new unit is installed right side up.

10   Remove all traces of old gasket material or sealant from the mating surfaces of the thermostat housing and the intake manifold.

11   Place the new thermostat in position.

12   Apply a bead of silicone sealant to the mating surface of the thermostat housing **(see illustration)**.

13   Install the thermostat housing and tighten the bolt and stud to the specified torque.

14   Reconnect the vacuum T-fitting to the vacuum hose.

15   Reattach the smaller coolant hose to the underside of the throttle body.

16   Reattach the radiator hose to the thermostat housing.

17   Place the intake air duct and mass air flow sensor assembly into position and plug in the electrical connector to the bottom of the mass air flow sensor. Install the two retaining clips to the sides of the mass air flow sensor and tighten the air intake duct hose clamp securely.

18   Fill the cooling system.

19   Connect the cable to the negative terminal of the battery.

20   Start the engine, bring the cooling system up to operating temperature and check the thermostat housing and hoses for leaks.

1992 and later models

Refer to illustrations 3.24, 3.25 and 3.27

21   Drain coolant from the radiator, until the coolant level is below the thermostat housing (about 1 gallon).

22   Remove the air intake duct (see Chapter 4).

23   Disconnect the radiator hose from the thermostat cover. **Note:** *Most models do not have a thermostat housing gasket, and it isn't necessary to remove the hose unless it is being replaced.*

24   Remove the bolts and lift the cover off **(see illustrations)**. It may be necessary to tap the cover with a soft-face hammer to break the gasket seal.

25   Note how it's installed, then remove the thermostat **(see illustration)**. Be sure to use a replacement thermostat with the correct opening temperature (see this chapter's Specifications).

26   Install the thermostat and make sure the correct end faces out - the spring is directed toward the engine.

27   A traditional gasket is not used, but

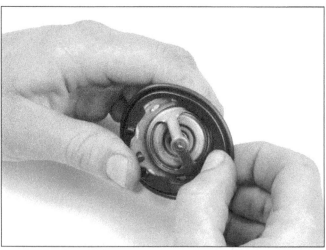

3.27 Install a new rubber seal around the thermostat

4.3 The cooling fan temperature switch is located in the right cylinder head just behind the engine oil dipstick. Unplug the electrical connector (arrow) and ground it and the fan should come on

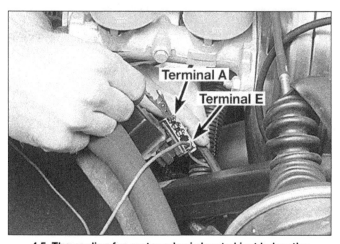

4.5 The cooling fan motor relay is located just below the master cylinder

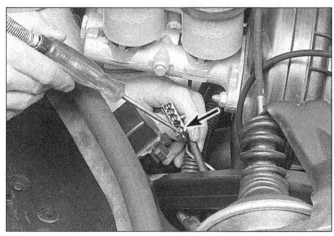

4.6 Check for battery voltage to the relay at terminal E (arrow) with a test light

rather a rubber ring around the thermostat. Replace this ring and install the thermostat cover without gasket sealant **(see illustration)**.

28   Carefully position the cover and install the bolts. Tighten them to the torque listed in this chapter's Specifications - do not over-tighten the bolts or the cover may crack or become distorted.

29   Reattach the radiator hose to the cover and tighten the clamp - now may be a good time to check and replace the hoses and clamps (see Chapter 1).

30   Refer to Chapter 1 and refill the system, then run the engine and check carefully for leaks.

---

4   Electric cooling fan motor and circuit - testing

1984 models

Refer to illustrations 4.3, 4.5 and 4.6

1   The cooling fan temperature switch is located on the right cylinder head, just behind the dipstick. Disconnect the electrical connector from the switch.

2   Turn the ignition key to the On position.

3   Ground the cooling fan temperature switch connector **(see illustration)**. The fan should come on. **Note:** *If the fan comes on, but has not been coming on when the engine is operating at above normal temperature, the thermo switch is faulty.*

4   If the fan does not come on, check the fan motor circuit fuse. Replace if necessary.

5   If the fuse is okay, unplug the cooling fan relay, located just below the master cylinder. Jump terminal A (black/red wire) to terminal E (red wire) **(see illustration)**. If the fan now comes on, the relay is faulty. Replace it.

6   If the fan does not come on after either test, check the battery voltage to the red wire with a test light **(see illustration)**.

7   If battery voltage is present, either the fan motor is faulty, the wire between the relay and the fan motor is faulty or the wire between the motor and ground is broken or disconnected.

8   To check the wire between the relay and

the fan motor, bypass the relay as in Step 5, attach a test light to ground and probe the black/red wire at the fan motor connector. If the light does not come on, there is an open in this wire. Repair or replace it.

9   To check the fan motor, connect a fused jumper wire from the battery directly to the fan motor hot terminal. If the motor does not come on, check the ground wire.

10   To check the fan motor ground wire, hook up a test light to a voltage source and probe the black wire at the fan motor connector. If the light does not come on, there is an open in the ground wire. Repair or replace it.

11   Aside from these checks, further diagnosis of the electric cooling fan motor and circuit is beyond the scope of the home mechanic because the fan is ECM-controlled during operation of the air conditioning system.

1985 through 1987 models

Refer to illustrations 4.13a, 4.13b and 4.15

Note: *Most models use an auxiliary cooling fan, typically mounted ahead of the radiator. The auxiliary cooling fan relay is usually*

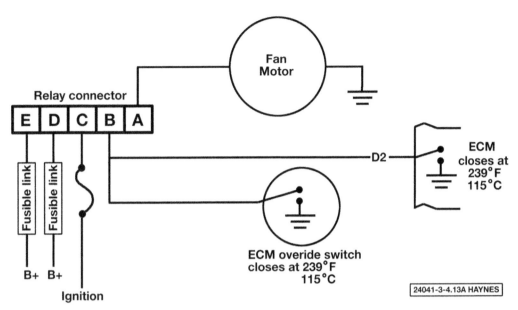

4.13a 1985 through 1987 Cooling fan wiring diagram - Some models use cavity E or D as the battery feed through the fusible link

4.13b 1985 through 1987 cooling fan relay location (arrow)

4.15 1985 through 1987 cooling fan fusible link location (arrow)

mounted on the left side of the radiator shroud. The same procedure for diagnosing the primary cooling fan can be used to diagnose the auxiliary cooling fan.

12   Ground the ALCL terminals A to B, (Chapter 6) turn the ignition on, engine not running. The cooling fan should be running.

13   If the fan is inoperative Locate the cooling fan relay mounted at the left side of the engine compartment by the brake booster. With the ignition key in the ON position ground the cooling fan relay connector terminal "B" (see illustrations). If the fan operates the problem is in the ECM control circuit. Check the wire for opens, shorts and grounds. If the wiring appears to be OK the problem may be in the ECM itself and you may want to have a professional repair shop perform further diagnostics.

14   If the fan runs check for voltage and ground at the inoperative fan. If power and ground are OK replace the defective fan.

Caution: The problem can be a poor connection at the cooling fan. the cooling fans may unexpectedly begin to run as you probe for power at the fan connectors. Keep hands and tools clear of the cooling fans at all times.

15   If there is no power at the inoperative fan(s). Remove the fan relay (located near the brake booster area) and check for battery voltage at the relay connector terminal "D" to ground. If there is less than battery voltage or no voltage, check the fusible links at the junction block located behind the battery and repair as necessary (see illustration).

16   If battery voltage is available at the relays check for battery voltage with the ignition key ON at the relay connector terminal "C" If no voltage check the ignition fuses and wiring and repair as necessary.

17   If the above checks are as specified, jump the cooling fan relay connector terminals "A" to "D", the fan should run. If it runs, replace the relay.

18   If cooling fan runs with the ignition off, replace the relay.

19   If the cooling fan problem hasn't been isolated it may be necessary to take the vehicle to a dealer or professional mechanic to diagnose the computer control of the cooling fan.

1988 and later models

Refer to illustrations 4.22 and 4.24

Note: 1995 and 1996 models utilize three cooling fan relays and use various combinations of these relays to obtain several cooling fan speeds. Due to the complexity of this circuitry the relay circuit must be diagnosed using equipment normally not available to the home mechanic. These vehicles should be diagnosed by a professional repair shop.

20   Most models use two cooling fans, a primary and a secondary. during normal operation the ECM will command the primary cooling fan on when engine coolant tempera-

4.22 On 1988 and later models, the cooling fan relays are located on the left side of the radiator shroud

4.24 1988 and later cooling fan fusible link location (arrow)

ture reaches 226-degrees F, when the air conditioning is on or if a cooling system or coolant sensor diagnostic code is occurring. The ECM will command the secondary cooling fan on when the primary fan is running and the air conditioning is on if the vehicle speed is below 55 MPH, if the engine temperature exceeds 235-degrees F or if a cooling system or coolant sensor diagnostic code is occurring.

21   Ground the ALCL terminals A to B, (see Chapter 6) turn the ignition on, engine not running. Both cooling fans should be running.

22   If one or both fans are inoperative, locate the cooling fan relays mounted at the left side of the radiator shroud **(see illustration)**. With the ignition key in the ON position ground the inoperative cooling fan relay connector terminal "F". If the fan operates the problem is in the ECM control circuit. Check the wire for opens, shorts and grounds. If the wiring appears to be OK the problem may be in the ECM itself and you may want to have a professional repair shop perform further diagnostics.

23   If no fans or only one fan runs check for voltage and ground at the inoperative fan(s). If power and ground are OK replace the defective fan. **Caution:** *The problem can be a poor connection at the cooling fan. the cooling fans may unexpectedly begin to run as you probe for power at the fan connectors. Keep hands and tools clear of the cooling fans at all times.*

24   If there is no power at the inoperative fan(s), remove the fan relays and check for battery voltage at the relay connector terminal "E" to ground. If there is less than battery voltage or no voltage, check the fusible links at the junction block located behind the battery and repair as necessary **(see illustration)**.

25   If battery voltage is available at the relays check for battery voltage with the ignition key ON at the relay connector terminal "D". If there's no voltage check the ignition fuses and wiring and replace as necessary.

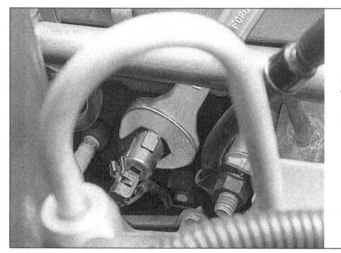

5.3 The cooling fan temperature switch is mounted in the right cylinder head, just behind the engine oil dipstick

26   If the above checks are good, jump the cooling fan(s) relay connector terminals "A" to "E", the fans should run. If they do, replace the relay(s).

27   If either of the cooling fans run with the ignition off, replace the relays.

28   If the cooling fan problem hasn't been isolated it may be necessary to take the vehicle to a dealer or professional mechanic to diagnose the computer control of the cooling fans.

5   Cooling fan temperature switch - replacement

Refer to illustration 5.3

1   Disconnect the cable from the negative terminal of the battery.

2   Disconnect the electrical connector from the cooling fan temperature switch.

3   Remove the switch **(see illustration)**.

4   Wrap the threads of the new switch with Teflon tape.

5   Install the new switch and tighten it securely.

6   Reconnect the electrical connector.

7   Reconnect the cable to the negative terminal of the battery.

8   Start the engine, warm it up to operating temperature and verify that the new switch is operating properly. That is, it is now turning on the fan motor when the cooling system overheats.

6   Electric cooling fans - removal and installation

Refer to illustration 6.4

1   Disconnect the cable from the negative terminal of the battery.

2   Loosen the hose clamp at the air intake duct and pop off the retaining clips on the side of the mass air flow sensor body. Pull the air flow sensor away from the air cleaner housing enough to disconnect the electrical connection on the bottom of the sensor, then remove the air intake duct and mass air flow sensor and set them aside. **Caution:** *The mass air flow sensor is delicate - handle it carefully.*

3   Remove the two locking knobs on the top of the air cleaner housing and remove the

6.4 Engine-side (primary) cooling fan - 1984 model shown

7.5 Use a flare nut wrench to disconnect the transmission cooler fittings from the radiator side tank

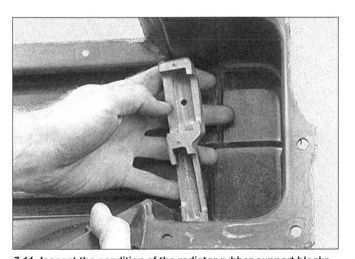

7.11 Inspect the condition of the radiator rubber support blocks - replace them if they are cracked or deteriorated

8.3a The water pump has two weep holes, one on the top and one on the bottom (arrow)

air cleaner housing.

4   Remove the four cooling fan motor frame-to-shroud bolts, lift the fan motor frame up far enough to disconnect the electrical connector from the underside of the fan motor and remove the fan motor frame assembly **(see illustration)**. **Note:** *Early models have one fan in front of the radiator (the auxiliary fan) and one on the engine side (the primary fan). Later models have both cooling fans on the engine side of the radiator.*

5   Remove the fan retaining nut and remove the fan from the motor.

6   Remove the four motor-to-frame mounting bolts and remove the motor from the frame.

7   Installation of the new motor is the reverse of the removal procedure.

7   Radiator - removal and installation

Refer to illustrations 7.5 and 7.11

1   Disconnect the cable from the negative terminal of the battery.

2   Drain the coolant (Chapter 1).

3   Remove the fan motor frame assembly (see Section 6).

4   Disconnect the upper and lower radiator hoses and the overflow hose.

5   Disconnect the transmission cooler fittings **(see illustration)**.

6   Remove the 12 bolts and 6 screws that attach the upper shroud to the lower one.

7   Remove the power steering reservoir-to-shroud bolt and loosen the rear bolt, then rotate the reservoir to the left far enough to clear the radiator shroud.

8   Pop the fan motor wire harness retaining clip loose from the shroud.

9   Push the accumulator bracket aside and lift the shroud from the engine compartment.

10   Lift the radiator straight up and remove it.

11   Installation is the reverse of the removal procedure. **Note:** *Be sure to inspect the condition of the radiator rubber support blocks before installing the radiator* **(see illustration)**. *Replace them if they are cracked or deteriorated.*

8   Water pump - check

Refer to illustrations 8.3a and 8.3b

1   Water pump failure can cause overheating of and serious damage to the engine. There are three ways to check the operation of the water pump while it is installed on the engine. If any one of the three following quick checks cast suspicion on the water pump, it should be replaced immediately.

2   Start the engine and warm it up to normal operating temperature. Squeeze the upper radiator hose. If the water pump is working properly, you should feel a pressure surge as the hose is released.

3   A seal protects the water pump impeller shaft bearing from contamination by engine coolant. If this seal fails, weep holes in the top and bottom of the water pump snout **(see illustrations)** will leak coolant onto the ground underneath when the vehicle is parked. If the weep hole is leaking, shaft bearing failure will follow. Replace the water pump immediately.

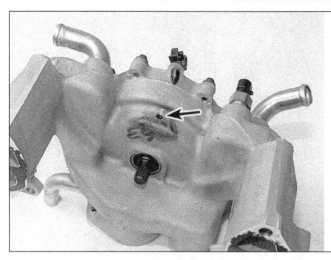

8.3b Check the weep hole (arrow) for leakage (1992 and later models)

4   Besides contamination by coolant after a seal failure, the water pump impeller shaft bearing can also be prematurely worn out by an improperly tensioned drivebelt. When the bearing wears out, it emits a high pitched squealing sound. If such a noise is emanating from the water pump during engine operation, the shaft bearing has failed. Replace the water pump immediately.

5   To identify excessive bearing wear before the bearing actually fails, grasp the water pump pulley and try to force it up and down or from side to side. If the pulley can be moved either horizontally or vertically, the bearing is nearing the end of its service life. Replace the water pump.

9   Water pump - removal and installation

1991 and earlier models

Refer to illustrations 9.5, 9.9, 9.11, 9.16, 9.17 and 9.20

Removal

1   Relieve fuel system pressure (Chapter 4), then disconnect the cable from the negative terminal of the battery.
2   Drain the coolant (Chapter 1).
3   Loosen the hose clamp at the air intake duct and pop off the retaining clips on the side of the mass air flow sensor body. Pull the air flow sensor away from the air cleaner housing enough to disconnect the electrical connection on the bottom of the sensor, then remove the air intake duct and mass air flow sensor and set them aside. **Caution:** *The mass air flow sensor is delicate - handle it carefully.*
4   Remove the drivebelt (Chapter 1).
5   Disconnect the electrical connector and vacuum line from the air switching valve. Disconnect the electrical connector and vacuum

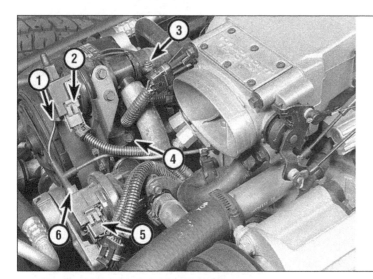

9.5 Once the air intake duct and mass airflow sensor are out of the way, the following items must be disconnected

1   Vacuum fitting at the air switching valve
2   ECM electrical connector at the air switching valve
3   Air hose-to-exhaust manifold
4   Air hose-to-exhaust pipe between reducing and oxidizing catalysts
5   ECM electrical connector at the diverter valve
6   Vacuum fitting at the air diverter valve

9.9 Remove the air switching valve bracket bolts (arrows)

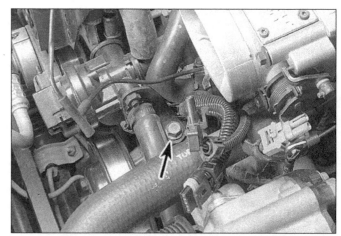

9.11 Remove the diverter valve bracket bolt (arrow)

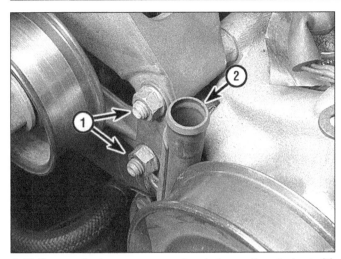

9.16 Remove the two compressor bracket mounting stud nuts (1) and the diverter pipe (2)

9.17 Disconnect both fuel fittings (arrows) - be sure to use a backup wrench on each fitting to prevent kinking the fuel lines

9.20 Using a screwdriver wedged between the hub and a bolt to prevent the pulley from turning, remove the pulley mounting bolts, then remove the two mounting studs (1) and the mounting bolts (2) from the water pump

line from the diverter valve **(see illustration)**.

6   The hose between the air injection reactor (AIR) pump and the diverter valve is routed right under the radiator hose. Loosen the hose clamp at the AIR pump end of the hose and disconnect the hose from the pump.

7   Disconnect the exhaust manifold AIR hose from the rear of the air switching valve.

8   Disconnect the exhaust pipe AIR hose from the underside of the air switching valve.

9   Remove the air switching valve bracket bolts **(see illustration)**.

10   Remove the small coolant hose from the underside of the throttle body.

11   Remove the diverter valve bracket bolt **(see illustration)**.

12   Remove the air management valves (the switching and diverter valves). **Note:** *If the throttle position sensor (TPS) switch wire harness has been routed in such a manner that it is in the way, disconnect it.*

13   Remove all four air conditioning accumulator bracket bolts (Section 15).

14   Loosen but do not remove the rear air conditioning compressor bracket bolt and remove the nut from the exhaust manifold

stud to which the compressor bracket is attached. Swing the bracket reinforcement arm free.

15   The bracket bolt beneath the lower edge of the compressor clutch pulley must be backed out from the cylinder head. **Note:** *It is not necessary to actually remove this bolt for this procedure - it can remain with the compressor/bracket assembly during removal and installation.*

16   Remove the two compressor bracket nuts from the water pump mounting studs just inboard of the tensioner pulley and remove the diverter valve pipe **(see illustration)**.

17   Disconnect both fuel fittings **(see illustration)**.

18   Lift the accumulator out of the way and slide the compressor and bracket off the bracket mounting studs. Wire the compressor out of the way so that it clears the water pump.

19   Loosen the hose clamps for the lower radiator hose and heater hose and disconnect the two hoses from the water pump. If you are replacing the water pump, remove the heater hose extension pipe for installation

on the new pump.

20   Prevent the pulley from turning by placing a screwdriver between the pulley hub and a bolt, loosen each bolt and remove the pulley bolts **(see illustration)**.

21   Remove the two mounting bolts from the left side of the water pump and the two mounting studs from the right side of the water pump **(see illustration 9.20)**. Note that the two studs on the right side of the pump must not be switched with the bolts on the left side of the pump because the air conditioning compressor bracket is mounted on these two studs.

22   Remove the water pump.

Installation

23   If you are replacing the water pump, remove the heater hose pipe from the old pump and install it on the new unit. Use Teflon sealing tape to prevent leakage.

24   Clean the sealing surfaces on both the block and the water pump, install new water pump-to-block gaskets and tighten the two bolts and the two studs to the specified torque.

25   Install the pulley and tighten the pulley bolts securely.

26   Install the lower radiator hose, heater hose and hose clamps. Tighten the hose clamps securely.

27   Slide the air conditioning compressor and its bracket onto the bracket mounting studs and install the bracket reinforcement arm onto the exhaust manifold stud. Tighten the compressor bracket mounting nuts, the compressor rear bracket bolt and the bracket reinforcement arm retaining nut securely.

28   Place the accumulator back in position, install the bracket bolts and tighten them securely.

29   Attach both fuel fittings and tighten them securely.

30   Install the air management valves (the switching and diverter valves). If the throttle position sensor (TPS) switch wire was disconnected, reconnect it. Install the diverter

9.46 Use a soft-faced hammer to break the gasket seal
on the pump

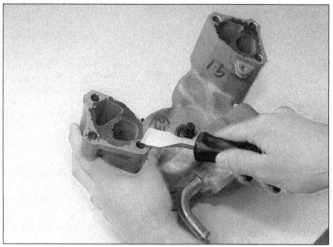

9.47 Remove all traces of old gasket material - use care to avoid
gouging the soft aluminum

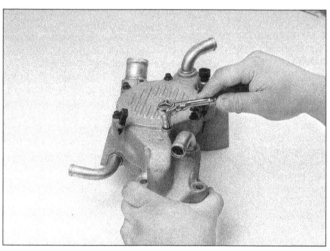

9.48 The front cover can be removed from the water pump
housing on V8 engines to allowing pressing in a new impeller and
bearings - the cover is O-ring sealed

9.50a The water pump is mechanically driven by this splined shaft
from the timing cover - held here is the splined coupler that joins
the shaft to the splines on the back of the pump

valve bracket bolt and tighten it securely.

31 Install the small coolant hose to the underside of the throttle body and tighten the hose clamp securely.

32 Install the air switching valve bracket bolts and tighten them securely.

33 Connect the exhaust pipe air injection reactor (AIR) hose to the underside of the air switching valve and tighten the hose clamp securely.

34 Connect the exhaust manifold AIR hose to the rear of the air switching valve and tighten the hose clamp securely.

35 Connect the hose between the AIR pump and the diverter valve to the pump and tighten the hose clamp securely.

36 Connect the electrical connector and vacuum line to the air switching valve. Connect the electrical connector and vacuum line to the diverter valve.

37 Install the drivebelt (Chapter 1).

38 Place the air intake duct and mass air flow sensor assembly into position and connect the electrical connector to the underside

of the sensor. Pop the retaining clips onto both sides of the sensor and tighten the air intake duct hose clamp securely.

39 Add coolant (Chapter 1).

40 Connect the cable to the negative terminal of the battery.

41 Start the engine and check the water pump and hoses for leaks.

1992 and later models

Refer to illustrations 9.46, 9.47, 9.48, 9.50a and 9.50b

Removal

42 Drain the coolant (see Chapter 1).

43 Remove the serpentine drivebelt (see Chapter 1).

44 Remove the upper and lower radiator hoses and heater hoses from the water pump.

45 Disconnect the electrical connector from the coolant temperature sensor on top of the water pump (see Chapter 6), then unclip its wire harness from the two clips

attached to the face of the water pump.

46 Remove the water pump mounting bolts and detach it. If necessary, strike the pump with a soft-face hammer, block of wood or wooden hammer handle to break the gasket seal **(see illustration)**. Do not pry between the pump and the block.

Installation

47 Clean the sealing surfaces of all gasket material on both the water pump and block **(see illustration)**. Wipe the mating surfaces with a rag saturated with lacquer thinner or acetone.

48 The water pump housing can be reused, and by removing a cover on the front, access is gained to the replaceable impeller and bearings, which must be pressed in at an automotive machine shop **(see illustration)**.

49 Apply a thin layer of RTV sealant to both sides of the new gasket and install the gasket on the water pump.

50 Pull the water pump driveshaft coupling **(see illustration)** from the driveshaft (the

9.50b Whenever the water pump is removed, replace the O-ring seals - one at the pump splines, one at the drive splines - use something tapered to slip the seal evenly around the shafts

10.4 The heater blower motor assembly

1   *Motor cooling tube*   2   *Electrical connectors*

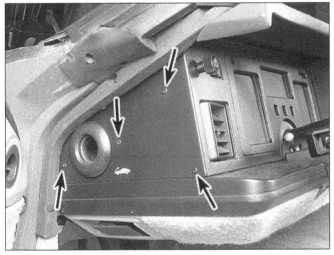

11.2a Remove the four retaining screws (arrows) on the left side of the instrument cluster bezel . . .

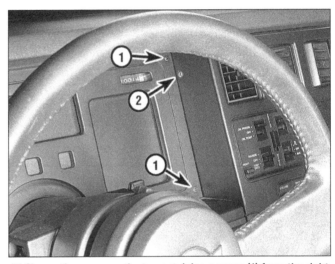

11.2b . . . then remove the two retaining screws (1) from the right end of the instrument cluster bezel and screw (2), which retains the upper left side of the accessory trim plate

short, splined shaft coming out of the block above the distributor). Use something tapered like a marking-pen top and slip lightly-oiled, new O-ring seals over the splines at the driven-gear end and the splines at the back of the water pump **(see illustration)**.

51   Place the water pump in position and install the bolts finger tight. Use caution to ensure that the gasket doesn't slip out of position. Install the water pump driveshaft coupling over the driveshaft and align the pump's splines with the coupling when putting the pump into place on the engine.

52   The remainder of the installation procedure is the reverse of removal.

53   Add coolant to the specified level (see Chapter 1) and start the engine and check for the proper coolant level and the water pump and hoses for leaks.

10   Heater blower motor - removal and installation

Refer to illustration 10.4

1   Disconnect the cable from the negative terminal of the battery.

2   Remove the weatherstripping retaining screw (at the top outside corner of the front wheel house rear panel) and peel back the wheel house weatherstripping.

3   Remove all 14 fasteners (five Torx screws and nine hex screws) and remove the front wheel house rear panel.

4   Remove the heater blower motor cooling tube **(see illustration)**.

5   Disconnect both electrical connectors.

6   Remove all five retaining screws and remove the motor.

7   Installation is the reverse of removal.

11   Heater and air conditioning control assembly - removal and installation

Warning: *Some of the models covered by this manual are equipped with Supplemental Inflatable Restraints (SIR), more commonly known as airbags. Always disable the airbag system before working in the vicinity of any airbag system components to avoid the possibility of accidental deployment of the airbag, which could cause personal injury (see Chapter 12).*

1989 and earlier models

Refer to illustrations 11.2a, 11.2b, 11.3, 11.5a and 11.5b

1   Disconnect the cable from the negative terminal of the battery.

2   Remove the instrument cluster bezel

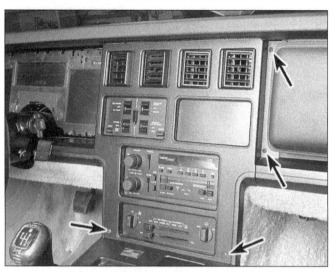

11.3 Remove the retaining screws (arrows) and the accessory trim plate

11.5a Unscrew the two rear retaining screws on the shifter trim plate and rotate it out of the way, then remove the four mounting screws (arrows) from the air conditioning and heater control assembly

retaining screws **(see illustrations)** and remove the bezel.

3   Remove the accessory trim plate retaining screws **(see illustration)** and remove the accessory trim plate.

4   Remove the two rear retaining screws from the shifter trim plate and lift the plate up and rotate it out of the way.

5   Remove the four air conditioning and heater control assembly retaining screws **(see illustration)** and pull the assembly out from the dash far enough to get at the backside **(see illustration)**.

6   Disconnect the temperature control cable, vacuum hose cluster, the big electrical connector and the two smaller electrical connectors.

7   Remove the heater and air conditioning control assembly.

8   Installation is the reverse of removal.

1990 and later models

9   Disconnect the cable from the negative battery terminal.

10   Remove the console trim plate by first removing the shifter button, automatic transmission shifters are retained by a snap ring, manual shifters are retained by a set screw (see Chapter 7).

11   Remove the console trim plate screws and the hidden screw under the ashtray (see Chapter 12).

12   Disconnect the electrical connectors for the cigarette lighter and the instrument harness, then remove the console plate.

13   Remove the accessory trim plate by first removing the two center outlet retaining screws (see Chapter 12).

14   Remove the accessory trim plate to instrument cluster screws and remove the trim plate.

15   Remove the upper trim pad screws and screws next to the center air conditioning outlet. Pull the trim pad out to get to the right-hand mounting screws on the air conditioning

11.5b Pull the assembly out far enough to disconnect the temperature control cable, the vacuum hose cluster and the electrical connectors

control assembly.

16   Remove the air conditioning control assembly mounting screws from the instrument panel, slide the air conditioning control assembly out and disconnect the electrical connectors.

17   Installation is the reverse of removal.

---

12   Air conditioning system - check and maintenance

Air conditioning system

Warning: *The air conditioning system is under high pressure. Do not loosen any hose fittings or remove any components until the system has been discharged. Air conditioning refrigerant should be properly discharged into an EPA-approved recovery/recycling unit by a dealer service department or an automotive air conditioning repair facility. Always wear eye protection when disconnecting air conditioning system fittings.*

1   The following maintenance checks

should be performed on a regular basis to ensure that the air conditioner continues to operate at peak efficiency:

a) *Inspect the condition of the compressor drivebelt. If it is worn or deteriorated, replace it (see Chapter 1).*

b) *Check the drivebelt tension and, if necessary, adjust it (see Chapter 1).*

c) *Inspect the system hoses. Look for cracks, bubbles, hardening and deterioration. Inspect the hoses and all fittings for oil bubbles or seepage. If there is any evidence of wear, damage or leakage, replace the hose(s).*

d) *Inspect the condenser fins for leaves, bugs and any other foreign material that may have embedded itself in the fins. Use a fin comb or compressed air to remove debris from the condenser.*

e) *Make sure the system has the correct refrigerant charge.*

2   it's a good idea to operate the system for about ten minutes at least once a month. This is particularly important during the winter months because long term non-use can

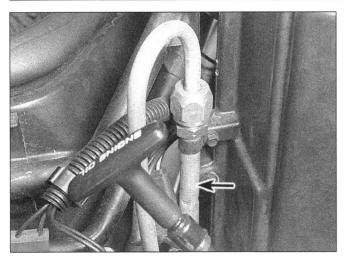

12.9 To determine whether the refrigerant charge is adequate, feel the evaporator inlet pipe between the orifice (arrow) and the evaporator, lay your other hand on the accumulator (not visible) and note the temperature of each

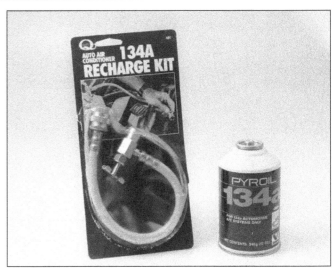

12.12 A basic charging kit for R-134a systems is available at most auto parts stores

cause hardening, and subsequent failure, of the seals.

3   Leaks in the air conditioning system are best spotted when the system is brought up to operating temperature and pressure, by running the engine with the air conditioning ON for five minutes. Shut the engine off and inspect the air conditioning hoses and connections. Traces of oil usually indicate refrigerant leaks.

4   Because of the complexity of the air conditioning system and the special equipment required to effectively work on it, accurate troubleshooting of the system should be left to a professional technician.

5   If the air conditioning system doesn't operate at all, check the A/C fuse in the fuse panel and the air conditioning relay, located in the engine compartment.

6   The most common cause of poor cooling is simply a low system refrigerant charge. If a noticeable drop in cool air output occurs, the following quick check will help you determine if the refrigerant level is low. For more complete information on the air conditioning system, refer to the Haynes Automotive Heating and Air Conditioning Manual.

Checking the refrigerant charge

Refer to illustration 12.9

7   Warm the engine up to normal operating temperature.

8   Place the air conditioning temperature selector at the coldest setting and put the blower at the highest setting. Open the doors (to make sure the air conditioning system doesn't cycle off as soon as it cools the passenger compartment).

9   With the compressor engaged - the clutch will make an audible click and the center of the clutch will rotate. After the system reaches operating temperature, feel the evaporator inlet pipe between the orifice and the evaporator (see illustration). Place your

other hand on the accumulator surface.

10   If both surfaces feel about the same temperature and if both feel cooler than the ambient temperature, the refrigerant charge is probably sufficient.

11   If the evaporator inlet pipe has frost accumulation or feels considerably cooler than the accumulator surface, the refrigerant charge is low.

Adding refrigerant

Refer to illustrations 12.12 and 12.15

Caution: *Refrigerant use has changed from the use of R-12, through 1993 models, to the environmentally friendly R-134a used in 1994 and later models. The two refrigerants are NOT compatible. Even after purging and evacuating an R-12 system, there is enough residual oil and refrigerant in the hoses and components that simply filling the system with R-134a cannot be done. Special fittings and manifold gauge sets are used on the different refrigerant types so that an accidental hook-up of the two systems cannot be made. When replacing entire components, additional refrigerant oil should be added equal to the amount that is removed with the component being replaced. Refrigerant oils, just like refrigerant R-12 vs. R-134a, are not compatible. Be sure to read the can before adding any oil to the system, to make sure it is compatible with the type of system being repaired.*

Note: *Because of Federal regulations by the Environmental Protection Agency, R-12 refrigerant is not available for home-mechanic use, however, cans of R-134 refrigerant are commonly available in auto parts stores. Models with R-12 systems will have to be serviced at a dealership or air conditioning shop.*

12   Buy an automotive charging kit at an auto parts store. A charging kit includes a 14-ounce can of R-134a refrigerant, a tap valve and a short section of hose that can be attached between the tap valve and the sys-

tem low side service valve (see illustration). Because one can of refrigerant may not be sufficient to bring the system charge up to the proper level, it's a good idea to buy a couple of additional cans. Try to find at least one can that contains red refrigerant dye. If the system is leaking, the red dye will leak out with the refrigerant and help you pinpoint the location of the leak.

13   Connect the charging kit by following the manufacturer's instructions.

14   Back off the valve handle on the charging kit and screw the kit onto the refrigerant can, making sure first that the O-ring or rubber seal inside the threaded portion of the kit is in place. **Warning:** *Wear protective eye wear when dealing with pressurized refrigerant cans.*

15   Remove the dust cap from the low-side charging port and attach the quick-connect fitting on the kit hose (see illustration). **Warning:** *DO NOT hook the charging kit hose to the system high side! The fittings on the charging kit are designed to fit only on the*

12.15 A typical aftermarket air conditioner recharge kit hooked up to the LOW SIDE charging port (R-134a only)

13.2 Disconnect the electrical connector (1) from the top of the air conditioning compressor and remove the line fitting bolt (2) from the backside of the compressor

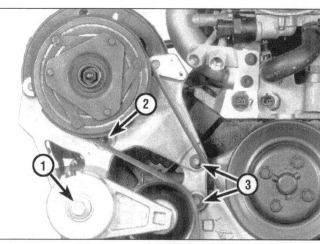

13.4 Remove the drivebelt, the tensioner pulley bolt (1), the air conditioning compressor bracket bolt (2), rotate the compressor clutch pulley until either notch is aligned with the bolt and remove the bolt, then remove the bracket mounting nuts (3)

low side of the system.

16   Warm the engine to normal operating temperature and turn on the air conditioner. Keep the charging kit hose away from the fan and other moving parts.

17   Turn the valve handle on the kit until the stem pierces the can, then back the handle out to release the refrigerant. You should be able to hear the rush of gas. Add refrigerant to the low side of the system until both the accumulator surface and the evaporator inlet pipe feel about the same temperature. Allow stabilization time between each addition. **Warning:** *Never add more than two cans of refrigerant to the system. The can may tend to frost up, slowing the procedure. Wrap a shop towel wet with hot water around the bottom of the can to keep it from frosting.*

18   If you have an accurate thermometer, you can place it in the center air conditioning duct inside the vehicle to monitor the air temperature. A charged system that is working properly, should output air down to approximately 40 degrees F.

19   When the can is empty, turn the valve handle to the closed position and release the connection from the low-side port. Replace the dust cap.

20   Remove the charging kit from the can and store the kit for future use with the piercing valve in the UP position, to prevent inadvertently piercing the can on the next use.

Heating systems

21   If the air coming out of the heater vents isn't hot, the problem could stem from any of the following causes:

a) *The thermostat is stuck open, preventing the engine coolant from warming up enough to carry heat to the heater core. Replace the thermostat (see Section 3).*

b) *A heater hose is blocked, preventing the flow of coolant through the heater core. Feel both heater hoses at the firewall. They should be hot. If one of them is*

cold, there is an obstruction in one of the hoses or in the heater core, or the heater control valve is shut. Detach the hoses and back flush the heater core with a water hose. If the heater core is clear but circulation is impeded, remove the two hoses and flush them out with a water hose.

c) *If flushing fails to remove the blockage from the heater core, the core must be replaced (see Section 16).*

22   If the blower motor speed does not correspond to the setting selected on the blower switch, the problem could be a bad fuse, circuit, switch, blower motor resistor or motor.

23   If there isn't any air coming out of the vents:

a) *Turn the ignition ON and activate the fan control. Place your ear at the heating/air conditioning register (vent) and listen. Most motors are audible. Can you hear the motor running?*

b) *If you can't (and have already verified that the blower switch and the blower motor resistor are good), the blower motor itself is probably bad.*

24   If the carpet under the heater core is damp, or if antifreeze vapor or steam is coming through the vents, the heater core is leaking. Remove it and install a new unit (most radiator shops will not repair a leaking heater core).

25   Inspect the drain hose from the evaporator case and make sure it is not clogged.

---

13   Air conditioning compressor - removal and installation

Refer to illustrations 13.2, 13.4 and 13.7
Warning : *The air conditioning system is under high pressure. Do not loosen any hose fittings or remove any components until the system has been discharged. Air conditioning refrigerant should be properly discharged into*

an EPA-approved recovery/recycling unit by a dealer service department or an automotive air conditioning repair facility. Always wear eye protection when disconnecting air conditioning system fittings.

1987 and earlier models
Removal

1   Have the air conditioning system discharged (see **Warning** above). Disconnect the cable from the negative terminal of the battery.

2   Disconnect the electrical connector from the top of the air conditioning compressor and remove the line fitting bolt from the backside of the compressor **(see illustration)**.

3   Remove the drivebelt (refer to Chapter 1).

4   Remove the drivebelt tensioner pulley bolt and the tensioner pulley assembly **(see illustration)**.

5   Loosen the compressor bracket bolt beneath the compressor pulley. A couple of notches 180 degrees apart in the edge of the compressor pulley clutch face have been provided for the removal of this bolt. Rotate the pulley until a notch is aligned with the bracket bolt and remove the bolt.

6   Remove the two compressor bracket nuts from the water pump mounting studs just inboard of the tensioner pulley.

7   Remove the rear air conditioning compressor bracket bolt and remove the nut from the exhaust manifold stud to which the compressor bracket reinforcement arm is attached. Remove the bracket reinforcement arm **(see illustration)**.

8   Slide the compressor and its bracket off the bracket mounting studs and lift the compressor and bracket off together.

9   Remove the remaining compressor bracket bolt from the lower backside of the compressor and separate the compressor from the bracket.

13.7 Remove the compressor bracket bolt (1) and nut (2), then remove the bracket reinforcement arm

14.3 Disconnect both air conditioning line fittings, using a backup wrench to prevent kinking

Installation

Note: *To assure proper alignment of bolt and stud holes, do not tighten any compressor mounting nuts or bolts until all fasteners have been installed finger tight.*

10   Install the last compressor bracket bolt removed first (see Step 9 above).
11   Slide the compressor and its bracket onto the bracket mounting studs.
12   Install the bracket reinforcement arm between the exhaust manifold and the backside of the compressor. Install the rear air conditioning compressor bracket bolt and the exhaust manifold stud nut.
13   Install the two compressor bracket nuts to the water pump mounting studs just inboard of the tensioner pulley.
14   Rotate the compressor pulley until one of its notches is aligned with the compressor bracket bolt just underneath the pulley and install the bolt.
15   Install the tensioner pulley assembly and the tensioner pulley assembly retaining bolt.
16   Once all the compressor mounting nuts and bolts are installed, tighten them securely.
17   Install the drivebelt (Chapter 1).
18   Connect the electrical connector to the top of the compressor. Install the line fitting bolt to the backside of the compressor and tighten it securely.
19   Connect the cable to the negative terminal of the battery.
20   Take the vehicle to a dealer service department or an air conditioning specialist and have the air conditioning system evacuated and recharged.

1988 and later models

21   Have the air conditioning system discharged (see **Warning** above). Disconnect the cable from the negative terminal of the battery.
22   Remove the serpentine drive belt.
23   Disconnect the compressor clutch electrical connector on the top of the compressor.

24   Remove the refrigerant line fitting bolt disconnect the refrigerant lines and cap the open fittings to prevent the entry of dirt or moisture.
25   Remove the compressor to exhaust manifold brace.
26   Remove the compressor mounting bolts and remove the compressor.
27   Installation is the reverse of removal.
28   Take the vehicle to a dealer service department or an air conditioning specialist to have the system evacuated, recharged and leak checked.

---

14   Air conditioning condenser - removal and installation

Refer to illustration 14.3
Warning: *The air conditioning system is under high pressure. Do not loosen any hose fittings or remove any components until the system has been discharged. Air conditioning refrigerant should be properly discharged into*

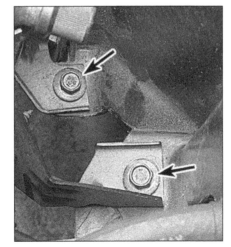

15.3a The accumulator front bracket bolts (arrows)

an EPA-approved recovery/recycling unit by a dealer service department or an automotive air conditioning repair facility. Always wear eye protection when disconnecting air conditioning system fittings.

1   Have the air conditioning system discharged (see **Warning** above). Disconnect the cable from the negative terminal of the battery.
2   Remove the radiator shroud (Section 7).
3   Disconnect the air conditioning line fittings **(see illustration)**. Be sure to use a backup wrench to prevent damage to the fittings.
4   Lift out the condenser.
5   Installation is the reverse of removal.

---

15   Air conditioning accumulator - removal and installation

Refer to illustrations 15.3a and 15.3b
Warning: *The air conditioning system is under high pressure. Do not loosen any hose fittings or remove any components until the system has been discharged. Air conditioning refrigerant should be properly discharged into an EPA-approved recovery/recycling unit by a dealer service department or an automotive air conditioning repair facility. Always wear eye protection when disconnecting air conditioning system fittings.*

Removal

1   Have the air conditioning system discharged (see **Warning** above). Disconnect the cable from the negative terminal of the battery.
2   Disconnect the accumulator inlet and outlet connections. Cap or plug the open lines immediately.
3   Remove the accumulator attaching bolts **(see illustrations)** and remove the accumulator assembly.

15.3b The accumulator rear bracket bolts (arrows)

17.3 Loosen the clamps and remove the heater hoses from the heater control valve (arrow)

Installation

4   Check the amount of oil in the old accumulator and install this amount of fresh 525 viscosity refrigerant oil into the new accumulator, plus two ounces.
5   Place the new accumulator into position, install the four mounting bolts and tighten them securely.
6   Install the inlet and outlet connections, using clean 525 viscosity refrigerant oil on new O-rings.
7   Connect the cable to the negative terminal of the battery.
8   Have the air conditioning system evacuated and recharged by a dealer service department or an air conditioning specialist.

16 Heater core -removal and installation

Warning: *Some of the models covered by this manual are equipped with Supplemental Inflatable Restraints (SIR), more commonly known as airbags. Always disable the airbag system before working in the vicinity of any airbag system components to avoid the possibility of accidental deployment of the airbag, which could cause personal injury (see Chapter 12).*

1985 through 1989 models

1   Drain the cooling system (Chapter 1).
2   Disconnect the cable from the negative battery cable.
3   Remove the instrument cluster bezel, tilt wheel lever and instrument panel pad.
4   Remove the air conditioning duct, disconnect the flex hose and remove the right side hush panel.

5   Remove the right side window defroster duct hose, window to heater cover screws then disconnect the extension.
6   Remove the temperature control cable and bracket at the heater cover and disconnect the heater door shaft.
7   Disconnect the ECM wiring connectors and remove the ECM (see Chapter 6).
8   Remove the support brace from the door pillar to aluminum instrument panel reinforcement brace.
9   Remove the heater core cover attaching screws.
10   Remove the heater pipe and water control valve bracket and screws.
11   Carefully cut the heater hoses at the heater core pipes, then measure the hoses and replace with new hoses when installing the heater core.
12   Remove the heater core.
13   Installation is the reverse of removal.
14   Refill the cooling system and check for leaks.

1990 and later models

15   Drain the cooling system(Chapter 1).
16   Remove the instrument panel upper trim pad.
17   Remove the in vehicle temperature aspirator hose and disconnect the temperature sensor.
18   Remove the floor heat deflector screws, the right hand knee bolster brace the floor heat deflector brace and the relays from the multi- use relay bracket.
19   Remove the nuts holding the wiring harness to the radio control head and remove the harness from the control head.
20   Remove the multi-use harness.
21   Remove the passenger knee bolster brace attachments.

22   Remove the side window defroster duct clip and hose from the knee bolster brace, then pull the carrier back and remove the knee bolster brace.
23   Disconnect the electrical connector from the multi-use relay bracket, the cruise module and the radio control head.
24   Remove the screws holding the side window defroster duct to the rear of the heater case.
Remove the fuse block, the vacuum hose from the actuator and the vacuum hose retaining tape from the heater housing.
25   Remove the harness from the retainer clip on the bottom of the rear heater case.
26   Remove the side window center defrost duct extension from the heater case.
27   Remove the rear heater case screw and heater case vent.
28   Remove the high fill reservoir, the heater hoses from the heater core and remove the core from the heater case.
29   Installation is the reverse of removal.
30   Fill the cooling system and check for leaks.

17 Heater water control valve replacement

Refer to illustration 17.3
1   Drain the cooling system (Chapter 1)
2   Disconnect the heater hoses from the control valve **(see illustration)**.
3   Remove the control valve vacuum line.
4   Remove the screws securing the control valve.
5   Remove the heater control valve.
6   Installation is the reverse of removal.
7   Fill cooling system and check for leaks.

Chapter 4
Fuel and exhaust systems

Contents

Specifications

1984 Crossfire Injection (TBI) models

Fuel pressure	9 to 13 psi
Idle Air Control (IAC) dimension A	1-1/8 inch
Maximum shoulder-to-ball clearance	0.200 inch
Minimum throttle rod-to-casting clearance	0.040 inch

Torque specifications

	Ft-lbs (unless otherwise noted)
Idle Air Control (IAC) assembly	13
Fuel meter body screws	35 in-lbs
Fuel inlet and return fittings	21
Fuel connecting tube nuts	24
Fuel meter cover screws	28 in-lbs
TBI mounting bolts	8 to 14
Intake cover plate bolts and stud	15 to 25
Intake manifold bolts	35
Exhaust pipe U-bolt nuts	40
Exhaust manifold-to-exhaust pipe bolts	17

1985 and later Tuned Port Injection (TPI) models

Fuel pump flow test	1/2 pint or more in 15 seconds
Fuel pressure	
1985 through 1991	34 to 39 psi
1992 and later	41 to 47 psi
Minimum idle speed idle stop screw adjustment	
automatic transmission	400 rpm (in Drive)
manual transmission	450 rpm (in Neutral)
Throttle position sensor (TPS) adjustment	0.54 volt plus or minus 0.075 volt
Idle Air Control (IAC) dimension A	1-1/8 inches

Torque specifications

	Ft-lbs (unless otherwise noted)
Intake manifold bolts	35
Fuel filter fittings	22
Fuel feed and return line-to-fuel rail fittings	22
Fuel rail-to-intake manifold bolts	15 to 22
Fuel pressure connection assembly	88 in lbs
Throttle body-to-plenum bolts	8 to 18
Idle air control (IAC) valve	13
Idle air control/coolant cover assembly bolts	27 in lbs
EGR valve bolts	14
EGR control solenoid nut	17
Exhaust manifold-to-exhaust pipe flange bolts	17
Exhaust pipe U-bolt nuts	
1985	40
1986-on	45

1 General information

Fuel system

The fuel system consists of a rear mounted fuel tank, an electrically operated in-tank fuel pump, a fuel injection system, and the fuel feed and return lines between the fuel injection system and the tank.

The Crossfire Fuel Injection system on 1984 models consists of two Throttle Body Injection (TBI) units mounted diagonally on a common intake manifold. In 1985, the Crossfire system was replaced by a Tuned Port Injection system with individual injectors for each cylinder.

Exhaust system

Every model covered by this manual is equipped with an exhaust system consisting of two four-into-one exhaust headers connected to exhaust pipes that meet behind the transmission, just in front of the catalytic converter. Behind the converter, the exhaust splits into two pipes again, leading to dual mufflers. The exhaust system was equipped with a single, three-stage catalytic converter until 1986, when two additional reducing converters were added, one to each exhaust pipe between the exhaust manifold and the main converter.

2 Fuel pressure relief

Warning: *Gasoline is extremely flammable, so take extra precautions when you work on any part of the fuel system. Don't smoke or allow open flames or bare light bulbs near the work area, and don't work in a garage where a natural gas-type appliance (such as a water heater or a clothes dryer) with a pilot light is present. Since gasoline is carcinogenic, wear latex gloves when there's a possibility of being exposed to fuel, and, if you spill any fuel on your skin, rinse it off immediately with soap and water. Mop up any spills immediately and do not store fuel-soaked rags where they could ignite. The fuel system is under constant pressure, so, if any fuel lines are to be disconnected, the fuel pressure in the system must be relieved first. When you perform any kind of work on the fuel system, wear safety glasses and have a Class B type fire extinguisher on hand.*

Note: *After the fuel pressure has been relieved. it's a good idea to lay a shop towel over any fuel connection to be disassembled, to absorb the residual fuel that may leak out when servicing the fuel system.*

1   The fuel system remains under pressure even after the engine has been shut off for an extended time. Therefore it is necessary to relieve the pressure in the fuel system before any work is done on fuel injection components or lines.

2   Remove the cover from the fuse block located in the right end of the dashboard.

3   Pull the fuel pump fuse from the fuse block to disable the pump. The fuel pump fuse is identified by the letters FP.

4   Start the engine and let it run until it dies from lack of fuel, then crank the engine over for several seconds with the starter to insure that all pressure has been eliminated from the system.

5   Be sure to replace the fuel pump fuse when work is completed on the fuel system.

3 Fuel filter - replacement

Refer to illustration 3.3

Note: *On 1992 and later vehicles the right catalytic converter is located next to the fuel filter. It will be necessary to allow the vehicle to cool for a minimum of one hour before attempting to service the fuel filter. The battery must be disconnected to prevent shorting the starter and the starter electrical components also located near the fuel filter. Expect minimum fuel spillage of up to one quart during filter service.*

1   Relieve fuel pressure (Section 2). See the **Warning** in Section 2.

2   Raise the vehicle and support it securely on jackstands.

3   Using a backup wrench, disconnect the fuel line-to-fuel filter fittings **(see illustration)**.

4   Loosen the fuel filter clamp bolt at the right frame rail.

5   Remove the filter.

6   Place the new filter in position and tighten the filter clamp bolt securely. Make sure the arrow on the filter points toward the engine.

7   The remaining steps of the installation procedure are the reverse of removal.

4 Fuel pump - check

Refer to illustrations 4.2, 4.6, 4.12, 4.13a, 4.13b and 4.15

1   Before testing the fuel pump, check the fuel pump fuse (FP in the fuse box on the right end of the dashboard). If the fuse is blown, replace it before proceeding.

3.3 To remove the fuel filter (located on the right frame rail), disconnect the fuel line fittings from both ends of the filter and loosen the clamp bolt enough to slide the filter out

4.2 The Assembly Line Communications Link (ALCL) diagnostic connector is located underneath the instrument panel, just to the left of the console - to energize the fuel pump, apply battery voltage to terminal G (arrow)

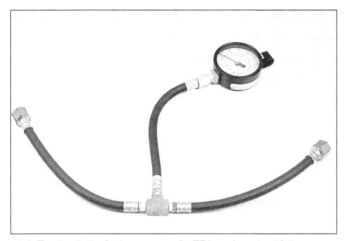

4.6 To check the fuel pressure of a TBI-equipped vehicle, use a gauge capable of reading between 9 and 13 psi and short lengths of hose with a plastic T-fitting and hose clamps

4.12 Measure fuel flow at the fuel feed line - it's necessary to remove the MAF sensor, intake ducting and air control valve to attach the hose to the fuel feed line

2   Remove the plastic cover from the Assembly Line Communications Link (ALCL) located underneath the instrument panel, just to the left of the console. Energize the fuel pump by applying battery voltage to terminal G **(see illustration)**. You should hear the pump come on inside the fuel tank. If the pump does not come on, either there is a loose connection or an open in the fuel pump circuit, or the fuel pump is faulty. Check the entire fuel pump electrical circuit (see Wiring Diagrams) before diagnosing the fuel pump as faulty. If the circuit checks out, replace the pump.
3   If the pump runs, check the fuel pressure.

TBI fuel pressure check

4   Remove the air cleaner and plug the Thermac intake vacuum port on the TBI unit.
5   Relieve fuel pressure (Section 2). See the **Warning** in Section 2. Remove the fuel line between the front and rear throttle bodies. Use a back-up wrench to prevent the fuel nut from turning when loosening the fuel line fittings.

6   Connect a fuel injection pressure gauge to two short sections of rubber hose with a T-fitting **(see illustration)** and install it between the two TBI balance line fittings. The gauge must be capable of reading fuel pressure between 9 and 13 psi.
7   Start the engine and observe the fuel pressure reading. It should be between 9 and 13 psi.
8   If there is no pressure (even though the fuse is okay and the pump is running), the inline fuel filter is probably clogged. Replace it and check the pressure again. If there is still zero pressure, the fuel feed line itself may be plugged. If the fuel feed line is clear, check the fuel pump inlet filter (Section 5).
9   If the pressure is below 9 psi, the inline fuel filter may be restricted. Replace it and check the pressure again. If it is still low, check for a loose fuel line fitting or a punctured fuel line. Also inspect both TBI injectors and the fuel meter cover for a leak. If nothing is leaking, disconnect the electrical connectors to both injectors, block the fuel return line by pinching the flexible hose portion and

apply 12 volts to the ALCL pump test terminal G and note the indicated pressure. If the pressure is now above 13 psi, check the line between the fuel pump and the test gauge. If the line is okay, replace the rear TBI fuel meter cover. If the pressure is still below 9 psi, replace the fuel pump.
10   If the pressure is above 13 psi, the problem is either a plugged fuel return line or a faulty fuel meter cover.
11   Remove the fuel pressure gauge and install the steel fuel line between the throttle bodies. Remove the plug covering the Thermac vacuum port on the TBI unit and install the air cleaner.

TPI fuel flow check

12   Relieve fuel pressure (Section 2). See the **Warning** in Section 2. Disconnect the fuel feed line from the fuel rail and connect a hose to the feed line. Run the other end of the hose into a suitable unbreakable container **(see illustration)**.
13   If you have an assistant to monitor the hose and container, the easiest way to acti-

4.13a Remove these two nuts (arrows) to remove the fuel pump relay from the firewall

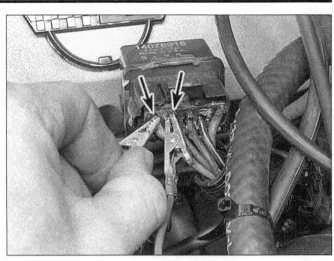

4.13b Jump terminals D and E (arrows) of the fuel pump relay plug to energize the fuel pump

vate the fuel pump is at the Assembly Line Communications Link (ALCL) underneath the dashboard, just to the left of the console. Remove the plastic cover from the ALCL diagnostic connector and apply battery voltage to terminal G of the fuel pump test connector **(see illustration 4.2)**. However, if you do not have an assistant, you can keep an eye on the container while energizing the fuel pump by using the fuel pump relay located on the engine compartment firewall. Remove both relay mounting nuts **(see illustration)**, disconnect the relay plug and thread it out from between its mounting bracket and the windshield wiper motor. Using a jumper wire with alligator clips on both ends, jump terminals D and E of the relay plug **(see illustration)** to apply battery voltage to the pump.

14   The fuel pump should supply 1/2 pint or more in 15 seconds. If the fuel system flow is below the minimum, inspect the fuel system for a restriction between the pump and the fuel rail. If there is no restriction, check the fuel pump pressure.

TPI fuel pressure check

1985 through 1991 models

15   Relieve fuel pressure (Section 2). See the **Warnings** in Section 2. Remove the cap from the Schrader valve on the right fuel rail and install a pressure gauge **(see illustration)**. The gauge must be capable of reading between 34 and 39 psi.

16   Start the engine and observe the fuel pressure reading. It should be between 34 and 39 psi.

17   If there is no pressure (even though the pump is running), the inline fuel filter is probably clogged. Replace it and check the pressure again. If there is still zero pressure, the fuel feed line may be clogged. If the fuel feed line is clear, check the fuel pump inlet filter (Section 5).

18   If the fuel pressure is below 34 psi, the inline fuel filter may be restricted. Replace it and check the pressure again. If it is still low,

check for a loose fuel line fitting or a punctured fuel line. Also inspect the pressure regulator, fuel rail, cold start valve and the injectors for a leak. If nothing is leaking, block the fuel return line and apply 12 volts to the fuel pump test connector, terminal G of the ALCL. Note the indicated pressure. If the pressure is now above 39 psi, check for a restriction in the fuel feed line between the fuel pump and the test gauge. If the pressure is still less than 34 psi, replace the fuel pump.

19   If the fuel pressure is above 39 psi, disconnect the fuel return line flexible hose and attach a 5/16-inch ID flexible hose to the pressure regulator side of the return line. Insert the other end in an approved gasoline container. Note the fuel pressure within two seconds after the ignition is turned to the On position. If pressure is still above 39 psi, replace the regulator. If the fuel pressure is now between 34 and 39 psi, the return line is restricted.

4.15 To determine the fuel pressure, attach a pressure gauge capable of reading between 34 and 39 psi to the Schrader valve on the right fuel rail

1992 and later models

20   Relieve fuel pressure (Section 2). See the **Warning** in Section 2. With the ignition off attach a fuel pressure gauge to the fuel pressure fitting on the fuel rail.

21   While observing the fuel gauge, turn the ignition on. The fuel pump will run approximately 2 seconds.

22   As the pump runs the pressure should be between 41 to 47 psi. When the pump stops running the pressure should drop slightly and hold steady.

23   If the pressure is within specification, start the engine and observe the fuel pressure. It should be 3 to 10 psi lower than in step 22.

24   If the fuel pressure is less than specified check the fuel quantity, the fuel filter and fuel lines for restrictions and leaks, if none are found replace the fuel pump.

25   If the fuel pressure is in excess of specified, check the fuel return line for restrictions, if none are found replace the pressure regulator.

---

5   Fuel pump - removal and installation

Warning: *Gasoline is extremely flammable, so take extra precautions when you work on any part of the fuel system. Don't smoke or allow open flames or bare light bulbs near the work area, and don't work in a garage where a natural gas-type appliance (such as a water heater or a clothes dryer) with a pilot light is present. Since gasoline is carcinogenic, wear latex gloves when there's a possibility of being exposed to fuel, and, if you spill any fuel on your skin, rinse it off immediately with soap and water. Mop up any spills immediately and do not store fuel-soaked rags where they could ignite. The fuel system is under constant pressure, so, if any fuel lines are to be disconnected, the fuel pressure in the system must be relieved first. When you perform*

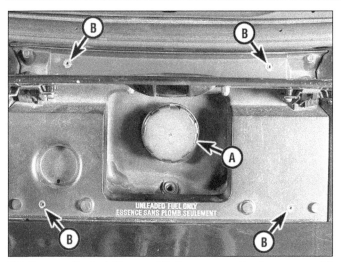

5.2 Remove the fuel filler cap (A), then remove the four screws (B) and the fuel filler door assembly

5.4 Peel the rubber fuel filler neck housing loose from the body, disconnect the overflow return hose (not visible) and remove the housing, then push the overflow hose aside

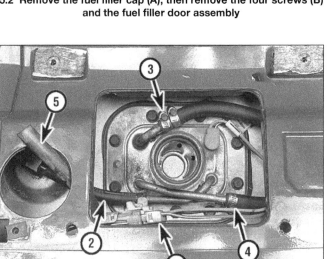

5.5 Before you can remove the fuel pump/fuel level sender assembly, disconnect the following items

1   Fuel pump/fuel level sender electrical connector
2   Fuel vapor hose
3   Fuel feed line
4   Fuel return line
5   Overflow hose (already disconnected) should be
    pushed aside

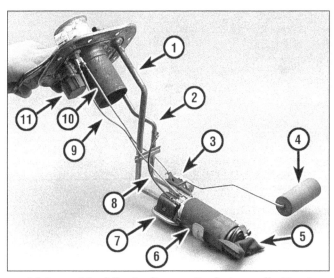

5.8 Inspect the following items on the fuel pump/fuel level sender assembly

1   Fuel feed line   7   Pulsation damper
2   Fuel return line   8   Fuel pump ground wire
3   Fuel level sender   9   Fuel pump hot wire
4   Fuel level float   10   Fuel level sender hot wire
5   Fuel pump intake strainer   11   Splash cup liquid vapor
6   Fuel pump   separator

any kind of work on the fuel system, wear safety glasses and have a Class B type fire extinguisher on hand.

Removal

Refer to illustrations 5.2, 5.4, 5.5 and 5.8

1   Disconnect the cable from the negative terminal of the battery.
2   Remove the fuel filler door (see illustration).
3   Remove the fuel cap.
4   Peel the fuel tank filler neck housing loose from the body (see illustration), disconnect the drain hose from the bottom, push the hose aside and remove the filler neck housing.

5   Disconnect the fuel pump/fuel level sender electrical connector. Loosen the fuel vapor, feed and return line hose clamps and disconnect all three hoses from the fuel pump/fuel level sender cover (see illustration).
6   Remove the screws that attach the fuel pump/fuel level sender assembly to the tank.
7   Remove the fuel pump/fuel level sender assembly and gasket.
8   If you are installing the same pump, inspect the pump intake filter strainer for tears, debris or deterioration. If it's damaged, replace it. Inspect the pump ground and hot wires and the fuel level sender unit hot wire for damage (see illustration). If damage to

any wire is evident, repair it before installing the pump.

Installation

Refer to illustration 5.11

9   The fuel pump/fuel level sender cover gasket must be installed with the notched corner at the right rear corner of the fuel tank opening.
10   Inspect the fuel pump/fuel level sender cover screw O-rings before installing the screws. If the O-rings are deteriorated, replace them.
11   Thread the fuel pump/fuel level sender cover screws in by hand until they are engaged with the threaded inserts in the fuel

tank bladder (liner), then tighten them **(see illustration)**.

12   The remaining installation steps are the reverse of the removal procedure.

6   Fuel tank - removal and installation

Warning: *Gasoline is extremely flammable, so take extra precautions when you work on any part of the fuel system. Don't smoke or allow open flames or bare light bulbs near the work area, and don't work in a garage where a natural gas-type appliance (such as a water heater or a clothes dryer) with a pilot light is present. Since gasoline is carcinogenic, wear latex gloves when there's a possibility of being exposed to fuel, and, if you spill any fuel on your skin, rinse it off immediately with soap and water. Mop up any spills immediately and do not store fuel-soaked rags where they could ignite. The fuel system is under constant pressure, so, if any fuel lines are to be disconnected, the fuel pressure in the system must be relieved first. When you perform any kind of work on the fuel system, wear safety glasses and have a Class B type fire extinguisher on hand.*

Removal

1   Disconnect the cable from the negative terminal of the battery.
2   Disconnect the fuel line from the fuel filter on the frame rail and drain the fuel tank. If the fuel will not drain through the pump, a siphon hose will have to be used to drain the tank. **Warning:** *Do not use your mouth to start the siphon. Use a hand pump or siphon pump designed specifically for gasoline.*
3   Remove the fuel tank filler door (Section 5).
4   Remove the fuel tank filler neck housing and drain hose (see Section 5).
5   Disconnect the fuel feed hose, fuel return hose, vapor hose and electrical connector from the fuel level sender and pump assembly (see Section 5).
6   Remove the license plate to gain access to the fascia-to-impact bar bolts.
7   Remove the two carriage bolts securing the fascia to the impact bar.
8   Raise the vehicle and place it securely on jackstands.
9   Remove the spare tire and tire carrier.
10   Remove the rear section of the exhaust system as a complete unit by disconnecting it from the rear mounting flange of the catalytic converter.
11   Detach both inner fender braces.
12   Remove both inner fender panels.
13   Disconnect the antenna ground strap.
14   Remove the bolt securing the fuel vapor line to the left side fuel tank strap.
15   Disconnect the fuel tank cables from the right and left rear stabilizer shaft brackets.
16   Remove all rear lights - side marker, tail, backup, license plate and spare tire (Chapter 12).

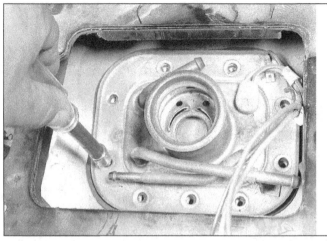

5.11 Thread the fuel pump/fuel level sender cover screws by hand until they are engaged with the threaded inserts in the fuel tank bladder (liner), then tighten them

17   Detach the bottom edge of the fascia from the energy absorber pad (Chapter 11).
18   Remove the three nuts securing each side of the fascia to the horizontal body retainer (Chapter 11).
19   Remove the nuts from the right and left vertical retainers securing the fascia to the body (Chapter 11).
20   Take out the four frame extension bolts (two per side).
21   Loosen but do not remove the four remaining frame extension bolts.
22   With the aid of an assistant, remove the remaining frame extension bolts and pull the fuel tank and frame assembly to the rear while pushing the cover out. Tilt the assembly down to clear the cover.
23   Detach the tank straps and remove the tank. If the fuel tank is being replaced, transfer the cables, the fuel level sender and pump and the fuel vapor connector to the new tank.

Installation

24   Position the tank on the frame extension and secure it to the tank supports with straps.
25   With the aid of an assistant, install the tank and frame assembly in the vehicle and secure it with the bolts.
26   Install the nuts on the right and left vertical retainers.
27   Thread the three nuts securing each side of the fascia to the horizontal body retainer (Chapter 11).
28   Install the screws securing the bottom edge of the fascia to the energy absorber pad (Chapter 11).
29   Replace all the rear lights - side marker, tail, backup, license plate and spare tire (Chapter 12).
30   Install the fuel tank cable to the stabilizer shaft brackets.
31   Install the bolt that secures the fuel vapor line to the left side fuel tank strap.
32   Connect the antenna ground strap.
33   Install both inner fender panels.
34   Replace both inner fender braces.
35   Install the exhaust system.
36   Install the spare tire and carrier.
37   Lower the vehicle.

38   Install the bolts securing the fascia to the impact bar.
39   Install the license plate.
40   Connect the fuel feed hose, fuel return hose, vapor hose and fuel level sender electrical connector.
41   Install the fuel tank filler neck housing and drain hose.
42   Install the fuel tank filler door.
43   Connect the battery ground cable.
44   Add fuel to the tank.

7   Fuel tank - cleaning and repair

1   Any repairs to the fuel tank or filler neck should be carried out by a professional who has experience in this critical and potentially dangerous work. Even after cleaning and flushing of the fuel system, explosive fumes can remain and ignite during repair of the tank.
2   If the fuel tank is removed from the vehicle, it should not be placed in an area where sparks or open flames could ignite the fumes coming out of the tank. Be especially careful inside garages where a natural gas appliance is located, because the pilot light could cause an explosion.

8   Fuel injection system - general information

Crossfire Throttle Body Injection (TBI)

The Corvette TBI assembly consists of two throttle bodies diagonally mounted on the intake manifold. In effect, the throttle body injection system is similar to but simpler than a conventional carburetor. Though superior to carbureted fuel induction systems, TBI is somewhat less efficient than the port injection used on 1985 and later Corvettes.

Each injector is an electrically-operated Bosch-type solenoid. System fuel pressure is maintained between 9 and 13 psi by a pressure regulator in the rear TBI unit. The Electronic Control Module (ECM) controls the duration of time that each injector is opened or closed. The open time determines the amount of fuel delivered to the engine.

Electronic control module

To determine how much fuel is required at any given moment, the Electronic Control Module processes signals from information sensors which monitor engine temperature, throttle position, vehicle speed, manifold vacuum and the exhaust oxygen content, updating the information every 1/10 second.

Idle Air Control (IAC) valve

The idle speed on the TBI is also controlled by the ECM, through an idle air control valve. When the engine is cold or when accessory loads, such as air conditioning, necessitate a higher idle speed, the ECM activates a "stepper motor," which opens the IAC valve to allow more air into the system. At the same time, the ECM increases opening time of the fuel injectors to maintain the proper fuel/air ratio.

Fuel pump

While the opening time of the injectors is controlled by the ECM, the availability of a sufficient quantity of fuel at the injectors is determined by the fuel pressure, which must be constantly maintained at 9 to 13 psi. This is accomplished through the use of an electric pump mounted in the fuel tank. The pump intake is in a special sump in the fuel tank so that neither a low fuel level, hard cornering, acceleration, deceleration or any combination of the above can uncover the pump inlet and cause a pressure drop in the fuel system.

Fuel pressure accumulator and regulator

Incoming fuel flow from the pump always exceeds injector needs, so excess fuel is directed through an accumulator in the first TBI unit to compensate for the momentary fuel pressure drop between the two units due to injector openings. Fuel from the accumulator is carried through the connecting fuel tube to the rear unit. Again, the incoming fuel flow exceeds the needs of the injector, so the excess fuel passes through a pressure regulator, which maintains a constant pressure in the system.

Tuned Port Injection

The Corvette multi-port fuel injection system utilizes an injector solenoid at each intake port. The Electronic Control Module (ECM) controls the opening time of the injectors, which determines the amount of fuel delivered to the engine.

All of the injectors operate simultaneously. Half of the fuel required by each cylinder is injected with each injector pulse. The pulses occur once every crankshaft revolution. Therefore, the full charge needed is available in the intake port for induction into the cylinder when the intake valve opens every other revolution.

Electronic control module

In determining how much fuel is required at any given moment, the Electronic Control Module processes signals from sensors recording engine coolant temperature, exhaust oxygen content, throttle position, intake air mass, engine rpm, vehicle speed and accessory load.

Mass Air Flow (MAF) sensor

The hot wire type Mass Air Flow (MAF) sensor measures the mass (weight) of the air entering the intake manifold, rather than the volume, which gives better control of the air/fuel mixture. The mass of any given volume of air is dependent upon the temperature of the air. Cold air weighs more than warm or hot air, and therefore the colder the air is, the more fuel is required to maintain the proper fuel/air ratio.

On the Corvette air mass sensor system a heated film is used to measure the intake air mass. The temperature of the incoming air is constantly monitored and the air mass sensor plate is maintained at exactly 75 degrees above the temperature of the air. Since the incoming air, passing over the plate, acts to cool the plate, the measurement of the amount of electrical energy needed to maintain that 75 degree temperature differential accurately indicates the mass of air entering the engine, and therefore the amount of fuel necessary to mix with that mass to provide the proper fuel/air ratio.

Idle air control

The idle speed on the port injection system is also controlled by the electronic control module, through an idle air control bypass channel. When the engine is cold or when accessory loads, such as air conditioning, necessitate a higher idle speed, the ECM, through a stepper motor, opens the bypass valve to provide more air into the system. At the same time, the ECM increases the fuel injector open time to maintain the proper fuel/air ratio.

Fuel pump

While the open time of the injectors is controlled by the electronic control module, the actual fuel delivery is determined by the fuel pressure, which must be maintained at a constant level. This is done by a positive displacement roller vane pump mounted in the fuel tank. The pump intake is mounted in a special sump in the fuel tank so that a low fuel level, combined with hard cornering, acceleration or deceleration, cannot uncover the pump inlet, causing a pressure drop.

Fuel pressure regulator

A fuel pressure regulator maintains the required pressure in the system by returning unneeded fuel to the tank. A fuel accumulator is used to dampen the pulses that would otherwise be set up in the fuel rail and fuel line when the injectors open and close.

9   TBI system - diagnosis, removal and installation

Diagnosis

If a malfunction in the TBI system is suspected, note the symptom(s), then study this Section carefully before disassembling or removing anything. If any of the following problems are noted, check, repair or replace the appropriate parts as required.

Flooding

Inspect the large and small fuel injector O-rings for damage, such as cuts, distortion, etc. Be sure that the back-up washer is located beneath the large (upper) O-ring. If, after disconnecting the fuel injector electrical connections, the fuel injector continues to supply fuel, replace the injector as required.

Hesitation

Inspect the fuel injector fuel filter to see if it is plugged or dirty. Clean or replace as necessary.

Check for incorrect installation of injectors (front versus rear TBI unit). Injectors must not be interchanged. Check the part number and always use the correct injector for the specific application.

If a fuel pressure check indicates incorrect fuel inlet or return pressure readings, check for restricted passages, incorrect fuel nut position (front versus rear TBI unit) or an inoperative fuel pressure regulator. Repair or replace as required.

Check for incorrect throttle synchronization. Refer to Throttle valve synchronization for the checking and adjusting procedure.

Hard Starting - poor cold operation

See the items listed under 'Hesitation' above.

Check for proper IAC motor actuation and positioning following vehicle shutdown.

Rough idle

Inspect the large and small fuel injector O-rings for damage, such as cuts, distortion, etc. Be sure that the steel back-up washer is located beneath the large (upper) O-ring.

Inspect the fuel injector filter for damage, dirt, etc. Clean or replace as necessary.

Check for incorrect installation of injectors. Front and rear TBI unit injectors must not be interchanged. Use only the correct injector for the specific application.

If, after disconnecting the fuel injector electrical connections, the fuel injector continues to supply fuel, replace the injector.

Check for improper throttle synchronization. Refer to Throttle valve synchronization for the checking and adjusting procedure.

Check for proper IAC motor actuation and positioning following vehicle shutdown.

Note: *The procedure for removing either of the TBI units from the assembly is similar. But there are some important differences. The*

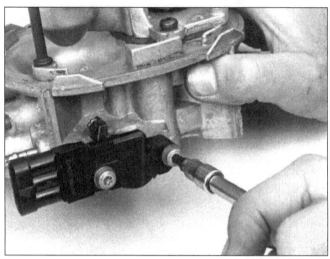

9.14 Scribe an alignment mark between the TPS and the throttle body housing, then remove the two Torx mounting screws, lockwashers and retainers

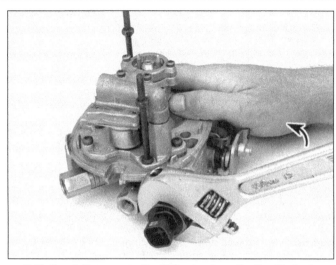

9.15 Carefully unscrew the IAC valve from the throttle body housing with a large wrench

major one is that while the rear TBI unit can be removed from the intake manifold by itself, the front unit, because of the manner in which the throttle linkages of the two units are connected, cannot be removed until the rear unit has already been removed. Practically speaking, this means that if you need to remove the front TBI unit, you will have to remove the rear unit as well. It is therefore recommended that you clean, inspect and overhaul both units as necessary any time they must both be removed. Be sure to read the following instructions carefully. Some steps refer to the front unit, some to the rear unit and some to both units. If you are removing only the rear unit, disregard references to removal of items from the front unit. Also, although the following procedure details the complete disassembly and reassembly of a TBI unit, components such as the fuel meter covers, the fuel injectors, the idle air control valves, the throttle position sensor or the fuel balance tube and fittings between the two TBIs can be removed/replaced with both TBI units installed on the intake manifold.

Removal

1   Relieve the pressure in the fuel system (Section 2). See the **Warning** in Section 2.
2   Disconnect the negative cable from the battery.
3   Remove the air cleaner assembly.
4   If you are removing either the front or both TBI units, disconnect the electrical leads from the fuel injectors and Idle Air Control (IAC) valves on both units. If you are removing only the rear unit, you may leave the above items connected to the front unit.
5   Disconnect the electrical connector from the throttle position sensor on the rear TBI unit.
6   Remove the throttle cable from the throttle linkage of the rear TBI unit (Section 11).
7   Disconnect the throttle rod retaining clip

from the front TBI throttle lever stud. A new clip, supplied with the service repair package, must be used when reconnecting the two throttle bodies. One throttle rod end bearing is permanently attached to the throttle lever stud of the rear unit and must not be removed from it. **Note:** *Should the throttle rod and bearing assembly need replacement, it will be necessary to replace the entire rear throttle body.*
8   If your vehicle is equipped with an automatic transmission and/or cruise control, disconnect the transmission detent and/or cruise control cables from the throttle linkage.
9   Label the location of all vacuum hoses leading to the TBI units and disconnect the hoses.
10   Remove the fuel line between the two TBI units. If you are removing the front TBI unit, disconnect the fuel inlet and return lines as well. Use a backup wrench to prevent the fuel nuts from turning on the TBI units.
11   Remove the short bolt, stud/bolt and stud from the rear TBI unit.
12   Lift the rear TBI unit off the manifold cover locator pins, then raise it above the front unit and slide the throttle rod end bearing off the front unit throttle lever stud. Remove the rear TBI unit.
13   If you are removing the front TBI unit, remove the short bolt, stud/bolt and stud. Then lift the front TBI unit and gasket off the manifold.

Disassembly

Refer to illustrations 9.14, 9.15, 9.17, 9.18a, 9.18b, 9.19a, 9.19b and 9.20
Note: *It is recommended that the TBI unit be mounted on a holding fixture during overhaul. Without the use of such a fixture, damage to the throttle plate may occur. If you do not have a suitable carburetor mounting fixture, avoid moving the throttle linkage when the TBI unit is laying flat on the workbench.*

Throttle Position Sensor (TPS)

14   If you are replacing the TPS or overhauling the rear TBI unit, scribe an alignment mark between the TPS and the throttle body housing, then remove the TPS Torx mounting screws, lockwashers and retainers and remove the TPS **(see illustration)**. Note the position of the retainers in relation to the TPS before removal. **Caution:** *Failure to scribe an alignment mark between the TPS and the throttle body housing will make a digital voltmeter necessary for realignment of the TPS.*

Idle Air Control (IAC) valve

15   Unscrew the Idle Air Control (IAC) assembly from the throttle body **(see illustration)**.

Fuel meter cover/accumulator (front), fuel meter cover/pressure regulator (rear)

16   Remove the fuel meter cover mounting screws and separate the cover from the body. Note the positions of the two shorter fuel meter cover screws - they must be reinstalled in the same positions. **Caution:** *Do not immerse the fuel meter cover in solvent. It might damage the pressure regulator diaphragm and gasket.*
17   The fuel meter cover, which includes the fuel pressure regulator (rear TBI unit) or accumulator (front unit), is serviced only as a complete assembly. If either the regulator/accumulator or cover requires replacement, the entire assembly must be replaced. **Warning:** *Both the fuel pressure regulator and the accumulator are under heavy spring tension. The four screws securing the pressure regulator/accumulator to the fuel meter cover* **(see illustration)** *should not be removed, as personal injury could result.*

Fuel injector

Note: *If you are removing both injectors, be sure to label each injector to ensure that it is*

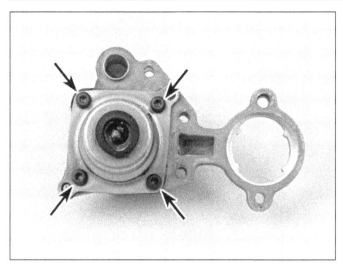

9.17 The fuel meter cover and the accumulator (front unit) or pressure regulator (rear unit) are an integral assembly that cannot be serviced separately - do not remove the four screws (arrows) under any circumstances

9.18a Place a screwdriver shank on top of the fuel meter cover gasket surface to serve as a fulcrum and, using another screwdriver, carefully pry the fuel injector from th fuel meter body

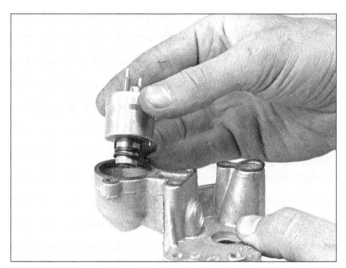

9.18b Note the position of the terminals on top and the dowel pin on the bottom of the injector in relation to the fuel meter body when you lift the injector out of the body - the injector has to be reinstalled in exactly the same position

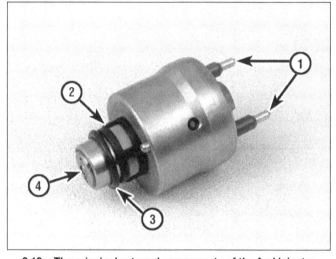

9.19a The principal external components of the fuel injector

1	Terminals	3	O-ring
2	Fuel filter	4	Nozzle

installed in the same fuel meter body from which it is removed.

18   Place a screwdriver shank on top of the fuel meter cover gasket surface to serve as a fulcrum and, using another screwdriver, carefully pry the fuel injector out of the fuel meter body **(see illustration)**. The old fuel meter cover gasket can be used on the fuel meter body to prevent the surface from being scratched by the screwdriver during the procedure. Note the position of the terminals on top and the dowel pin on bottom of the injector in relation to the fuel meter body when you lift the injector out of the body **(see illustration)**. **Caution:** *Do not push the injector out from underneath as this could damage the injector tip.*

19   Remove the small O-ring from the nozzle end of the injector **(see illustration)**. Carefully rotate the injector fuel filter back

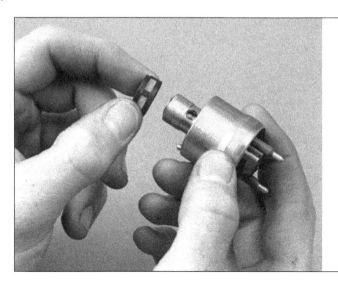

9.19b Carefully rotate the fuel filter back and forth and remove it from the injector

9.20 If you are removing the fuel meter body from the throttle body housing, first remove the fuel inlet and return fittings

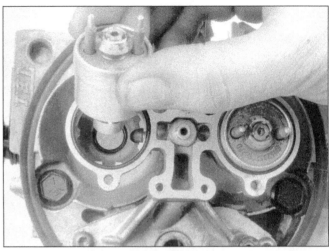

9.32 Make sure the lug is aligned with the groove in the bottom of the fuel injector cavity, then push firmly down

and forth and remove the filter from the base of the injector **(see illustration)**. Gently clean the filter in solvent and allow it to drip dry. It is too small and delicate to dry with compressed air. **Caution:** *The fuel injector itself is an electrical component. Do not immerse it in any type of cleaning solvent.*

Fuel meter body

20   Remove the fuel inlet and return line fittings and gaskets from the fuel meter body, marking their locations for proper reassembly **(see illustration)**.
21   Remove the three Torx screws and lockwashers from the fuel meter body, then remove the fuel meter body and insulator gasket from the throttle body assembly.
22   Remove the old gasket from the fuel meter cover and discard it. Remove the large O-ring and steel back-up washer from the upper counterbore of the fuel meter body injector cavity.

Reassembly

Refer to illustrations 9.32, 9.33, 9.34, 9.39 and 9.74
23   Clean all metal parts in a cold, immersion-type cleaner and blow them dry with compressed air. **Caution:** *The throttle position sensor, idle air control assembly, fuel meter cover/fuel pressure regulator, fuel injector, fuel injector filter and all diaphragms and other rubber and plastic parts should not be immersed in cleaner, as it will cause swelling, hardening and distortion, and in some cases might even dissolve the part. Clean these parts by hand with clean shop rags. Make sure all air and fuel passages are clear. Any thread locking compound residue remaining on the idle air control assembly mounting threads after cleaning should be allowed to remain.*
24   Inspect the TBI casting mating surfaces for damage that could affect gasket sealing.

Fuel meter body

25   Install a new fuel meter body insulator gasket on the throttle body assembly. The

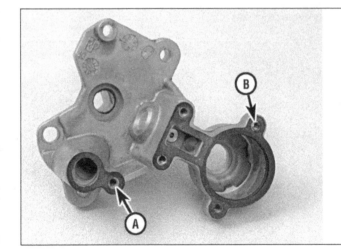

9.33 Place a new fuel outlet passage gasket (A) and a new fuel meter cover gasket (B) on the fuel meter cover

gasket is installed correctly when cutout portions of the gasket match the opening in the throttle body assembly.
26   Install the fuel meter body onto the insulator gasket. Apply a thread-locking compound such as Loctite 262 to the threads of the three fuel meter body screws in accordance with the instructions on the package. Install the fuel meter body screws and lockwashers and tighten them to the specified torque.
27   Install the fuel inlet and return fittings, with new gaskets, in the same locations at which you marked them before removal, and tighten them to the specified torque.

Fuel injector

28   Install the new fuel injector nozzle filter on the end of the fuel injector with the larger end of the filter facing the injector so that the filter covers the raised rib at the base of the injector. Use a twisting motion to position the filter against the base of the injector.
29   Lubricate a new small O-ring with automatic transmission fluid. Push the O-ring onto the nozzle end of the injector until it presses against the injector fuel filter.
30   Insert the steel back-up washer into the

top counterbore of the fuel meter body injector cavity.
31   Lubricate a new large O-ring with automatic transmission fluid and install it directly over the back-up washer. Be sure that the O-ring is properly seated in the cavity and is flush with the top of the fuel meter body casting surface. **Caution:** *The back-up washer and large O-ring must be installed before the injector or improper seating of the large O-ring could cause fuel to leak.*
32   Install the injector in the cavity in the fuel meter body. If you have removed both injectors, make sure the injector goes back into the same fuel meter body from which it was removed. Align the raised lug on the injector base with the cast-in notch in the fuel meter body cavity. Push straight down on the injector with both thumbs **(see illustration)** until it is fully seated in the cavity. **Note:** *The electrical terminals of the injector should be approximately parallel to the throttle shaft.*
33   Install a new fuel outlet passage gasket and a new fuel meter cover gasket on the fuel meter cover **(see illustration)**.
34   Install a new dust seal into the recess on the fuel meter body **(see illustration)**.
35   Attach the fuel meter cover to the fuel

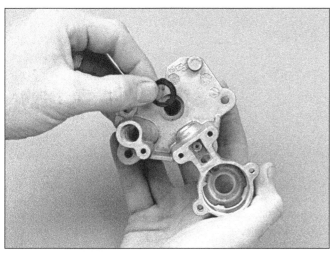

9.34 Install a new dust seal into the recess on the fuel meter body

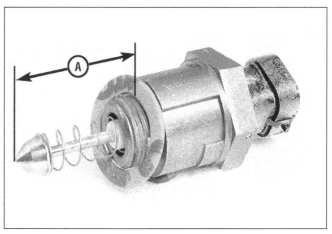

9.39 Distance (A) should be less than 1-1/8 inch for either type of Idle Air Control valve - if it isn't, determine which kind of valve you have and adjust it accordingly

meter body, making sure that the pressure regulator dust seal and cover gaskets are in place.

36 Apply thread locking compound to the threads of the fuel meter cover screws. **Note:** *Service repair kits include a small vial of thread compound with directions for use. If material is not available, use Loctite 262 or equivalent. Do not use a higher strength locking compound than recommended, as this may prevent subsequent removal of the screws or cause the screw head to break should removal become necessary. Install the screws (the two short screws go next to the injector) and tighten them to the specified torque.*

Idle Air Control (IAC) valve

37 Three different designs are used for the IAC pintle valve. The first design is a single taper, the second is a dual taper and the third is a blunt valve. Be sure to use the correct design when replacing an IAC valve.

38 Before installing the IAC valve, measure the distance that the valve is extended. The measurement should be made from the motor housing to the end of the cone. The distance should be no greater than 1-1/8 inch (28 mm). If the cone is extended too far, damage to the valve may occur when it is installed.

39 If the distance is too large, the pintle must be retracted the requisite amount to bring the above distance within specification. But first determine if the assembly is a Type I, with a collar at the electric terminal end of the valve, or a Type II, without a collar **(see illustration)**.

40 Retract the pintle on a Type I by pushing firmly on the end of the pintle until it is sufficiently compressed. A slight side-to-side motion may be helpful.

41 Retract the pintle on a Type II by pushing it in while turning it clockwise. Return the spring to its original position with the straight portion of the spring end aligned with the flat surface of the valve.

42 Install a new gasket on the IAC valve, screw the valve into the throttle body housing and tighten it to the specified torque. **Note:**

No further adjustment of the IAC valve is necessary. Once the TBI assembly is fully operational, the ECM will reset idle speed the first time that the vehicle is driven at 35 mph.

Preliminary installation steps

Note: *This section refers only to the reinstallation of both the original TBI units on the original manifold cover.*

43 Clean all old gasket material from the intake manifold mating surface, then install new front and rear throttle body mounting gaskets.

44 Place the front throttle body on the manifold cover locator pins, install the mounting bolts and tighten them to the specified torque.

45 Holding the rear unit above the front unit, install the throttle rod end bearing on the front TBI unit throttle lever stud.

46 Install the rear TBI unit onto the manifold cover locator pins, install the mounting bolts and tighten them to the specified torque.

47 Using needle-nose pliers, carefully install a new throttle rod and bearing assembly retaining clip (supplied in the service kit).

48 Proceed to the Throttle valve synchronization and Throttle rod alignment check procedures outlined below.

Replacement of TBI unit(s)

Note: *This section must be followed when replacing one or both throttle bodies with new ones. Replacement of one or more of these assemblies requires throttle valve synchronization and checking of throttle rod alignment.*

49 If tamper resistant plugs over the throttle stop screws have been installed, remove them in accordance with the following procedure. **Note:** *Replacement throttle bodies do not have throttle stop screw plugs installed.*

 a) *Use a punch to establish the location, over the centerline of the throttle stop screw, for the hole to be drilled.*

 b) *Drill a 5/32-inch (4 mm) diameter hole through the throttle body casting to the hardened plug.*

 c) *Use a 1/16-inch (2 mm) diameter punch to drive through the drilled hole and knock out the plug.*

50 If the front TBI unit throttle synchronizing screw has a welded retaining collar, grind off the weld. (A replacement throttle body may have a throttle synchronizing screw with a non-welded collar).

51 Block movement of the throttle lever by relieving the force of the heavy spring against the throttle synchronizing screw to prevent the levers from coming into contact. An Allen wrench can be inserted under the stop screw plate to do this. **Caution:** *If the lever is not blocked before the throttle synchronizing screw is removed, the screw may be damaged and reinstallation will be accomplished only with great difficulty.*

52 Remove the synchronizing screw and collar, discard the collar and reinstall the screw. Remove the block from the throttle lever. **Caution:** *The collar must be removed to prevent possible interference with the air cleaner.*

53 Install the throttle body mounting gaskets.

54 With either throttle body in place, install the throttle rod end bearing on the front unit throttle lever stud.

55 Install the other throttle body on the manifold cover locator pins.

56 Tighten all mounting bolts to the specified torque. Do not install the throttle rod and bearing assembly retaining clip yet.

Throttle valve synchronization - preliminary adjustment

Note: *The following adjustment procedure may be performed with the entire assembly on the vehicle or as a bench procedure.*

57 Turn both front and rear unit throttle stop screws counterclockwise enough to break contact with the throttle lever tangs.

58 Adjust the throttle synchronizing screw to allow both throttle valves to close. The throttle rod end bearing will move freely on the front unit throttle lever stud when both valves are closed.

59   Turn the front throttle stop screw clockwise slowly until it makes contact with the throttle lever tang. Turn it clockwise an additional 1/4-turn.
60   Turn the rear throttle stop screw clockwise slowly until it makes contact with the throttle lever tang. Turn it clockwise an additional 1/2-turn.
61   Final adjustments are made with the assembly installed on the vehicle and with the engine running.

Throttle rod alignment check

62   Actuate the rear TBI unit throttle lever to bring both units to the wide open throttle position by rotating the throttle valves and load them in the direction shown. Move the throttle rod toward the front unit casting boss.
63   Check the clearance between the shoulder of the stud and the side of the ball surface. Maximum clearance must not exceed 0.200-inch (5.0 mm). If the clearance is greater, the replacement assembly (manifold cover or throttle body) must be exchanged and the preliminary throttle valve synchronizing adjustment must be repeated.
64   Use needle-nose pliers to carefully install a new throttle rod and bearing assembly retaining clip (supplied in the service package).
65   Move both the front and rear throttle levers through the total throttle travel while simultaneously loading the throttle valves in the forward direction.
66   Check the clearance between the throttle rod and front throttle body casting boss. If the minimum clearance is less than 0.040-inch (1.02 mm), the replacement assembly must be exchanged. If the minimum clearance is at least 0.040-inch (1.02 mm), proceed to the Final installation steps.
67   If the assemblies were replaced in the previous Step, repeat the Throttle valve synchronization - preliminary adjustment and the Throttle rod alignment check Steps to ensure that both clearances are still correct.

Final installation steps

68   Reinstall the connecting fuel tube between the throttle body assemblies and tighten it to the specified torque. Use a back-up wrench to prevent the fuel nuts on the throttle bodies from turning.
69   Reconnect the fuel inlet and return line fittings to the front TBI assembly, using a back-up wrench on the fuel nuts to prevent them from turning. Make sure that the fuel line O-rings are not nicked, cut or damaged before connecting the lines.
70   Reattach the throttle cable, transmission detent and, if equipped, the cruise control linkage.
71   Reinstall the related vacuum hoses.
72   Connect the electrical connectors to the IAC valves, the fuel injectors and the TPS switch. Reinstall the wiring harness on the throttle bodies.
73   With the engine off, depress the accelerator pedal to the floor and release it, checking for free return of the pedal.

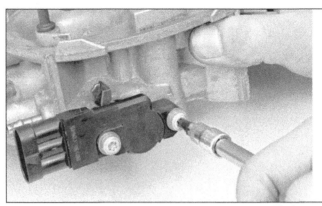

9.74 Make sure that you use the new TPS switch attaching screws (Torx), lockwashers and retainers that are included in the service kit and that the TPS pickup lever is located above the tang on the throttle actuator lever

74   With the throttle plate in the closed (idle) position, install the TPS switch onto the throttle body with the retainers and the two new screws and lockwashers which came in the service kit **(see illustration)**. **Caution:** *Do not reuse the old screws. Be sure to use the thread-locking agent supplied in the service kit on the screw threads.*
75   Make sure that the TPS is lined up in accordance with the alignment mark you made prior to removing it and that the throttle position sensor pickup lever is located above the tang on the throttle actuator lever. **Note:** *The following TPS adjustment procedure cannot be performed until the vehicle is running, and it requires the use of a digital voltmeter capable of reading 0.525 ± 0.075 volts. If you don't have this tool, the following procedure is impossible to perform. But you can drive the vehicle to a dealer immediately after reassembly of the TBI system and have the TPS adjusted by a dealer service department.*
76   Disconnect the TPS harness connector from the TPS switch. Use three jumper wires to reconnect the TPS connector to the TPS switch.
77   Hook up the digital voltmeter to terminals A and B and, with the ignition On but the engine not running, rotate the TPS switch to obtain a reading of 0.525 ± 0.075 volts at the closed throttle position. If you cannot obtain this reading, replace the TPS.
78   Tighten the screws, then recheck your reading to insure that the adjustment has not changed.
79   Turn the ignition Off, remove the jumper wires and reconnect the TPS harness connector to the TPS switch.
80   If the manifold cover or throttle bodies were replaced, adjust the minimum idle speed (Section 12) and synchronize the throttle valves (Steps 57 through 61).
81   Install the air cleaner. With the engine off, depress the accelerator pedal to the floor and release, checking for free return of the pedal.
82   Start the engine and check for fuel leaks.
83   Allow the engine to reach normal operating temperature.
84   On manual transmission equipped vehicles the idle speed will automatically be controlled when normal operating temperature is reached.

85   On vehicles with an automatic transmission, the idle air control unit will begin controlling idle speed when the engine is at normal operating temperature and the transmission is shifted into Drive.
86   If the idle speed is too high and does not regulate back to normal after a few moments, operate the vehicle at a speed of 45 mph (72 kph). At that speed the electronic control module will command the idle air control pintle to extend fully to the mating seat in the throttle body, which will allow the electronic control module to establish an accurate reference point with respect to pintle position. Proper idle regulation will result.

10   Tuned Port Injection system - removal and installation

Note: *Although the following procedure details the complete removal and installation of the principal components of the Tuned Port Injection system, the removal or replacement of some parts is independent from the rest of the system. Therefore, read the following procedures carefully.*

Removal

1985 through 1991 models

Refer to illustrations 10.4, 10.6, 10.8, 10.10, 10.11, 10.12, 10.15, 10.25, 10.26, 10.29, 10.36, 10.38, 10.39, 10.46, 10.48, 10.50, 10.51, 10.53 and 10.54

1   Relieve the fuel pressure (Section 2). See the **Warning** in Section 2.
2   Disconnect the negative cable from the battery.

Fuel pressure connection assembly

Note: *Removal and installation of the fuel pressure connection assembly is the only procedure in this Section that requires no preliminary removal of other fuel injection components. Unless you are only replacing the fuel pressure connection assembly seal, or the valve itself, skip the next three Steps and proceed to Step 6.*

3   Thoroughly clean the area around the fuel pressure connection assembly.
4   Remove the fuel pressure connection assembly. Note which end of the valve is threaded into the fuel rail **(see illustration)**.

10.4 Exploded view of the fuel pressure connection assembly - the ends of the valve are not the same so it is important to note which end threads into the fuel rail

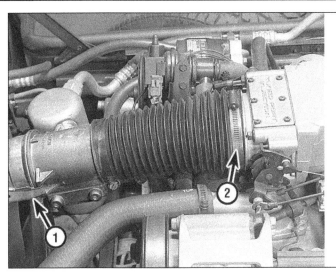

10.6 Disconnect the spring clips (1) attaching the MAF sensor to the air intake duct and loosen the hose clamp (2) attaching the flexible ducting to the throttle body

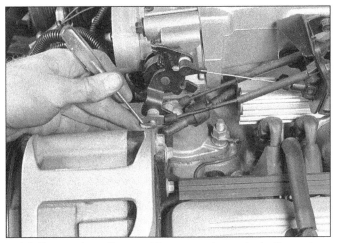

10.8 Two retainer clips must be removed - one locks the cruise control cable (the one being disconnected in the photo) and the other locks the throttle cable to the throttle shaft lever arm

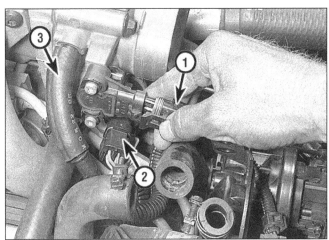

10.10 Disconnect the electrical connectors from the TPS (1) and the IAC valve (2), then detach the PCV hose (3) from the throttle body

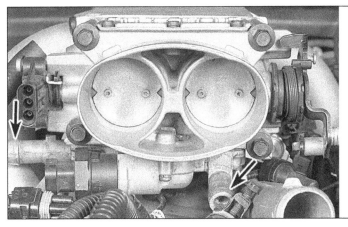

10.11 Loosen the hose clamps and detach the coolant hoses (arrows) from the throttle body coolant cover

Discard the old seal.

5   Install a new seal on the valve, install the fuel pressure connection assembly and tighten it to the specified torque.

Mass Air Flow (MAF) sensor

6   If you are only removing the MAF sensor and flexible air intake ducting to gain access to other fuel injection components, loosen the hose clamp which attaches the ducting to the throttle body and pop off the retaining clips on the side of the MAF sensor body **(see illustration)**. If you are replacing the MAF sensor itself, do not disconnect the ducting from the throttle body - loosen the hose clamp attaching the ducting to the rear of the MAF sensor and separate the ducting from the MAF sensor.

7   Pull the MAF sensor away from the air cleaner housing enough to disconnect the electrical connection on the bottom of the sensor, then remove the ducting and MAF sensor as an assembly and set them aside. **Caution:** *The MAF sensor is delicate - if you are going to reinstall the existing unit, handle it carefully.*

Throttle body

8   Detach the cruise control, accelerator and transmission TV cables from the throttle body **(see illustration)**.

9   Remove the three cable bracket bolts and separate the cable bracket from the plenum.

10   Unplug the wiring harness connectors from the Idle Air Control (IAC) and Throttle Position Sensor (TPS). Disconnect the PCV hose, which is attached to the throttle body right behind the TPS **(see illustration)**.

11   Drain the cooling system (Chapter 1), then loosen the hose clamps and detach the two coolant hoses from the throttle body fittings **(see illustration)**.

12   There is a vacuum hose between the left

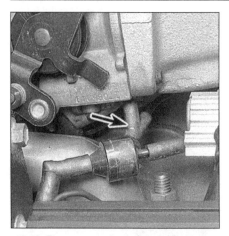

10.12 The vacuum hose (arrow) between the throttle body and the vacuum damper valve (the small black plastic cylinder) must be disconnected

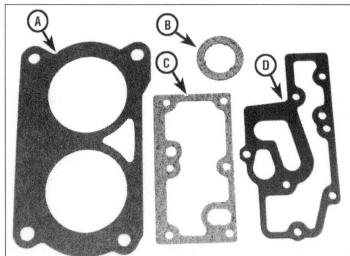

10.15 All of the throttle body gaskets should be replaced

A   Throttle body gasket
B   Idle Air Control (IAC) valve gasket
C   Clean air cover gasket
D   Idle air/vacuum signal housing gasket

rear underside of the throttle body and a vacuum damper valve situated at the left front corner of the plenum **(see illustration)**. Disconnect this hose from the damper valve.

13   If you are removing or replacing the throttle body or the plenum, remove the throttle body mounting bolts, then remove the throttle body and gasket from the front of the plenum. If you are removing/replacing an intake runner gasket, the fuel rail or any component attached to the fuel rail (pressure regulator or a fuel injector), do not remove the throttle body from the plenum.

14   The throttle body is not serviceable. If the throttle shaft or either throttle plate is damaged or worn, the throttle body must be replaced.

15   The four gaskets for the IAC valve, the IAC valve coolant cover, the clean air cover, the throttle body itself, the IAC valve or the TPS may be replaced if necessary **(see illustration)**. The following procedure outlines the procedure for disassembly of the throttle body.

16   If you intend to reinstall the same Throt-

tle Position Sensor (TPS) switch, scribe alignment marks between the top and bottom edges of the TPS and the throttle body to ensure proper reinstallation. **Caution:** *Failure to scribe alignment marks for the proper reinstallation of the TPS will necessitate adjustment with a digital voltmeter - if you don't have this tool, you must drive the vehicle to a dealer immediately after completing reassembly of the fuel injection system.*

17   Remove the TPS mounting screws, lockwashers, retainers and the TPS.

18   Unscrew and remove the IAC valve and gasket.

19   Remove the idle air/vacuum signal housing and coolant cover assembly mounting screws, the cover and the gasket.

20   Remove the clean air cover screws, the clean air cover and the clean air cover gasket.

21   The throttle body may be cleaned in a cold immersion-type cleaner. **Caution:** *The TPS, IAC valve and all four gaskets must not be soaked in liquid solvent or they will be damaged. If the TPS or the IAC valve are mounted in the throttle body, do not immerse*

the throttle body.

22   Clean all metal parts thoroughly and blow them dry. Be sure that all air passages are free of burrs and dirt.

23   Inspect the casting mating surfaces for damage that could affect gasket sealing.

24   Reassembly is the reverse of disassembly. Use new gaskets and tighten all screws securely. Do not install the IAC and TPS until the throttle body has been bolted onto the plenum.

Plenum (1985 through 1991 models only)

25   Label then disconnect the two vacuum lines from the right rear corner of the plenum **(see illustration)**.

26   Unscrew the power brake booster vacuum line threaded fitting at the left rear corner of the plenum **(see illustration)**.

27   Unplug the wiring harness electrical connector from the Manifold Air Temperature (MAT) sensor protruding down from the underside of the plenum, just forward of the pressure regulator.

28   Disconnect the PCV fresh air intake assembly from the right rocker arm cover, disconnect it from the PCV pipe assembly

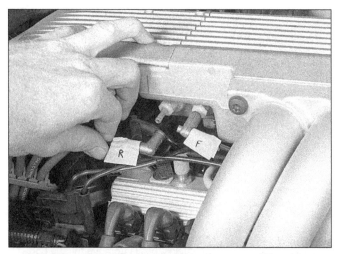

10.25 The vacuum lines at the right rear corner of the plenum should be labeled and disconnected from the plenum

10.26 Unscrew the power brake booster vacuum line threaded fitting (arrow) and pull it back slightly from the plenum

10.29 There are four Torx bolts attaching the upper end of each intake runner assembly to the plenum (only three are visible in this photo of the right runner)

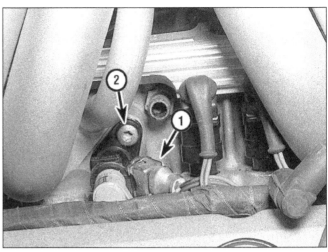

10.36 To remove the cold start valve from the intake manifold, disconnect the fuel injector wiring harness electrical connector (1) from the cold start valve electrical terminal and remove the cold start valve flange bolt (2) . . .

and remove it. Remove the two flanged hex screws from the right fuel injector wire harness shroud and remove the shroud.

29 Remove the four Torx bolts that attach the upper end of the right intake runner assembly to the plenum (see illustration) and the five Torx bolts that attach the lower end of the runners to the intake manifold. One of the bolts holding the runners to the manifold can only be removed by reaching under the plenum from the left side of the engine.

30 Remove the four Torx bolts that attach the upper end of the left intake runner assembly to the plenum.

31 Remove the plenum.

Intake runner assemblies and cold start valve assembly

32 Remove the right rocker arm cover upper nuts to free the PCV pipe brackets. Remove the PCV pipe to provide clearance for the intake runner flanges. Remove the

right intake runner assembly.

33 Label, then disconnect, the vacuum hoses at the EGR valve and fuel pressure regulator.

34 The left runner has one Torx bolt that is accessible only from the opposite side of the engine. It's located just below the pressure regulator. Remove it.

35 Disconnect the PCV valve assembly from the left rocker arm cover and the PCV pipe assembly and remove it. Remove the two flanged hex screws from the left fuel injector wire harness shroud and remove the shroud.

36 Disconnect the wire harness electrical connector from the cold start valve (see illustration). Disconnect the connectors from the left bank of fuel injectors.

37 Remove the left rocker arm cover upper nuts to free the PCV pipe brackets. Remove the PCV pipe assembly to provide clearance for removal of the left intake runner assembly.

38 Using a backup wrench on the fuel rail fitting, disconnect the cold start valve fuel line fitting from the rear of the left side fuel rail (see illustration). Remove the cold start valve mounting bolt. Clean the area around the cold start valve connection then remove the cold start valve assembly.

39 To replace the cold start valve or check the O-ring between the cold start valve and the elbow fitting of the tube and body assembly, bend back the tab on the elbow fitting of the tube and body assembly and unscrew the valve from the tube (see illustration). Before disassembling the valve from the tube and body assembly, note the relationship of the electrical connector to the tube and body assembly. This is important because the connector must be in the same position once the new valve is adjusted in order for the cold start valve assembly to fit properly.

40 Inspect the O-ring between the cold start valve and the elbow fitting of the tube

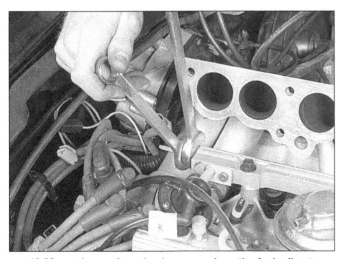

10.38 . . . then, using a backup wrench on the fuel rail nut, unscrew the cold start valve line threaded fitting from the left fuel rail

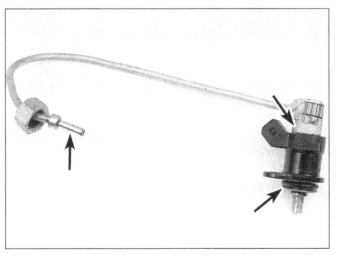

10.39 Cold start valve assembly - arrows indicate areas where O-rings should be replaced

10.46 To disconnect the fuel feed and return lines from the intake manifold, unscrew both threaded fittings (1 and 2), then remove both bracket bolts (arrows)

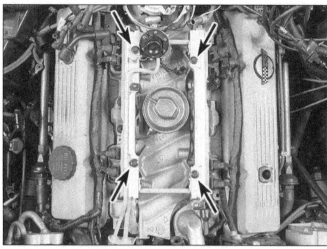

10.48 To remove the fuel rail assembly from the intake manifold, remove the four fuel rail mounting bolts (arrows)

10.50 Each fuel injector is locked to the fuel rail with a retainer clip (arrow)

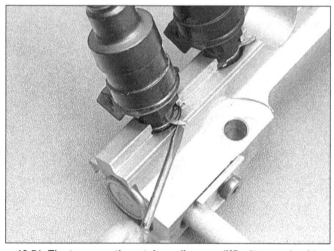

10.51 The tangs on the retainer clips are difficult to push with your finger - use a small screwdriver instead

and body assembly. Replace it if it's damaged.

41   To adjust the new cold start valve, screw it into the elbow fitting of the tube and body assembly until it bottoms. Be sure to lubricate the O-ring with engine oil.

42   Turn the valve back one full turn, until the electrical connector is at the top position you noted before separating the valve from the tube and body assembly.

43   Bend the tab on the side of the tube and body assembly elbow fitting forward to limit rotation of the valve to less than a full turn.

44   Remove the remaining four Torx bolts which attach the lower end of the left intake runner assembly to the intake manifold.

45   Remove the left intake runner assembly.

Fuel rail assembly, fuel injectors and fuel pressure regulator

46   On 1985 through 1991 models, disconnect fuel line fittings at the bracket on the right front of the intake manifold, then remove both bracket bolts from the manifold **(see**

illustration). On 1992 and later models, disconnect fuel line at the right rear of the intake manifold using a special fuel line quick disconnect tool.

47   Disconnect the wiring harness electrical connectors from the fuel injectors. Label the injector connector with its corresponding cylinder number.

48   Remove the four mounting bolts from the fuel rail **(see illustration)** and remove the fuel rail and fuel injectors as an assembly.

49   Before disassembly or removal of any component(s), the fuel rail assembly should be thoroughly cleaned with a spray type cleaner. **Caution:** *Do not immerse the fuel rail assembly or its components in liquid solvent.*

Fuel injectors

Note: *Whether you are replacing a malfunctioning fuel injector, a leaking injector O-ring seal or the fuel pressure regulator assembly, it is essential that all O-ring seals at both ends of all injectors be inspected. In view of the relative inaccessibility of these items, it is a*

good idea to replace all of them, since you have to remove the injectors to inspect the O-rings anyway.

50   Each port injector is located and held in position by a retainer clip that must be rotated to release and/or lock the injector **(see illustration)**.

51   Rotate the injector retaining clips to the unlocked position **(see illustration)**.

52   Pull the injectors from the fuel rail with a slight twisting motion.

53   Inspect all injector O-ring seals for wear. Remove and discard damaged O-rings by prying them off either end of the injector with a small screwdriver **(see illustration)**.

Fuel pressure regulator

Note: *The fuel pressure regulator is factory adjusted. Do not attempt to remove the regulator cover. If the regulator is malfunctioning, it must be replaced.*

54   Remove either front crossover tube retainer screw and crossover tube retainer **(see illustration)**.

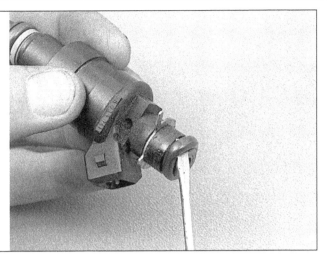

10.53 The O-rings on the ends of the injectors should be inspected and, if damaged, should be replaced

Fuel rail

79   Lubricate the nozzle ends of all fuel injectors with engine oil, place the fuel rail assembly into position above the intake manifold and carefully push down on the rails until all injectors are fully seated.
80   Install the fuel rail assembly mounting bolts and tighten them to the specified torque.

Left intake runner

81   Using new intake runner-to-intake manifold gaskets, place the left runner assembly in position and install the four outboard intake runner Torx mounting bolts. From the right side of the engine, install the fifth bolt just below the pressure regulator. Tighten all bolts securely.

Cold start valve assembly

82   Inspect the large O-ring between the cold start valve and the intake manifold and the small O-ring between the threaded fitting of the tube and body assembly and the rear of the right fuel rail. If there is evidence of damage to either O-ring, replace it.
83   Lubricate both O-rings with engine oil and install the cold start valve assembly.
84   Install the threaded fitting to the rear of the left fuel rail and the cold start valve flange mounting bolt to the intake manifold. Tighten the threaded fitting and the mounting bolt securely.
85   Install the PCV pipe, retaining brackets and left upper rocker arm cover nuts. Tighten the nuts securely.
86   Place the fuel injector wiring harness in position and connect the electrical connectors to the left bank of fuel injectors. Also connect the electrical connector to the cold start valve.
87   Place the fuel injector wiring harness shroud in position, install the flanged hex bolts and tighten them securely.
88   Install the PCV valve assembly into the left rocker arm cover and connect the hose to the PCV pipe assembly.

55   Remove the rear retainer-to-regulator base screw and crossover tube retainer.
56   Separate the left and right fuel rail assemblies.
57   Remove the regulator bracket-to-right fuel rail screw (just below the fuel pressure connection), the bracket-to-pressure regulator base screws and the pressure regulator mounting bracket.
58   Remove the fuel outlet tube rear retaining bracket stud.
59   Remove the pressure regulator base-to-right fuel rail mounting screw (on the side of the regulator base, just in front of the rear crossover tube).
60   Remove the pressure regulator and base assembly from the right fuel rail assembly.
61   Carefully rotate and pull on the regulator to remove it from the fuel outlet tube.
62   Remove the regulator base-to-fuel rail connector.
63   Remove the remaining front and rear crossover tube retainers and separate the crossover tubes from the fuel rails.
64   Inspect all O-rings on both ends of the connector, the fuel outlet tube and the crossover tubes for any damage that could prevent proper sealing. Replace any damaged O-ring seals. **Caution:** *The fuel rail assembly O-rings are not all the same size. When removing O-rings, be sure to note their location and size to ensure proper replacement.*

Installation

Refer to illustration 10.114

Fuel pressure regulator

65   Lubricate all O-rings with engine oil and install them on the connector, the fuel outlet tube and the front and rear crossover tubes.
66   Insert the front and rear crossover tubes into their respective holes in the left fuel rail assembly. Install the tube retainers and screws and tighten them securely.
67   Install the base-to-right fuel rail connector in the pressure regulator base.
68   Push the regulator and base assembly onto the fuel outlet tube.

69   Rotate the regulator and base assembly to position the base properly against the right fuel rail assembly.
70   Install and tighten securely the pressure regulator base-to-right fuel rail mounting screw.
71   Install the pressure regulator and base assembly mounting bracket, the bracket-to-base mounting screws and the bracket-to-fuel rail mounting screw.
72   Install and tighten the fuel outlet tube rear retaining bracket stud.
73   Reconnect the left fuel rail assembly and previously installed crossover tubes to the right fuel rail assembly by gently pushing the crossover tube ends into the right fuel rail.
74   Install the rear crossover tube-to-pressure regulator base retainer and mounting screw. Tighten the screw securely.
75   Install the front crossover tube-to-right fuel rail assembly retainer and mounting screw. Tighten the screw securely.

Fuel injectors

76   Lubricate all injector O-ring seals with engine oil.
77   Push the injectors back into the fuel rail.
78   Rotate the injector retainer clips to the locked position.

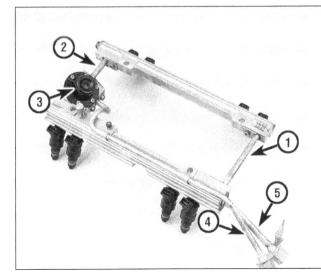

10.54 Components of the fuel rail assembly

1   *Front crossover tube*
2   *Rear crossover tube*
3   *Fuel pressure regulator*
4   *Fuel outlet pipe*
5   *Fuel inlet pipe*

4-18

Chapter 4 Fuel and exhaust systems

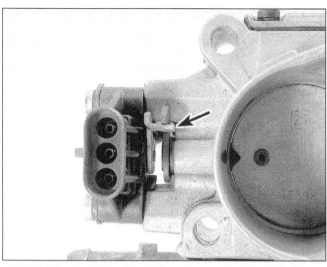

10.114 When installing the TPS, make sure that the TPS pickup lever (arrow) lines up with the tang on the throttle actuator lever

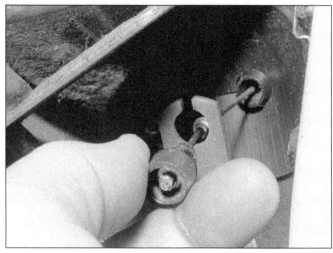

11.5 To detach the throttle cable from the accelerator pedal, pull the small plastic retainer loose from the upper end of the pedal lever, the lift the cable up and out of the groove in the lever

Plenum and right intake runner assemblies

89 Place the right intake runner assembly and runner-to-intake manifold gaskets in position and loosely tighten the four outboard Torx mounting bolts. Don't forget to install the fifth, inboard bolt located behind the second runner from the front. It must be installed from the opposite side of the engine.

90 Holding the plenum in position, connect the wiring harness electrical connector to the MAT sensor protruding from the bottom of the plenum.

91 Slide the new runner-to-plenum gaskets into place and bolt the left and right intake runner assemblies to the plenum. Tighten all bolts securely. Don't forget to tighten the bolts that attach the lower end of the right intake runner assembly to the intake manifold.

92 Install the PCV pipe, brackets and right rocker arm cover upper nuts. Tighten the nuts securely.

93 Lay the right fuel injector wiring harness in position and reconnect the electrical connectors to the fuel injectors.

94 Install the right fuel injector wiring harness shroud and the flanged hex bolts. Tighten the bolts securely.

95 Install the PCV fresh air intake assembly into the top of the right rocker arm cover and connect it to the PCV pipe assembly.

96 Reconnect the vacuum hoses at the EGR valve and the fuel pressure regulator.

97 Reattach the power brake booster vacuum line threaded fitting at the left rear corner of the plenum.

98 Reattach the two vacuum lines at the right rear corner of the plenum.

Throttle body

99 Install the throttle body and a new throttle body gasket on the front of the plenum and tighten the throttle body mounting bolts

to the specified torque.

100 Connect the vacuum hose between the throttle body and the vacuum damper valve beneath the lower left front corner of the plenum. Connect the PCV hose to the right side of the throttle body and tighten the hose clamp securely.

101 Attach the coolant hoses to the IAC valve coolant cover assembly. Tighten the hose clamps securely.

102 Attach the cruise control, accelerator and transmission TV cables to the throttle body linkage.

103 Install the cable bracket to the plenum and tighten the bracket bolts securely.

104 Add coolant to the cooling system.

105 Hook up the cable to the negative terminal of the battery.

106 If you are installing a new IAC or TPS, they must be adjusted. The new IAC can be adjusted on the bench, but the TPS must be adjusted during installation on the throttle body.

IAC adjustment

107 Measure the distance from the gasket-mounting surface of the IAC valve assembly to the tip of the pintle (see illustration 9.39).

108 If this distance is greater than 1-1/8 inch (28 mm), reduce it as follows:

109 If the IAC valve assembly has a collar around the electrical connector end, retract the pintle by applying firm hand pressure (a slight side-to-side motion may help).

110 If the IAC valve assembly has no collar, compress the pintle retaining spring toward the body of the IAC and try to turn the pintle clockwise.

111 If the pintle turns, continue turning it until the specified distance is attained. Return the spring to its original position, with the straight part of the spring end lined up with the flat surface under the pintle head. **Note:** *If the pintle will not turn, use firm hand pressure to retract it.*

112 Install a new gasket onto the IAC valve and screw the IAC valve and gasket into the idle air/vacuum signal housing of the coolant cover assembly. Tighten the IAC valve to the specified torque.

113 Reconnect the wiring harness electrical connector to the IAC valve electrical terminal. **Note:** *No further adjustment of the IAC valve is necessary. It will reset itself after the first time the engine is started and the ignition is turned off.*

TPS adjustment

114 With the throttle valve in the closed idle position, install the TPS on the throttle body assembly, making sure that the TPS lever lines up with the tang on the throttle actuator lever **(see illustration)**.

115 Install the retainers and the TPS mounting screws.

116 Install three jumper wires between the TPS and the harness connector.

117 With the ignition turned to On, and a digital voltmeter connected to the terminals marked A and B on the connector (the top and middle terminals), adjust the TPS to obtain a reading of 0.54 ^ 0.075 volts.

118 Tighten the TPS mounting screws and recheck your reading to insure that the adjustment has not changed.

119 Turn the ignition Off, remove the jumper wires and connect the wiring harness electrical connector to the TPS.

120 Connect the wiring harness electrical connector to the terminal on the underside of the MAF sensor and install the MAF sensor/flexible ducting assembly between the air intake assembly and the throttle body. Snap the spring clips into place to lock the MAF sensor to the air intake and tighten the hose clamp that attaches the flexible ducting to the throttle body.

121 Start the engine and then turn it off. The IAC valve will automatically reset itself.

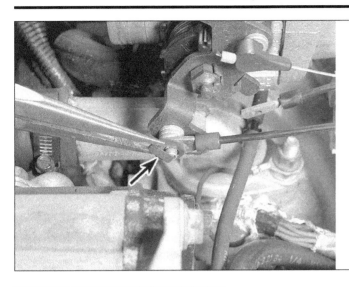

11.10 Remove the clip to disengage the TPI end of the cable from the throttle body lever

11 Throttle cable - removal and installation

Crossfire (TBI) system

Refer to illustration 11.5

1   Remove the throttle cable retaining clip from the rear throttle body throttle shaft lever arm.

2   Push the rear throttle body throttle shaft lever forward slightly and disconnect the cable from the throttle linkage.

3   Disconnect the plastic cable retainer from the bracket on the intake manifold.

4   Peel away the body sealing compound from the retainer at the firewall plate. Pinch the tabs on the top and bottom of the retainer and disconnect the retainer from the plate.

5   Disconnect the throttle cable from the accelerator pedal **(see illustration)**.

6   Remove the cable.

7   Installation is the reverse of the removal procedure.

8   There are no linkage adjustments because the replacement cable is a specific length. Therefore, it is imperative that the correct cable be installed. Only the specific replacement cable will work properly on this vehicle.

9   After assembly, the accelerator linkage must operate freely without binding between the closed and wide open positions. Check for correct opening and closing by operating the accelerator pedal and, if any binding is present, check the routing of the cable and make sure there are no kinks in the cable.

TPI system

Refer to illustration 11.10

10   Remove the clip that secures the throttle cable to the throttle body lever **(see illustration)**.

11   Push the throttle lever to the rear and disconnect the cable.

12   Pinch the tabs on the top and bottom of the plastic retainer attaching the cable to the cable bracket. Slide the retainer through the bracket.

13   Unclip the plastic cable bracket attached to the side of the plenum. Lift the cable out of the bracket.

14   Peel away the body sealing compound from the retainer at the firewall plate. Pinch the tabs on the top and bottom of the retainer and free the retainer from the plate.

15   Disconnect the throttle cable from the accelerator pedal.

16   Remove the throttle cable.

17   Installation is the reverse of removal.

18   No linkage adjustment is necessary because the replacement cable is a specific length. Therefore, it is imperative that the correct cable be installed. Only the specific replacement cable will work properly on this vehicle.

19   After assembly, the throttle linkage must operate freely without binding between the closed and wide open positions. Check for correct opening and closing by operating the accelerator pedal and, if any binding is present, check the routing of the cable and make sure there are no kinks in the cable.

12 Minimum idle speed - adjustment

Crossfire (TBI) system

Note: *The following procedure cannot be performed without two special tools. One is a balance manometer and the other is an idle air valve plug set.*

1   The throttle position of each throttle body must be balanced so that the throttle plates are synchronized and open simultaneously.

2   Adjustment should be performed only when a manifold cover or throttle body has been replaced, or if there is any indication of tampering with the throttle valve synchronizing screw on the front unit.

3   Remove the air cleaner and air cleaner-to-throttle body gaskets. Plug the Thermac vacuum port on the rear throttle body.

4   If in place, remove the tamper resistant plugs covering both throttle stop screws. Refer to Section 9 if necessary.

5   Block the drive wheels and apply the parking brake.

6   Connect a tachometer.

7   Disconnect the Idle Air Control (IAC) valve electrical connector.

8   Plug the idle air passages of each throttle body with a special plug available at most auto supply stores. Be certain that the plugs are fully seated and that no air leaks exist.

9   Start the engine and allow engine rpm to stabilize at normal operating temperature.

10   Place the transmission in Drive. Engine rpm should decrease below curb idle speed. If engine rpm does not decrease, check for a vacuum leak.

11   Remove the cap from the ported tube on the rear throttle body and connect the water manometer.

12   Adjust the rear unit throttle stop screw to obtain approximately six inches of water on the manometer. If unable to adjust to this level, be sure that the front unit stop screw is not limiting throttle travel.

13   Remove the manometer from the rear throttle body and install the cap on the ported tube.

14   Remove the cap from the ported tube on the front throttle body and connect the manometer. The reading should also be approximately six inches of water. If adjustment is required, proceed as follows:

a) *Locate the throttle synchronizing screw and collar on the front throttle body. The screw retaining collar is welded to the throttle lever to discourage tampering with this adjustment.*

b) *If the collar is in place, grind off the weld from the screw collar and throttle lever.*

c) *Block movement of the throttle lever by relieving the force of the heavy spring against the throttle synchronizing screw to prevent the levers from coming into contact.* **Caution:** *If the lever is not blocked before the throttle synchronizing screw is removed, the screw may be damaged and reinstallation will be done only with great difficulty.*

d) *Remove the screw and collar. Discard the collar.*

e) *Reinstall the throttle synchronizing screw, using thread locking compound (supplied in the service repair package) in accordance with the directions. If the compound is not available, use Loctite 262 or an equivalent.*

f) *Immediately adjust the screw to obtain approximately six inches of water on the manometer.*

15   Remove the manometer from the front throttle body and install the cap on the ported tube.

16   Adjust the rear throttle stop screw on the rear unit to obtain 475 rpm.

17   Turn the ignition to Off and place the transmission in Neutral.

18   Adjust the front throttle stop screw to obtain 0.005-inch clearance between the front throttle stop screw and the throttle lever tang.

12.30 To remove the idle stop screw plug from the throttle body of the TPI system, pierce it with an awl and apply leverage to remove it

13.2 Before the exhaust system can be removed, it is necessary to disconnect the rubber hose from the AIR check valve and remove the valve from the metal line which leads to the catalytic converter

19   Remove the idle air passage plugs and reconnect the IAC valves.

20   Start the engine. It may run at high rpm at first, but the rpm will decrease when the IAC valves close the air passages. Stop the engine when the idle rpm has decreased.

21   Check the TPS voltage and adjust if necessary.

22   Install the air cleaner gasket, connect the vacuum line to the throttle body and install the air cleaner.

TPI system

Refer to illustration 12.30

Note: *Before attempting to adjust the minimum air rate the throttle body and throttle plate must be clean and free of debris.*

23   With the engine off, remove the MAF sensor (if equipped) and the air inlet duct. Hold the throttle wide open and spray a generous quantity of aerosol carburetor cleaner on the back of the throttle plate and around the inside of the throttle body. Allow this to soak for several minutes.

24   Soak a corner of a soft rag with carburetor cleaner and scrub the throttle plate and the inside of the throttle body.

25   Using aerosol carburetor cleaner, spray the remaining carbon deposits from the throttle plate and throttle body.

26   Wipe any remaining cleaner from the throttle body.

27   Reinstall the MAF sensor and air inlet duct. Start the engine and allow it to idle.

28   Road test the vehicle and re-evaluate the need to adjust the minimum air rate. Often, cleaning the throttle body and throttle plate will restore minimum air flow and further adjustments are not necessary. If the minimum air must be adjusted proceed as follows:

29   The idle stop screw, which regulates the minimum idle speed of the engine, is adjusted at the factory and covered with a plug to discourage unnecessary adjustment.

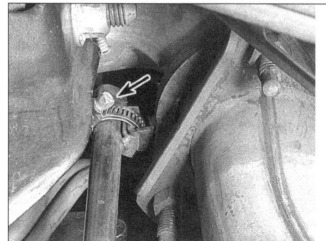

13.4 To free the AIR system metal pipe, loosen this hose clamp (arrow) and lift the pipe off the mounting bracket welded to the exhaust manifold flange

However, if it is necessary to gain access to the idle stop screw assembly, proceed as follows.

30   Pierce the idle stop screw plug with an awl and apply leverage to remove it **(see illustration)**.

31   With the IAC valve connected, ground the diagnostic terminal of the ALCL connector under the left side of the dash.

32   Turn the ignition to On but do not start the engine. Wait at least 30 seconds.

33   With the ignition On, disconnect the IAC electrical connector.

34   Remove the ground from the diagnostic lead and start the engine. Run the engine until it reaches normal operating temperature.

35   Adjust the idle screw to obtain either 400 rpm in Drive with an automatic transmission or 450 rpm in Neutral with a manual transmission.

36   Turn the ignition to Off and reconnect the IAC valve connector.

37   Following the procedure outlined in Section 10, adjust the TPS to 0.54-volts plus-or-minus 0.075-volts. Secure the TPS and

recheck the voltage.

38   Start the engine and check for proper idle operation.

13   Exhaust system - removal and installation

Refer to illustrations 13.2, 13.4 and 13.8

1   Disconnect the cable from the negative terminal of the battery.

2   Loosen the hose clamp on the Air Injection Reaction (AIR) system check valve hose adjacent to the right exhaust manifold **(see illustration)** and, using a backup wrench on the metal pipe fitting, remove the check valve.

3   Raise the vehicle and place it securely on jackstands.

4   From underneath the vehicle, loosen the AIR pipe retaining clamp bolt and push the clamp up and off the exhaust manifold flange bracket **(see illustration)**.

5   Disconnect the oxygen sensor pigtail electrical connector.

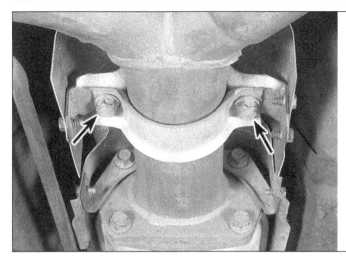

13.8 Remove the two bolts (arrows) and the clamp ahead of the converter

6   There are three exhaust manifold-to-exhaust pipe flange nuts per flange - remove all six nuts.

7   Support the exhaust system with a floor jack and a block of wood positioned underneath the intersection of the left and right exhaust pipes (the 'Y" located ahead of the catalytic converter).

8   Remove the clamp bolts and the clamp just in front of the catalytic converter **(see illustration)**.

9   There is one hanger bolt per muffler.

Remove both of them.

10   With the help of an assistant, carefully lower the floor jack and remove the exhaust system.

11   Installation is the reverse of removal.

---

14   Exhaust system back pressure check

1   If the exhaust system is suspected of being restricted, exhaust system back pres-sure can be checked by either of the two ways outlined in the following steps.

Vacuum gauge method

2   Attach a vacuum gauge to any intake manifold vacuum source (a source that indicates vacuum at idle such as the fuel pressure regulator hose or the brake vacuum booster hose).

3   Start the engine and allow it to idle. Observe the vacuum reading at idle.

4   Accelerate the engine to 2000 rpm and while observing the vacuum gauge, release the throttle.

5   The vacuum reading on the gauge should increase momentarily as the throttle is released, then return to the prior idle reading.

6   If the vacuum reading does not increase as the throttle is released the exhaust system may be restricted.

Fuel pressure gauge method

7   Disconnect one of the AIR check valves from the AIR pipe.

8   Attach a fuel pressure gauge with an appropriate adapter to the AIR pipe.

9   Start the engine and accelerate and maintain 2000 rpm, observe the fuel pressure gauge.

10   Pressure should not exceed 3 psi at 2000 rpm. If pressure is above specification the exhaust system may be restricted.

Notes

Chapter 5
Engine electrical systems

Contents

Specifications

Firing order	1-8-4-3-6-5-7-2
Distributor rotation	clockwise
Ignition coil resistance	
Primary resistance	0.3 to 1.3 ohms
Secondary resistance	5,000 to 7,000 ohms

1 Ignition system - general information and precautions

The ignition system consists of the ignition switch, the battery, the coil, the primary (low tension) and secondary (high tension) wiring circuits, the distributor and the spark plugs.

High Energy Ignition (HEI) distributor (1991 and earlier models)

A High Energy Ignition (HEI) distributor with Electronic Spark Timing (EST) is used on 1991 and earlier models. The ignition coil is housed in the distributor cap and is connected through a resistance brush to the rotor. The HEI distributor has an internal magnetic pick-up assembly that contains a permanent magnet, a pole piece with internal teeth and a pick-up coil.

All spark timing changes in the HEI/EST distributor are carried out electronically by the Electronic Control Module (ECM), which monitors data from various engine sensors, computes the desired spark timing and signals the distributor to change the timing accordingly. A back-up spark advance system is incorporated to signal the ignition module in case of ECM failure. No vacuum or mechanical advance is used.

Distributor Ignition (DI) system (1992 and later models)

The 1992 and later models are equipped with a Distributor Ignition (DI) system that incorporates the Opti Spark ignition system. The Opti Spark ignition system consists of the distributor housing, cap and rotor, optical position sensor, sensor disc, pick-up assembly, distributor drive shaft, ignition module, ignition coil, primary and secondary wiring, spark plugs and the necessary control circuits (wiring harness) for the entire system. Although the distributor is composed of components, the entire assembly must be replaced as a single unit.

The Opti Spark distributor is mounted on the front cover and it is driven directly by the camshaft. The cap and rotor directs the spark from the ignition coil to the proper spark plug secondary wire. The distributor cap is marked with the corresponding cylinder numbers for easy routing of the plug wires.

The distributor contains an optical pick-up system that provides actual crankshaft position (in degrees) to the ECM. The system uses two infrared optical sensors and a flat disc with two rows of notches (slots) cut around the circumference. One row has 360 notches (one-degree widths) and the other row has eight notches. When the optical pick-up turns, it produces two modulated digital signals. The first row (360-degrees) produces the signal for the timing while the second row (8 notches) produces the signal for the RPM reference.

Electronic Spark Control (ESC)

All models are equipped with an Electronic Spark Control (ESC), which uses a knock sensor in connection with the ECM to control spark timing to allow the engine to

2.2a To get at the battery, first remove the three bolts (arrows) that attach the wheelhouse panel to the fender skirt . . .

have maximum spark advance without spark knock. This improves driveability and fuel economy.

Secondary (spark plug) wiring

The secondary (spark plug) wires are a carbon-impregnated cord conductor, encased in an 8 mm (5/16-inch) diameter rubber jacket with an outer silicone jacket. This type of wire will withstand very high temperatures and provides an excellent insulator for the high secondary ignition voltage. Silicone spark plug boots form a tight seal on the plug. The boot should be twisted 1/2-turn before removing (for more information on spark plug wiring, refer to Chapter 1). **Warning:** *Because of the very high voltage generated by the ignition system, extreme care should be taken whenever an operation involving ignition components is performed. This not only includes the distributor, coil, control module and spark plug wires, but related items such as the plug connections, tachometer and testing equipment.*

2.2b . . . then remove the bolt at the upper rear corner of the skirt . . .

2   Battery - removal and installation

Refer to illustrations 2.2a, 2.2b, 2.2c, 2.4 and 2.5

1   Disconnect the cable from the negative terminal of the battery.
2   Remove the three bolts along the rear edge of the left front wheelhouse, the bolt at the upper rear corner of the fender skirt and the bolt at the lower rear corner of the skirt **(see illustrations)**.
3   Remove the fender skirt.
4   Remove the battery hold down clamp bolt and clamp from the carrier **(see illustration)**, located at the rear bottom edge of the battery.
5   Remove the battery from the vehicle **(see illustration)**. **Warning:** *Do not tilt the battery when moving it.*
6   Installation is the reverse of the removal procedure. **Note:** *The bottom bolt hole is slotted so that you can slide the skirt in or out to align the skirt with the door. Before tightening the upper and lower rear fender skirt*

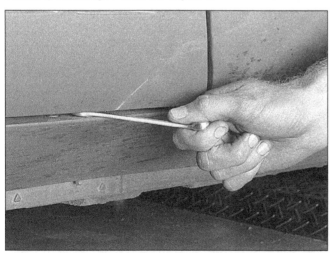

2.2c . . . and loosen the bolt at the lower rear corner enough to release the slotted tab and slide the skirt free

2.4 Remove this battery hold down bolt and the hold down clamp from the carrier

2.5 When removing the battery, always keep it level to avoid spilling battery acid

5.2 To test the HEI system with a calibrated ignition tester, pull off any plug boot, push it onto the tester, clip the tester to a metal bracket, crank the engine and see if a spark jumps from the center electrode of the tester to the cylinder surrounding it

bolts, make sure the trailing edge of the skirt is properly aligned with the leading edge of the left door by carefully opening the door. If the door rubs the skirt, pull the skirt out slightly and retighten it. Repeat this Step until the skirt and door are properly aligned.

3 Battery - emergency jump starting

Refer to the Booster battery (jump) starting procedure at the front of this manual.

4 Battery cables - check and replacement

1   Periodically inspect the entire length of each battery cable for damage, cracked or burned insulation and corrosion. Poor battery cable connections can cause starting problems and decreased engine performance.
2   Check the cable-to-terminal connections at the ends of the cables for cracks, loose wire strands and corrosion. The presence of white, fluffy deposits under the insulation at the cable terminal connection is a sign that the cable is corroded and should be replaced. Check the terminals for distortion, missing mounting bolts and corrosion.
3   If only the positive cable is to be replaced, be sure to disconnect the negative cable from the battery first.
4   Disconnect and remove the cable. Make sure that the replacement cable is the same length and diameter.
5   Clean the threads of the starter or ground connection with a wire brush to remove rust and corrosion. Apply a light coat of petroleum jelly to the threads to ease installation and prevent future corrosion.
6   Attach the cable to the starter or ground connection and tighten the mounting nut securely.
7   Before connecting the new cable to the

battery, make sure that it reaches the terminal without having to be stretched.
8   Connect the positive cable first, followed by the negative cable.

5 Ignition system - check

Warning: *Because of the very high voltage generated by the secondary ignition system, extreme care should be taken whenever an operation is performed involving ignition components. This not only includes the distributor, coil, control module and spark plug wires, but related items such as the electrical connections, tachometer and any test equipment.*

HEI system (1991 and earlier models)
Refer to illustrations 5.2 and 5.6
1   If the engine will not start even though it turns over, check for spark at the spark plug by installing a calibrated ignition system tester to one of the spark plug wires. **Note:** *The tool is available at most auto parts stores. Be sure to get the correct tool for your particular ignition system (HEI).*
2   Connect the clip on the tester to a ground such as a metal bracket or valve cover bolt, crank the engine and watch the end of the tester for bright blue, well defined sparks **(see illustration)**.
3   If sparks occur, sufficient voltage is reaching the plugs to fire the engine. However the plugs themselves may be fouled, so remove and check them as described in Chapter 1 or replace them with new ones.
4   If no sparks occur, the distributor cap, rotor or spark plug wires may be defective. Remove the distributor cap and check the cap, rotor and spark plug wires as described in Chapter 1. Check the carbon button in the center of the cap for proper contact with the rotor. Replace defective parts as necessary.
5   If moisture is present in the distributor cap, use WD-40 or something similar to dry

out the cap and rotor, then reinstall the cap and repeat the spark test at the spark plug wire.
6   If no sparks occur, check the primary wire connections at the distributor cap to make sure they are clean and tight. Check for voltage to the ignition coil (BAT terminal on the distributor cap connection) **(see illustration on next page)**. Battery voltage should be available to the ignition coil with the ignition switch ON. **Note:** *If the reading is 7 volts or less, repair the primary circuit from the ignition switch to the ignition coil.*
7   If there's still no spark, check the ignition module (the ignition module requires special test equipment to test, many auto parts stores will perform this test for you - see Section 7 for removal). Check the ignition pick-up coil (see Section 8) and the ignition coil (see Section 9). Replace the defective parts as necessary.

Distributor Ignition (DI) system (1992 and later models)
Refer to illustrations 5.12, 5.15, 5.17, 5.18a and 5.18b
8   Check all ignition wiring connections for tightness, cuts, corrosion or any other signs of a bad connection. A faulty or poor connection at a spark plug could also result in a misfire. Also check for carbon deposits inside the spark plug boots. Remove the spark plugs, if necessary, and check for fouling.
9   Check for ignition and battery supply to the ECM. Check the ignition fuses (see Chapter 12).
10   Use a calibrated ignition tester to verify adequate available secondary voltage (25,000 volts) at the spark plug **(see illustration 5.2)**. Using an ohmmeter, check the resistance of the spark plug wires). Each wire should measure less than 30,000 ohms.
11   Check to see if the fuel pump and relay are operating properly (see Chapter 4). The fuel pump should activate for two seconds when the ignition key is cycled ON. Install an

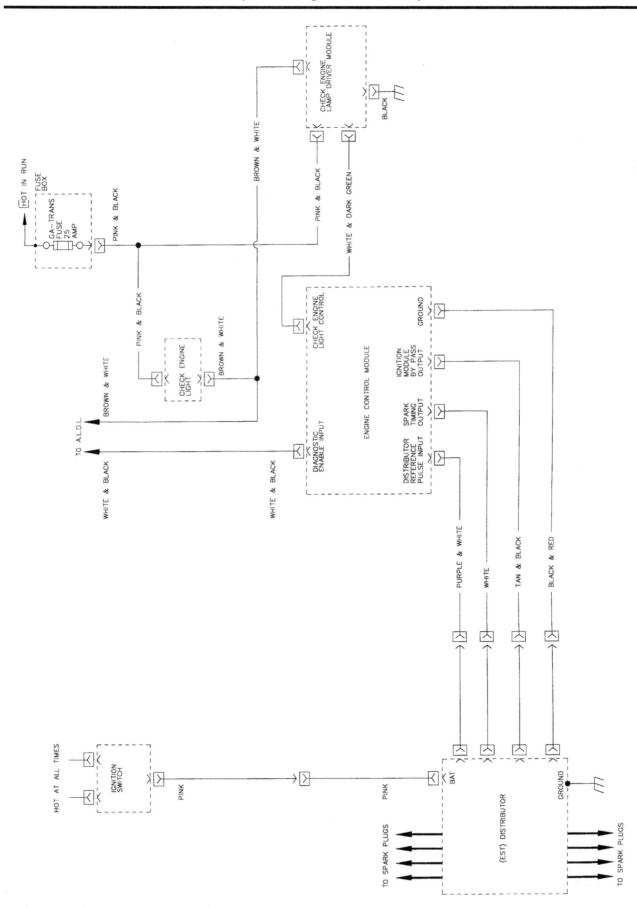

5.6 Ignition schematic for the HEI ignition system

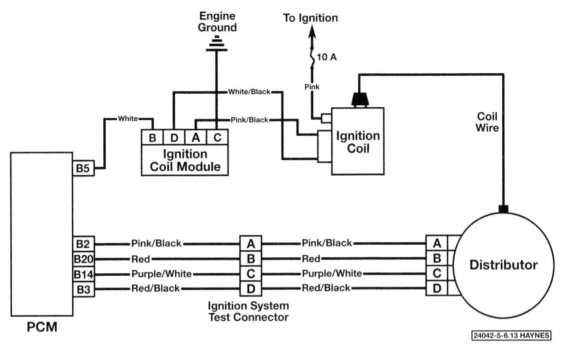

5.12 Ignition schematic for the Distributor Ignition (DI) system

injector test light and monitor the blinks as the injector voltage signal pulses (see Chapter 4, Fuel Injection System - check).

No start condition

12   Refer to the accompanying schematic for terminal pin designations for testing the Distributor Ignition (DI) ignition system **(see illustration)**.

13   Check for spark at the spark plugs, checking at least two or more spark plug wires **(see illustration 5.2)**. On a NO START condition, check the fuel pump relay and fuel pump systems for fuel delivery problems in the event the ignition spark is available.

14   Check for spark at the coil wire with a spark tester while an assistant is cranking the engine. A spark indicates that the problem

must be the distributor unit. This test separates the ignition wires from the ignition coil. **Note:** *A few sparks followed by no spark is the same condition as no spark at all.*

15   If there is NO spark, check the ignition coil (see Section 7). If the coil checks are correct, disconnect the ignition module connector, turn the ignition key ON (engine not running) and check for battery voltage at the terminals A and D. Normally, there should be battery voltage at the A and D terminals **(see illustration)**. Low voltage indicates an open or high resistance circuit from the distributor to the coil or ignition switch (primary ignition circuit). Also, check the OBD self diagnosis system for codes that may indicate an ignition system failure (see Chapter 6). **Note:** *If Code 16 or 36 is present, the distributor unit*

may be defective, have the system checked by a by a dealership or other properly equipped repair facility.

16   Switch the voltmeter to the A/C scale and measure the voltage on terminal B while cranking the engine over. Have an assistant turn the engine over while observing the voltmeter. This voltage signal from the computer acts as reference voltage for ignition control. This test will eliminate the PCM as a source of the problem. If there is a voltage signal present, a no start condition will most likely be caused by a faulty module. Continue testing. If there is no voltage signal from the PCM, replace the computer (see Chapter 6).

17   Turn the ignition key OFF and install a LED type test light onto the battery positive terminal (+). Observe that the test light is ON

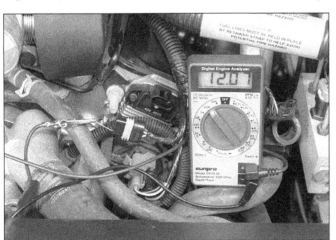

5.15 Disconnect the ignition module connector and check for battery voltage on the pink/black wire (terminal A)

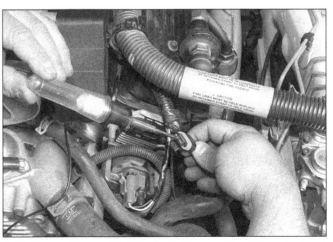

5.17 Connect the test light to the battery positive (+) terminal and probe terminal C. The light should illuminate to indicate a good ground circuit

5.18a Check for a pulsing voltage signal from the module on the red wire (+) by backprobing the distributor electrical connector

5.18b Remove the coil secondary wire and ground the coil to the engine using a suitable jumper wire

6.4 To disconnect the retaining ring from the distributor cap, push the locking latches away from the coil cap and pull up on the ring

when terminal C on the ignition module connector (harness side) is probed. This will check for a complete ground circuit from the ignition module. The test light should illuminate when touched to the connector terminal C **(see illustration)**. If there is no light, check the wiring harness for an open circuit.

18　Remove the coil wire and ground it to the engine using a suitable jumper wire. Disconnect the ignition coil harness connector, connect a LED type test light to the positive (+) battery terminal and check for a pulsing signal voltage at the white/black wire while the engine is cranked over. This test will check for the pulsing signal voltage from the ignition distributor after it has been triggered by the PCM. Have an assistant crank over the engine while observing the LED test light. Be sure to remove the coil secondary wire and ground the coil to the engine using a suitable jumper wire **(see illustrations)**. There should be a steady and obvious flashing LED light.

19　If all the tests results are correct except Step 18 and there still is a NO START condition, it will be very difficult to distinguish a faulty distributor unit from a faulty ignition module. In most cases, the ignition module, when defective, will not produce a pulsing voltage signal with symptoms of a definite start or no-start condition while a faulty DI unit will produce intermittent ignition failures when the engine is running. Take the vehicle to a dealer service department or other qualified repair shop for diagnosis.

20　DI ignition systems are prone to water entering the distributor unit through the venting system. Thoroughly check the vent system for cracks, leaks or damaged components:

a) *Clean the air supply hose from the air intake duct to the distributor.*
b) *Check the vacuum supply hose and check valve from the intake manifold.*
c) *Check the venting system harness from the lower section of the distributor to the check valve filter.*

d) *Make sure the air intake duct is not plugged or damaged.*

6　Distributor - removal and installation

1991 and earlier models

Removal

Refer to illustrations 6.4 and 6.5

1　Disconnect the cable from the negative terminal of the battery.
2　If your vehicle is equipped with Crossfire throttle body injection (1984), remove the air cleaner.
3　If your vehicle is equipped with Tuned Port Injection (1985 through 1991), remove the three distributor shield Torx screws and the shield from the rear of the plenum.
4　Using your thumbs, push the latches on the spark plug wire retaining ring away from the coil cover **(see illustration)** to release the ring, then lift it slightly.
5　Push down on the spark plug wire boots **(see illustration)** to detach the boots from their respective holes in the ring. Remove the ring. Note that the holes for the plug wires and the wires themselves are numbered. The wires must be installed in exactly the same holes from which they were removed.
6　Detach the spark plug wires from the distributor cap.
7　Unscrew the four screws attaching the distributor cap to the distributor base and rotate the cap slightly to move the electrical terminal on the underside of the cap away from the firewall.
8　Disconnect the ignition switch battery feed wire, the tachometer lead and the coil electrical connector (three-wire plug) from the cap.
9　Remove the distributor cap.
10　Using a socket and breaker bar on the crankshaft damper bolt, rotate the crankshaft until the piston in the number one cylinder is at top dead center (TDC). Refer to Chapter 2 if necessary.

11　Mark the alignment of the distributor housing to the manifold. Mark the position of the rotor to the distributor housing.
12　Remove the hold down clamp bolt and the clamp.
13　Lift the distributor up slightly and disconnect the four-wire electrical connector.
14　Remove the distributor. **Note:** *Do not turn the crankshaft once you have removed the distributor or the engine will have to be re-timed.*

Installation

15　Insert the distributor into the engine in exactly the same relationship to the block in which it was removed. **Note:** *If the crankshaft was turned with the distributor removed, Position the engine with number one cylinder at TDC (see Chapter 2A) and install the distributor with the rotor pointing at the number one spark plug wire terminal.*
16　Reconnect the four-wire electrical connector before proceeding to the next Step (you will not be able to get at this connector once the distributor is properly seated).
17　To mesh the helical gears on the

6.5 To remove the ring, pop the eight spark plug wire boots loose by pushing them out of their holes in the ring

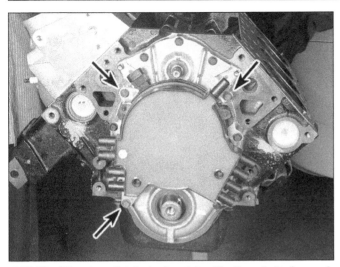

6.37 The DI system distributor is retained by three bolts (arrows)

7.4 To remove the module, remove the two mounting screws (1) and unplug the two electrical connectors (2)

camshaft and the distributor, it may be necessary to turn the rotor slightly. If the distributor does not seat completely against the manifold, it's because the lower end of the distributor shaft is not mating properly with the oil pump shaft.

18 Recheck the alignment marks between the distributor housing and the manifold to make sure that the distributor is aligned with the manifold in the same position it was in before removal. Also check the rotor to make sure that it is aligned with the mark you made on the housing to make sure that it too is in the same position that it was in before you removed it.

19 Using a socket and a breaker bar on the crankshaft damper bolt, turn the engine in its normal direction of rotation. Because the gear on the distributor shaft is engaged with the gear on the camshaft, their relationship to one another will not change as long as the distributor is not lifted from the engine. Therefore, the rotor will turn but the oil pump shaft will not, because the two shafts are not yet engaged. When the tang on the end of the distributor shaft and the slot on the end of the oil pump shaft are aligned, the distributor will drop down over the oil pump shaft and the distributor housing will seat properly against the manifold.

20 With the base of the distributor properly seated against the intake manifold, turn the distributor housing to align the marks made on the distributor base and the manifold.

21 With the distributor marks aligned, the rotor should be pointing at the alignment mark you made on the distributor housing before removal.

22 Place the hold down clamp in position and loosely install the bolt.

23 Install the distributor cap and the four screws that attach the cap to the housing. Tighten them securely.

24 Plug in the three-wire electrical connector to the coil.

25 Reconnect the ignition switch battery feed wire and tachometer lead to the distributor cap.

26 Reattach the spark plug wires to the distributor cap. The wires are numbered. Make sure that they are reconnected to the same towers from which they were removed.

27 Push the spark plug wire retaining ring back onto the spark plug wire boots. A small screwdriver is helpful for pushing the tops of the boots back into the holes in the retaining ring.

28 Push the locking tabs on the retaining ring until they snap into the cover.

29 If your vehicle is equipped with Tuned Port Injection (1985 on), install the three distributor shield Torx screws and the shield onto the rear of the plenum.

30 If your vehicle is equipped with Crossfire throttle body injection (1984), install the air cleaner.

31 Connect the cable to the negative terminal of the battery.

32 Check the ignition timing (Chapter 1) and tighten the distributor hold down bolt.

1992 and later

Removal
Refer to illustration 6.37

33 Disconnect the cable from the negative battery terminal.

34 Remove the water pump (see Chapter 3).

35 Remove the drivebelt pulley from the crankshaft (see Chapter 2A).

36 Remove the spark plug wires from the distributor.

37 Remove the distributor retaining bolts **(see illustration)**.

38 Remove the distributor from the engine.

Installation

39 Clean the water pump gasket surfaces on the engine block and water pump. Glue new gaskets to the water pump, do not install the pump at this time.

40 Insert the distributor unit onto the splined (or doweled, depending on year) shaft exactly the same relation to the block in which it was removed. To mesh the splines, it may be necessary to turn the distributor unit slightly and wiggle the unit until it slides completely onto the spline and meets the engine block. **Note:** *The splined shaft can be removed for cleaning. In the event the splined shaft gets inadvertently rearranged, it can be inserted into the engine block either direction without assembly trouble.* **Caution:** *Do not install the distributor bolts and tighten the unit in such a way as to pull (press) the distributor gears onto the spline. This will damage the distributor assembly.*

41 Install the bolts securely.

42 Install the crankshaft pulley.

43 Install the water pump, hoses and drive belt (see Chapter 3).

44 Install the distributor ignition wires and coil wire.

45 Connect the cable to the negative terminal of the battery.

46 Refill the cooling system, start the engine, check for coolant leaks.

7 Ignition module - replacement

1991 and earlier models

Refer to illustration 7.4

Note: *It is not necessary to remove the distributor from the engine to replace the ignition module.*

1 Disconnect the cable from the negative terminal of the battery.

2 Remove the distributor cap (Section 6).

3 Mark the position of the rotor in relation to the distributor housing. Remove the rotor retaining screws and the rotor.

4 Remove both module mounting screws and detach the module from the distributor **(see illustration)**.

5 Disconnect both wire harnesses from the module. Note that they cannot be interchanged.

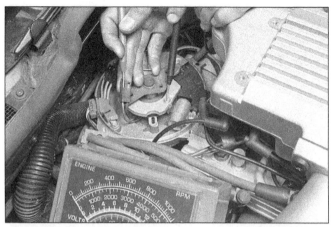

8.3 To test the pick-up coil, hook an ohmmeter lead to each of the terminals of the pick-up coil connector and ground the other lead to the distributor base

8.4 Place the leads in both terminals of the pick-up coil connector and flex the wires by hand to check for intermittent opens

8.7 Place the distributor in a bench vise and drive out the roll pin that locks the drive gear to the bottom of the shaft

8.10 Remove the C-clip retaining the pick-up coil/magnet/pole piece assembly and remove the assembly from the distributor

6   Do not wipe the grease from the module or the distributor base if the same module is to be reinstalled. If a new module is to be installed, a package of silicone grease will be included with it. Wipe the distributor base and the new module clean, then apply the silicone grease to the face of the module and the distributor base where the module seats **(see illustration)**. This grease is necessary for heat dissipation. **Note:** *The module cannot be tested without a module tester. If you suspect that the module is malfunctioning, have it checked by a dealer service department.*
7   Installation is the reverse of removal.

1992 and later models

8   The ignition module is located on the ignition coil mounting bracket at the front of the right cylinder head.
9   Disconnect the module electrical connector.
10   Remove the two module mounting screws and remove the module.
11   Installation is the reverse of removal. Be sure to apply a thin coat of silicone dielectric grease to the back of the module where it contacts the coil bracket.

8   Ignition pick-up coil - test and replacement

Refer to illustrations 8.3, 8.4, 8.7 and 8.10
Note: *This procedure applies to 1991 and earlier models only.*
1   Remove the distributor cap (Section 6).
2   Remove the rotor and disconnect the pick-up coil wires from the module.

Test

3   Connect an ohmmeter to each terminal of the pick-up coil wire and ground it to the distributor base **(see illustration)**. The ohmmeter reading should indicate infinite resistance. If it doesn't, the pick-up coil is defective.
4   Connect the ohmmeter between both terminals of the pick-up coil connector **(see illustration)**. Flex the wires by hand to check for intermittent opens. The ohmmeter reading should indicate one steady value within the 500 to 1500 ohm range as the wires are flexed. If it doesn't, the pick-up coil is defective.

5   If the pick-up coil fails either test, replace it.

Replacement

6   Remove the distributor. Mark the distributor shaft and gear so that they can be reassembled in the same position.
7   Secure the distributor shaft housing in a bench vise and drive out the roll pin with a hammer and punch **(see illustration)**.
8   Remove the gear and tanged washer, then pull the shaft from the distributor.
9   Remove the three attaching screws and detach the magnetic shield.
10   Remove the C-clip **(see illustration)** and detach the pick-up coil, magnet and pole piece.
11   Install the new pick-up coil, magnet and pole piece assembly.
12   Install the shaft. Make sure that it's clean and lubricated.
13   Install the tanged washer (with the tangs facing up), the drive gear and the roll pin.
14   Spin the shaft to ensure that the teeth on the distributor shaft do not touch the teeth on the pick-up coil pole piece.
15   If the teeth touch, loosen, adjust and

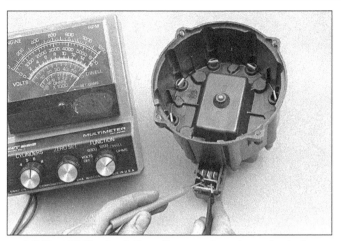

9.3 Test #1: To check the coil, connect an ohmmeter to the tachometer and battery lead terminals of the cap coil assembly - if the reading doesn't indicate zero or nearly zero ohms, replace the coil

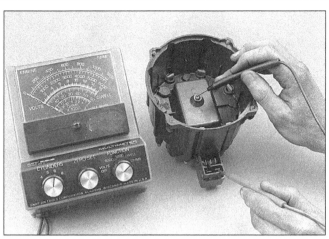

9.4a Test #2: Connect the ohmmeter between the tachometer terminal of the cap coil assembly and the center electrode of the distributor cap . . .

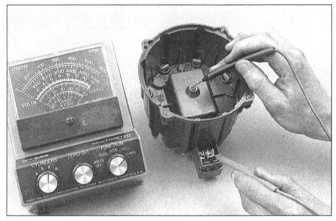

9.4b . . . then connect the ohmmeter between the ground terminal of the cap coil assembly and the center electrode of the distributor cap - if both readings indicate infinite resistance, replace the coil

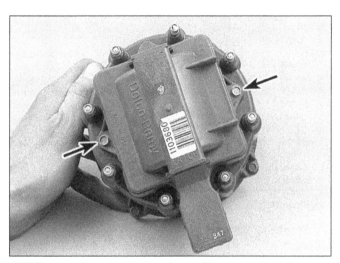

9.5 Remove these two screws (arrows) to detach the coil cover

retighten the pole piece to eliminate contact.
16   Install the distributor in the engine (Section 6).
17   Install the rotor and cap (Section 6).

9   Ignition coil - test, removal and installation

1991 and earlier models

Refer to illustrations 9.3, 9.4a, 9.4b, 9.5, 9.6 and 9.8

Test

1   Disconnect the cable from the negative terminal of the battery.
2   Remove the distributor cap (Section 6).
3   Place the cap upside down so that you can see the spade type coil terminals inside the hood protruding from the coil cap. Connect an ohmmeter to both terminals of the cap coil assembly **(see illustration)**. The ohmmeter reading should indicate zero resis-

9.6 Before disconnecting the coil assembly from the cap, push the spade connectors out of the hood with a small screwdriver

tance. If it doesn't, replace the coil.
4   Connect the ohmmeter between each of the terminals and the center contact of the distributor cap **(see illustrations)**. Be sure to use the high scale. If both ohmmeter readings indicate infinite resistance, replace the coil.

Removal and installation

5   Remove the coil cover attaching screws **(see illustration)** and lift off the cover.
6   Push the coil assembly leads through the top of the hood with a small screwdriver **(see illustration)**.

9.8 Remove the ignition coil arc seal from the bottom of the coil cavity in the distributor cap

11.2 A multimeter with a voltmeter mode is used for charging system tests

7   Remove the ignition coil mounting screws and lift the coil, with the leads, from the cap.

8   Remove the ignition coil arc seal **(see illustration)**.

9   Clean the coil housing and the rest of the cap with a soft cloth and inspect the cap for defects. Replace if necessary.

10   Install the arc seal.

11   Install the new coil and tighten the mounting screws securely.

12   Install the coil cover on the cap and tighten the screws securely.

13   Install the distributor cap on the distributor (Section 6).

14   Reconnect the cable to the negative terminal of the battery.

1992 and later models

Check

15   Check to make sure the coil is receiving battery voltage with the ignition key ON (engine not running).

16   Detach the primary electrical connector from the coil.

17   Using an ohmmeter on the low scale, connect the leads to the primary terminals on the ignition coil. The ohmmeter should indicate a very low resistance value. Refer to the Specifications listed in the beginning of this Chapter. If it doesn't, replace the coil.

18   Using the high scale, connect one lead to the ignition coil primary terminal and the other lead to the secondary terminal. The ohmmeter should not indicate an infinite resistance. Refer to the Specifications listed in the beginning of this Chapter. If it does, replace the coil.

Removal and installation

19   Detach the cable from the negative terminal of the battery.

20   Unplug the coil high tension wire and both electrical leads from the coil.

21   Remove both mounting nuts and remove the coil from the engine.

22   Installation of the coil is the reverse of the removal procedure.

10   Charging system - general information and precautions

The charging system includes the alternator, voltage regulator and battery. These components work together to supply electrical power for the ignition system, lights, radio, etc. The alternator is turned by a drivebelt at the front of the engine.

The purpose of the voltage regulator is to limit the alternator's voltage to a preset value. This prevents power surges, circuit overloads, etc., during peak voltage output. On all models with which this manual is concerned, the voltage regulator is contained within the alternator housing.

The charging system does not ordinarily require periodic maintenance. The drivebelt, electrical wiring and connections should, however, be inspected at the intervals suggested in Chapter 1.

Take extreme care when making circuit connections to a vehicle equipped with an alternator and note the following. When making connections to the alternator from a battery, always match correct polarity. Before using arc welding equipment to repair any part of the vehicle, disconnect the wires from the alternator and the battery terminal. Never start the engine with a battery charger connected. Always disconnect both battery leads before using a battery charger.

11   Charging system - check

Refer to illustrations 11.2 and 11.4

Note: *Two types of alternators are used on the models covered by this manual, depending on model year. The first type, used through 1985, is the SI-type alternator, and is serviceable. SI-type alternators are easily identified by the opening in the rectifier end frame for full fielding the regulator. It also has a cup type sealed rear bearing and a drive end frame external fan blade. The newer CS-type alternator used from 1986-on are identi-*fied by a sealed ball bearing in the rectifier end frame and the rotor axle can be seen in the center of the bearing. Most CS-type alternators have an internal drive end frame fan. The CS-type alternator can not be rebuilt and is serviced as a unit.

1   If a malfunction occurs in the charging circuit, do not immediately assume that the alternator is causing the problem. First check the following items:

a) *The battery cables where they connect to the battery. Make sure the connections are clean and tight.*

b) *Check the external alternator wiring and connections. They must be in good condition.*

c) *Check the drivebelt condition and tension (Chapter 1).*

d) *Make sure the alternator mounting bolts are tight.*

e) *Run the engine and check the alternator for abnormal noise.*

2   Using a voltmeter, check the battery voltage with the engine off. It should be approximately 12-volts **(see illustration)**.

3   Start the engine and check the battery voltage again. It should now be approximately 14-to-15 volts.

4   If equipped with an SI-type alternator. Locate the test hole in the back of the alternator and ground the tab that is located inside the hole by inserting a screwdriver blade into the hole and touching the tab and the case at the same time **(see illustration)**. **Caution:** *Do not run the engine with the tab grounded any longer than is necessary to obtain a voltmeter reading. If the alternator is charging, it is running unregulated during the test. This condition may overload the electrical system and cause damage to the components.*

5   The reading on the voltmeter should be 15-volts or higher with the tab grounded in the test hole.

6   If the voltmeter indicates low battery voltage, the alternator is faulty and should be replaced with a new one (Section 12).

7   If the voltage reading is 15-volts or higher and a "no charge" condition is present,

11.4 Ground the tab located inside the test hole on the backside of the alternator by inserting a screwdriver blade into the hole and touching the tab and the case at the same time

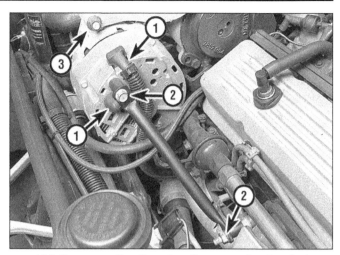

12.3 To remove the alternator, disconnect the electrical connectors (1), remove the bolts (2) and the bracket and remove the AIR pump-to-alternator bracket bolt (3)

12.8 Loosen the AIR pump-to-alternator bracket bolt at the AIR pump (arrow), swing the bracket up and out of the way and remove the alternator

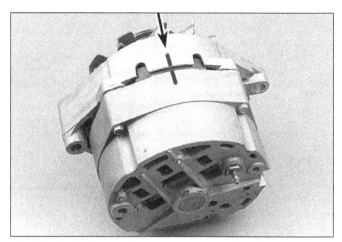

13.2 Mark the drive end frame assembly and rectifier end frame assembly with a scribe or paint before pulling the two halves apart

remove the alternator (Section 12) and have it checked further by a properly equipped repair facility. **Note:** *Many auto parts stores will bench check an alternator for you.*

12 Alternator - removal and installation

Refer to illustrations 12.3 and 12.8

1   Disconnect the cable from the negative terminal of the battery.
2   Remove the drivebelt (Chapter 1).
3   Disconnect the two electrical connectors from the alternator **(see illustration)**.
4   Remove the support bracket bolts from the alternator and the block, then remove the support bracket.
5   On 1991 and earlier models, remove the AIR pump-to-alternator bracket bolt from the alternator.
6   Remove the front lower bolt.
7   Remove the rear lower bolt.
8   On 1991 and earlier models, loosen the

AIR pump-to-alternator bracket bolt on the AIR pump and swing the bracket up and out of the way **(see illustration)**.
9   Remove the alternator.
10   Installation is the reverse of removal.

13 Alternator brushes - replacement

Refer to illustrations 13.2, 13.3, 13.6 and 13.8
Note: *This procedure applies to SI-type alternators only (see Section 11).*
1   Remove the alternator from the vehicle (see Section 12).
2   Scribe a match mark on the front and rear end frame housings of the alternator to facilitate reassembly **(see illustration)**.
3   Remove the four (4) throughbolts holding the front and rear end frames together, then separate the end frames **(see illustration)**.
4   Remove the nuts holding the stator to the rectifier bridge and separate the stator from the end frame.

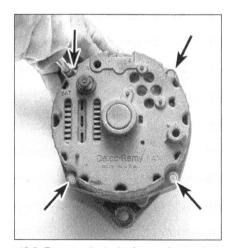

13.3 Remove the bolts (arrows) securing the rear end frame to the drive end frame

5   Remove the screw attaching the diode trio to the end frame and remove the diode trio.

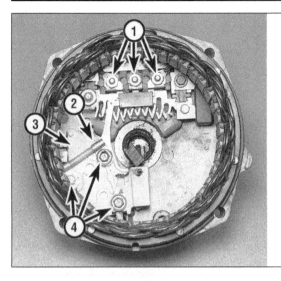

13.6 Brush replacement requires removal of the:

1   *Diode trio mounting nuts*
2   *Resistor (if equipped)*
3   *Voltage regulator*
4   *Brush holder mounting screws*

13.8 Insert a paper clip through the rear end frame to hold the brushes in place during disassembly and reassembly

6   Remove the screws retaining the brush holder and resistor (if equipped) to the end frame and remove the brush holder assembly **(see illustration)**.
7   Installation is the reverse of the removal procedure, noting the following.
8   After installing the brush holder assembly in the end frame, press the brushes into the brush holder and insert a section of wire such as a straightened paper clip through the rear of the end frame to hold both brushes in while reassembly is completed **(see illustration)**. The wire should not be removed until the front and rear end frames have been bolted together.

14   Starting system - general information

The function of the starting system is to crank the engine. The starting system is composed of a starting motor, solenoid and battery. The battery supplies the electrical energy to the solenoid, which then completes the circuit to the starting motor, which does the actual work of cranking the engine.

The electrical circuitry of the vehicle is arranged so that the starter motor can only be operated when the clutch pedal is depressed (manual transmission) or the transmission selector lever is in Park or Neutral (automatic transmission).

Never operate the starter motor for more than 30 seconds at a time without pausing to allow it to cool for at least two minutes. Excessive cranking can cause overheating, which can seriously damage the starter.

15   Starter motor - in-vehicle test

Note: *Before diagnosing starter problems, make sure that the battery is fully charged.*
1   If the starter motor does not turn at all when the switch is operated, make sure that the shift lever is in Park (automatic transmission) or that the clutch pedal is depressed

(manual transmission).
2   Make sure that the battery is charged and that all cables, both at the battery and starter solenoid terminals, are secure.
3   If the starter motor spins but the engine is not cranking, then the overrunning clutch in the starter motor is slipping and the starter motor must be removed from the engine and disassembled.
4   If, when the switch is actuated, the starter motor does not operate at all but the solenoid clicks, then the problem lies with either the battery, the main solenoid contacts or the starter motor itself.
5   If the solenoid plunger cannot be heard when the switch is actuated, the solenoid itself is defective or the solenoid circuit is open.
6   To check the solenoid, connect a jumper lead between the battery (+) and the S terminal on the solenoid. If the starter motor now operates, the solenoid is OK and the problem is in the ignition switch, Neutral start switch or the wiring.
7   If the starter motor still does not operate, remove the starter/solenoid assembly for disassembly, testing and repair.
8   If the starter motor cranks the engine at an abnormally slow speed, first make sure that the battery is charged and that all terminal connections are tight. If the engine is partially seized, or has the wrong viscosity oil in it, it will crank slowly.
9   Run the engine until normal operating temperature is reached, then disable the ignition system by disconnecting the wire from the BAT terminal on the distributor cap.
10   Connect a voltmeter positive lead to the starter motor terminal of the solenoid and then connect the negative lead to ground.
11   Actuate the ignition switch and take the voltmeter readings as soon as a steady figure is indicated. Do not allow the starter motor to turn for more than 30 seconds at a time. A reading of 9-volts or more, with the starter motor turning at normal cranking speed, is normal. If the reading is 9-volts or more but the cranking speed is slow, the motor is faulty. If the reading is less than 9-volts and

the cranking speed is slow, the solenoid contacts are probably burned.

16   Starter motor - removal and installation

Refer to illustrations 16.3, 16.5, 16.9 and 16.10
1   Disconnect the cable from the negative terminal of the battery.
2   Raise the vehicle and support it securely on jackstands.
3   Disconnect the electrical connectors from the starter motor solenoid. Note that the upper terminal (for the battery cable and electrical system hot wire) is larger in diameter than the terminal to the right (for the ignition switch wire), so it is impossible to con-

16.3 The solenoid terminals are unequally sized to help you remember which wire goes to which terminal

1   *Battery and electrical system hot wire terminal*
2   *Ignition switch terminal*
3   *Starter motor strap*

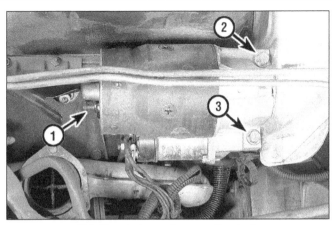

16.9 The starter motor mounting bolts

1   Nut (attaches the support bracket to the stud on the front of the starter)
2   Long starter-to-block mounting bolt
3   Short starter-to-block mounting bolt

16.5 Remove both knock sensor shield screws (arrows) and the shield, then disconnect the knock sensor electrical connector

fuse the wires when reattaching them to the solenoid terminals **(see illustration)**.
4   Remove the transmission cooler line bracket screw.
5   Remove the knock sensor shield screws and the shield **(see illustration)**. Disconnect the knock sensor electrical connector.
6   Remove the support bracket-to-starter nut.
7   Loosen the support bracket-to-engine bolt.
8   Swing the bracket out of the way.
9   Remove the two starter-to-block mounting bolts **(see illustration)**.
10   Slide the starter forward until the snout clears the bellhousing, then swing the snout end away from the vehicle and remove the

starter **(see illustration)**.
11   Installation is the reverse of removal.

17 Starter solenoid - removal and installation

Note: *This procedure applies to 1991 and earlier models only. On 1992 and later models the starter and solenoid must be serviced as a unit.*

Removal

Refer to illustration 17.5
1   Disconnect the cable from the negative terminal of the battery.
2   Remove the starter motor (Section 16).

3   Disconnect the strap from the solenoid terminal to the starter motor.
4   Remove the two screws that secure the solenoid to the starter motor.
5   Twist the solenoid in a clockwise direction to disengage the flange from the starter body **(see illustration)**.

Installation

6   Make sure that the return spring is in position on the plunger, then insert the solenoid body into the starter housing and turn the solenoid counterclockwise to engage the flange.
7   Install the two solenoid screws and connect the motor strap.

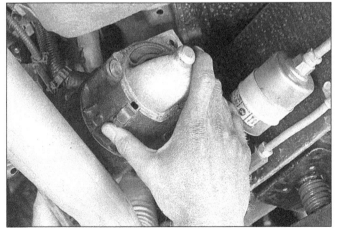

16.10 Slide the starter motor forward until the snout clears the bellhousing, then swing the starter out to the right, away from the engine

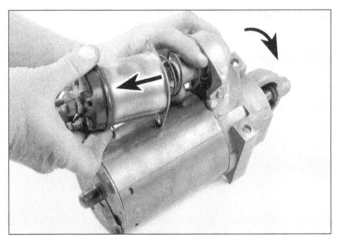

17.5 To disengage the solenoid from the starter, turn it in a clockwise direction

Notes

Chapter 6
Emissions and engine control systems

Contents

Specifications

Torque specifications

	Ft-lbs
Oxygen sensor	30
EGR valve-to-manifold bolts	14
EGR solenoid bracket nut	17

1 General information

Refer to illustration 1.7

To prevent pollution of the atmosphere from incompletely burned and evaporating gases, and to maintain good driveability and fuel economy, a number of emission control devices are incorporated. They include the:

Fuel control system
Electronic Spark Timing (EST)
Electronic Spark Control (ESC) system
Exhaust Gas Recirculation (EGR) system
Evaporative Emission Control System (EECS)
Positive Crankcase Ventilation (PCV) system
Thermostatic Air Cleaner (Thermac)
Air Injection Reaction (AIR) system
Catalytic converter

All of these systems are linked, directly or indirectly, to the On Board diagnostic (OBD) system.

The Sections in this Chapter include general descriptions, checking procedures within the scope of the home mechanic and component replacement procedures (when possible) for each of the systems listed above.

Before assuming that an emissions control system is malfunctioning, check the fuel and ignition systems carefully. The diagnosis of some emission control devices requires

1.7 The Vehicle Emission Control Information (VECI) label provides information regarding engine size, exhaust emission system used, engine adjustment procedures and specifications and an emission component and vacuum schematic diagram

specialized tools, equipment and training. If checking and servicing become too difficult or if a procedure is beyond the scope of your skills, consult your dealer service department.

This doesn't mean, however, that emission control systems are particularly difficult to maintain and repair. You can quickly and easily perform many checks and do most (if not all) of the regular maintenance at home with common tune-up and hand tools. **Note:** *The most frequent cause of emissions problems is simply a loose or broken vacuum hose or wiring connection, so always check the hose and wiring connections first.*

Pay close attention to any special pre-

cautions outlined in this Chapter. It should be noted that the illustrations of the various systems may not exactly match the system installed on your vehicle because of changes made by the manufacturer during production or from year-to-year.

A Vehicle Emissions Control Information label is located in the engine compartment **(see illustration)**. This label contains important emissions specifications and setting procedures, as well as a vacuum hose schematic with emissions components identified. When servicing the engine or emissions systems, the VECI label in your particular vehicle should always be checked for up-to-date information.

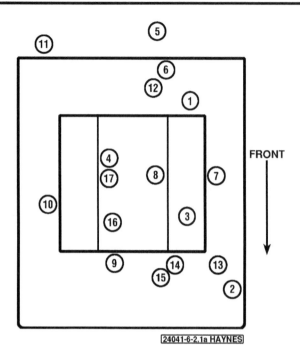

2.1a Underhood layout of emissions and engine control systems components - 1984 models

1 Base timing set connector	10 Knock sensor (ESC)
2 Vapor canister	11 Electronic Control module (ECM)
3 PCV valve	12 Fuel pump relay
4 EGR valve	13 Canister purge solenoid valve
5 Vehicle speed sensor	14 Air control solenoid valve
6 Manifold Pressure Sensor (MAP)	15 Air switching solenoid valve
7 Exhaust oxygen sensor	16 Idle air control
8 Throttle Position Sensor (TPS)	17 Exhaust gas recirculation solenoid valve
9 Coolant temperature sensor	

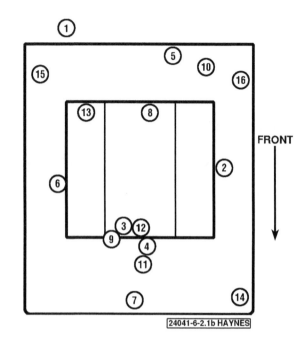

2.1b Underhood layout of emissions and engine control systems components - 1985 through 1991 models

1 Electronic Control Module (ECM)	10 Fuel pump relay
2 Exhaust oxygen sensor	11 Air control (divert) solenoid
3 Throttle position sensor	12 Air switching (catalytic converter) solenoid
4 Coolant temperature sensor	13 Exhaust Gas Recirculation vacuum solenoid
5 Vehicle speed sensor	14 Fuel vapor canister solenoid
6 Knock sensor (ESC)	15 Electronic Spark Control (ESC) module
7 Mass Air Flow sensor (MAF)	16 Engine coolant fan relay
8 Manifold air temperature sensor	
9 Idle air control motor	

2   On Board Diagnostic (OBD) system and trouble codes

Refer to illustrations 2.1a, 2.1b, 2.1c and 2.5
Note: *This Section applies to OBD-I equipped models only. Beginning in 1994 some models are equipped with OBD-II (On Board Diagnostics, second generation), this system is identified by the 16 terminal ALCL connector located to the right of the steering column under the knee bolster. The VECI label in the engine compartment also states "OBD-II". Models equipped with OBD-I systems, will use the 12 terminal ALCL connectors (see illustration 2.5). There is no provision to flash the trouble codes on an OBD-II system. The system can only be accessed with a diagnostic SCAN tool capable of interfacing with OBD class 2 serial data. If the CHECK ENGINE light illuminates on an OBD-II equipped model, take the vehicle to a dealer service department or other properly equipped repair facility for diagnosis.*

The On Board Diagnostic (OBD) system consists of an Electronic Control Module

(ECM) and information sensors which monitor various functions of the engine and send data back to the ECM **(see illustrations)**.

The OBD system is analogous to the central nervous system in the human body: The sensors (nerve endings) constantly relay information to the ECM (brain), which processes the data and, if necessary, sends out a command to change the operating parameters of the engine (body).

Here's a specific example of how one portion of this system operates: An oxygen sensor, located in the exhaust manifold, constantly monitors the oxygen content of the exhaust gas. If the percentage of oxygen in the exhaust gas is incorrect, an electrical signal is sent to the ECM. The ECM takes this information, processes it and then sends a command to the fuel injection system, telling it to change the air/fuel mixture. This happens in a fraction of a second and it goes on continuously when the engine is running. The end result is an air/fuel mixture ratio that is constantly maintained at a predetermined ratio, regardless of driving conditions.

One might think that a system using an

on-board computer and electrical sensors would be difficult to diagnose. This is not necessarily the case. The OBD system has a built-in diagnostic feature, which indicates a problem by flashing a Check Engine or Service Engine Soon light on the instrument panel. When this light comes on during normal vehicle operation, a fault in one of the information sensor circuits or the ECM itself has been detected. More importantly, the source of the malfunction is stored in the ECM's memory.

To retrieve this information from the ECM memory, you must use a short jumper wire to ground a diagnostic terminal. This terminal is part of a wiring connector known as the Assembly Line Communications Link (ALCL) **(see illustration)**. The ALCL is located underneath the dashboard, just below the instrument panel and to the left of the center console. To use the ALCL, simply remove the plastic cover by sliding it toward you. With the connector exposed to view, push one end of the jumper wire into the diagnostic terminal and the other end into the ground terminal.

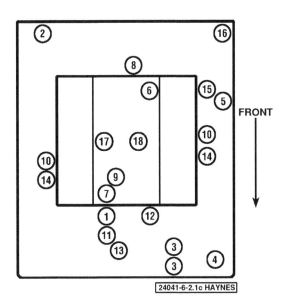

2.1c Underhood layout of emissions and engine control systems components - 1992 through 1996 models

1   Idle Air Control valve (IAC)
2   Fuel pump relay
3   Engine coolant fan relay
4   Electric Air Pump
5   Electric switching valve solenoid
6   EGR solenoid
7   Fuel vapor canister solenoid
8   Exhaust Gas Recirculation (EGR) valve
9   Manifold Absolute Pressure (MAP) sensor
10   Exhaust oxygen sensors
11   Throttle position sensor (TPS)
12   Coolant temperature sensor (CTS)
13   Intake air temperature sensor (IAT)
14   Knock (ESC) sensor(s)
15   Engine oil temperature sensor (EOT)
16   Electronic Control Module (ECM)
17   Ignition test connector
18   PCV valve

When the diagnostic terminal is grounded with the ignition on and the engine stopped, the system will enter the Diagnostic Mode. In this mode the ECM will display a "Code 12" by flashing the Check Engine light, indicating that the system is operating. A code 12 is simply one flash, followed by a brief pause, then two flashes in quick succession. This code will be flashed three times. If no other codes are stored, Code 12 will continue to flash until the diagnostic terminal ground is removed.

After flashing Code 12 three times, the ECM will display any stored trouble codes. Each code will be flashed three times, then Code 12 will be flashed again, indicating that the display of any stored trouble codes has been completed.

When the ECM sets a trouble code, the Check Engine light will come on and a trouble code will be stored in memory. If the problem is intermittent, the light will go out after 10 seconds, when the fault goes away. However, the trouble code will stay in the ECM memory until the battery voltage to the ECM is interrupted. Removing battery voltage for 10 seconds will clear all stored trouble codes. Trouble codes should always be cleared after repairs have been completed. **Caution:** *To prevent damage to the ECM, the ignition switch must be off when disconnecting power to the ECM.*

Following is a list of the typical trouble codes that may be encountered while diagnosing the Computer Command Control System. Also included are simplified trou-

2.5 A typical Assembly Line Communications Link (ALCL) connector - the two terminals you will be concerned with are the A (ground) and B (diagnostic) terminals

bleshooting procedures. If the problem persists after these checks have been made, more detailed service procedures will have to be done by a dealer service department.

Trouble codes
Note: *All codes apply to all models.*

Code	Circuit or system	Probable cause
13	Oxygen sensor circuit	Check the wiring and connectors from the oxygen sensor. Replace oxygen sensor (see Section 4).*
14	Coolant sensor circuit (high temperature indicated)	If the engine is experiencing overheating problems, the problem must be rectified before continuing (see Chapters 1 and 3). Check all wiring and connectors associated with the sensor. Replace the coolant sensor (see Section 4).*
15	Coolant sensor circuit (low temperature indicated)	See above. Also, check the thermostat for proper operation
16	Direct Ignition (DI) system fault	The ECM has detected a fault in the DI system (see Chapter 5).
21	TPS circuit (signal voltage high)	Check for sticking or misadjusted TPS. Check all wiring and connections at the TPS and at the ECM. Adjust or replace TPS (see Section 4).*
22	TPS circuit (signal voltage low)	See above.

Trouble codes (continued)
Note: *All codes apply to all models.*

Code	Circuit or system	Probable cause
23	Intake Air Temperature (IAT) sensor (low temperature indicated)	Check the resistance of the IAT sensor. Check the wiring and connections to the sensor. Replace the IAT sensor (see Section 4).*
24	Vehicle Speed sensor (VSS)	A fault in this circuit should be indicated only while the vehicle is in motion. Disregard code 24 if set when drive wheels are not turning. Check connections at the ECM. Check the TPS setting (see Section 4).
25	Intake Air Temperature (IAT) sensor (high temperature indicated)	Check the resistance of the IAT sensor. Check the wiring and connections to the sensor. Replace the IAT sensor (see Section 4).*
26	Quad Driver Module #1 circuit	Check the resistance of the EGR, Canister purge and the AIR pump relay coils all should be greater than 20 ohms. Replace any solenoid less than 20 ohms. Replace the ECM.*
27	Quad Driver Module #2 circuit	Check the 1 to 4 upshift relay (manual transmission only), Check the TCC solenoid. Resistance should be greater than 20 ohms. If less than 20 ohms replace the solenoid. Replace the ECM.*
28	Quad Driver Module #3 circuit	Check the resistance of the air conditioning clutch relay and the cooling fan relay coils. Replace any relay with less than 20 ohms. Replace the ECM.*
32	Exhaust Gas Recirculation circuit	Check the vacuum source and all vacuum lines. Check the system electrical connectors at the ECM and EGR valve. Replace the EGR valve or ECM as necessary (see Section 5).*
33 (1985 through 1990 models)	Mass Air Flow	Inspect the intake system for air and vacuum leaks. Check the MAF sensor connections. (MAF) sensor circuit. Replace the MAF relays. Replace the ECM.*
33 (1984 and 1991 through 1993 models)	Manifold Absolute Pressure (MAP) sensor circuit (signal voltage high)	Check the vacuum hose(s) from the MAP sensor. Check electrical sensor or circuit connections at the ECM. Replace MAP sensor (see Section 4).*
34 (1985 through 1990 models)	Mas Air Flow (MAF) sensor circuit	Inspect the intake system for air and vacuum leaks. Clean the throttle body. Check the MAF sensor connections. Replace the MAF relays. Replace the ECM.*
34 (1984 and 1991 through 1993 models)	Manifold Absolute Pressure (MAP) sensor circuit (signal voltage low)	Check vacuum hose(s) from MAP sensor. Check electrical sensor or circuit connections at the ECM. Replace MAP sensor (see Section 4).*
35	Idle Air Control (IAC) circuit	Idle RPM too low or too high. Check minimum idle speed (see Chapter 4), check fuel pressure, check for leaking injector and obstructions in the throttle body. Check for intake vacuum leaks. Replace the IAC valve.*
36 (1985 through 1990 models)	Mass Air Flow (MAF) burn-off circuit	Check the MAF sensor, power relay and burn-off relay electrical connections. Replace the MAF burn-off relay.*
36 (1992 and 1993 models)	Distributor Ignition (DI) system fault	The ECM has detected a fault in the DI system (see Chapter 5).
41 (1985 through 1991 models)	Cylinder select error	Check the connections at the ECM. Check for an incorrectly installed, defective or incorrect PROM/MEM-CAL (see Section 3).
41 (1992 and 1993 models)	Electronic Spark Timing (EST) Circuit	Check the wiring and connectors between the ignition module and the ECM. Replace the ignition module.* Replace the ECM.*

Code	Circuit or system	Probable cause
42	Electronic Spark Timing (EST) Circuit	Check the wiring and connectors between the ignition module and the ECM. Replace the ignition module.* Replace the ECM.*
43	Knock Sensor (KS) circuit	Check the ECM for an open or short to ground; if necessary, reroute the harness away from other wires such as spark plugs, etc. Replace the Knock Sensor (see Section 4).*
44	Lean exhaust	Check the wiring and connectors from the oxygen sensor to the ECM. Check the ECM ground terminal. Check the fuel pressure (Chapter 4). Replace the oxygen sensor (see Section 4).*
45	Rich exhaust	Check the evaporative charcoal canister and its components for the presence of fuel. Check for fuel contaminated oil. Check the fuel pressure regulator. Check for a leaking fuel injector. Check for a sticking EGR valve. Replace the oxygen sensor (see Section 4).*
46	Vehicle Anti-Theft System (VATS) circuit	The ECM has detected an error in the VATS system. Check for an open or short in the circuit from the VATS module to the ECM.
51	PROM/EEPROM error	Faulty or incorrect PROM/EEPROM. Diagnosis should be performed by a dealer service department or other qualified repair shop.
52	Engine oil temperature circuit (low temperature indicated)	Check the connections at the oil temperature switch. Replace the switch.*
53	System voltage high	Code 53 will set if the voltage at the ECM is greater than 17.1-volts or less than 10-volts. Check the charging system (see Chapter 5).
54	Fuel pump circuit (low voltage)	Check the fuel pump circuit for shorts or damage (see chapter 4).
55	Fuel Lean Monitor	Engine running lean during power acceleration indicating a possible fuel pump failure, fuel line restriction, etc. (see Chapter 4).
62	Engine oil temperature circuit (high temperature indicated)	Check the connections at the oil temperature switch. Replace the switch.*
63	Right bank oxygen sensor circuit (open circuit indicated)	Check wiring and connections from the oxygen sensor to ECM (see Section 4).*
64	Right bank oxygen sensor circuit (lean exhaust indicated)	Check the wiring and connectors from the oxygen sensor to the ECM. Check the ECM ground terminal. Check the fuel pressure (Chapter 4). Replace the oxygen sensor (see Section 4).*
65	Right bank oxygen sensor circuit (rich exhaust indicated)	Check the evaporative charcoal canister and its components for the presence of fuel. Check for fuel contaminated oil. Check the fuel pressure regulator. Check for a leaking fuel injector. Check for a sticking EGR valve. Replace the oxygen sensor (see Section 4).*
66	Air conditioning pressure sensor circuit	Check the sensor electrical terminal connections and possible short to ground or open circuit in the pressure sensor circuit sensor wiring. Replace the Air conditioning refrigerant pressure sensor.*
67	Air conditioning pressure sensor circuit	Check the pressure sensor or air conditioning clutch circuit for possible short to ground or open circuit pressure sensor circuit in the sensor wiring. Replace the air conditioning refrigerant pressure sensor.*

Trouble codes (continued)

Note: *All codes apply to all models.*

Code	Circuit or system	Probable cause
68	Air conditioning relay circuit	Check the air conditioning relay circuit and the relay for possible short to ground or open circuit (see chapter 3).
69	Air conditioning clutch circuit	Check the air conditioning clutch circuit and the relay or possible short to ground or open circuit (see chapter 3).
72	Gear selector switch	Check the connections at the TCC solenoid and Park Neutral switch. Replace the ECM or repair the transmission as needed.*

* Component replacement may not cure the problem in all cases. For this reason, you may want to seek professional advice before purchasing replacement parts.

3   Electronic Control Module/PROM removal and installation

Refer to illustration 3.6

1   The Electronic Control Module (ECM) is located under the instrument panel on 1984 through 1989, models and in the engine compartment above the battery on later models.

2   Disconnect the negative battery cable from the battery.

3   On 1984 through 1989 models remove the three right hand hush panel screws and detach the panel (under the right side of the dashboard). On later models remove the three nuts that attach the ECM to the ECM bracket in the engine compartment.

4   Disconnect the wire harnesses from the ECM. **Caution:** *The ignition switch must be turned to Off when removing or installing the ECM connectors.*

5   Remove the retaining bolts and carefully detach the ECM.

6   Turn the ECM so that the bottom cover is facing up **(see illustration)** and carefully place it on a clean work surface.

7   Remove the PROM access cover by removing the screws retaining it.

8   If you are replacing the ECM itself, the

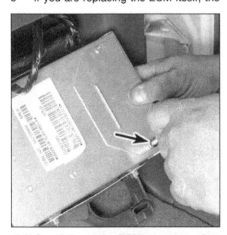

3.6 Carefully set the ECM on a clean, flat surface with the PROM access cover facing up, then remove the cover screws

new ECM will not contain a PROM. It will be necessary to remove the old PROM from the old ECM and install it in the new one.

9   Even if you are simply installing a new PROM in the old ECM, it will still be necessary to remove the old PROM from the old ECM and install a new PROM in its place.

10   On 1984 through 1989 models, remove the PROM using a PROM removal tool (available with a new PROM or a new ECM), grasp the PROM carrier at the narrow ends. Gently rock the carrier from end-to-end while applying firm upward force. The PROM and carrier should lift off the prom socket easily. **Caution:** *The prom carrier should only be removed with the special rocker type PROM removal tool. Removal without this tool or with any other type of tool may damage the PROM pr the PROM socket. On 1991 and later models, unlatch the PROM retaining clips at each end of the PROM. Firmly press down and outward on the clips using finger pressure only. The PROM should back out of the PROM socket.*

11   Note the reference end of the PROM carrier before setting it aside.

12   If you are replacing the ECM, remove the new ECM from its container and check the service number to make sure that it is the same as the number on the old ECM.

13   If you are replacing the PROM, remove the new PROM from its container and check the service number to make sure that it is the same as the number of the old PROM.

14   Position the PROM/carrier assembly squarely over the PROM socket with the small notched end of the carrier aligned with the small notch in the socket at the pin 1 end. Press on the PROM carrier until it seats firmly in the socket. If the PROM is new, make sure that the notch in the PROM is matched to the small notch in the carrier. **Caution:** *If the PROM is installed backwards and the ignition switch is turned on, the PROM will be destroyed. Using the tool, install the new PROM carrier in the PROM socket of the ECM. The small notch of the carrier should be aligned with the small notch in the socket. Press on the PROM carrier until it is firmly seated in the socket.* **Caution:** *Do not press*

on the PROM - press only on the carrier.

15   Attach the access cover to the ECM and tighten the two screws.

16   Install the ECM in the support bracket, plug in the electrical connectors to the ECM and install the hush panel.

17   Start the engine.

18   Enter the diagnostic mode by grounding the diagnostic lead of the ALCL (see Section 2). If no trouble codes occur, the PROM is correctly installed.

19   If Trouble Code 51 occurs, or if the Check Engine light comes on and remains constantly lit, the PROM is not fully seated, is installed backwards, has bent pins or is defective.

20   If the PROM is not fully seated, pressing firmly on both ends of the carrier should correct the problem.

21   It is possible to install the PROM backwards. If this occurs, and the ignition key is turned to On, the PROM circuitry will be destroyed and the PROM will have to be replaced.

22   If the pins have been bent, remove the PROM in accordance with the above procedure, straighten the pins and reinstall the PROM. If the bent pins break or crack when you attempt to straighten them, discard the PROM and replace it with a new one.

23   If carefully inspection indicates that the PROM is fully seated, has not been installed backwards and has no bent pins, but the Check Engine light remains lit, the PROM is probably faulty and must be replaced.

4   Information sensors

Refer to illustrations 4.1, 4.3, 4.4, 4.6 and 4.15

Engine coolant temperature sensor

1   The coolant sensor, mounted in the front of the intake manifold **(see illustration)** is a thermistor (a resistor which varies the value of its voltage output in accordance with temperature changes). A failure in the coolant

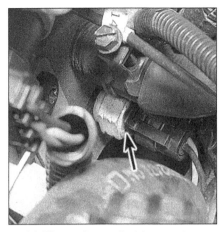

4.1 The engine coolant temperature sensor (arrow) is located in the front of the intake manifold on all vehicles - you will have to unbolt the air control bracket and move the assembly forward to get at the sensor

sensor circuit should set either a Code 14 or a Code 15. These codes indicate a failure in the coolant temperature circuit, so the appropriate solution to the problem will be either repair of a wire or replacement of the sensor.

2   To remove the sensor, remove the bracket bolts and slide the air control valve assembly forward until you have room to unscrew the sensor. Disengage the locking tab on the connector with a small screwdriver and unplug it from the sensor. Carefully unscrew the sensor itself. **Caution:** *Handle the coolant sensor with care. Damage to this sensor will affect the operation of the entire fuel injection system.*

3   Before installing the new sensor, wrap the threads with Teflon sealing tape to prevent leakage and thread corrosion **(see illustration)**. Installation is the reverse of removal.

Manifold Absolute Pressure (MAP) sensor (1984 and 1991 through 1994 models)

4   The Manifold Absolute Pressure (MAP)

4.3 Be sure to wrap the coolant temperature sensor's threads with Teflon tape to prevent leakage

sensor **(see illustration)** is located at the left rear corner of the intake manifold. The MAP sensor monitors the intake manifold pressure changes resulting from changes in engine load and speed and converts the information into a voltage output. The ECM uses the MAP sensor to control fuel delivery and ignition timing.

5   A failure in the MAP sensor circuit should set a Code 33 or a Code 34.

Intake Air Temperature (IAT) sensor (1985 and later models)

6   The Intake Air Temperature (IAT) sensor, located in the underside of the plenum **(see illustration)**, is a thermistor (a resistor which changes the value of its voltage output as the temperature changes). The ECM supplies a five-volt signal to the sensor through a resistor in the ECM and measures the actual voltage, which varies with the temperature, at the sensor. Thus the ECM "knows" the manifold air temperature. The ECM uses the MAT sensor signal to delay EGR until the manifold air temperature reaches 40-degrees F.

7   A failure in the MAT sensor circuit should set either a Code 23 or a Code 25.

Mass Air Flow (MAF) sensor (1985 through 1990, 1995 and 1996 models)

8   The Mass Air Flow (MAF) sensor, which is located between the fresh air intake and the flexible intake duct, measures the amount of air entering the engine. It is controlled by a "burn-off module" located behind the right hush panel, under the ECM. The burn-off module provides power to the MAF sensor and burns off (cleans) the sensor wire after engine shut down. The Bosch MAF sensor is a hot wire type. Current is supplied to the sensing wire to maintain a calibrated temperature but as air flow increases or decreases, the current will vary. This causes a voltage change within the circuitry of the MAF sensor directly proportional to the air mass. The voltage change, which varies from 0.5-volts at idle to about 4.7-volts at wide open throttle, is processed by the ECM for calculating fuel delivery.

9   A failure in the Bosch MAF sensor should set either a Code 33 or a Code 34.

10   The procedure for MAF sensor replacement is contained in Section 10 of Chapter 4.

Oxygen sensor

11   The oxygen sensor is mounted in the exhaust system where it can monitor the oxygen content of the exhaust gas stream.

12   By monitoring the voltage output of the oxygen sensor, the ECM will know what fuel mixture command to give the injector.

13   An open in the oxygen sensor circuit should set a Code 13. A low voltage in the circuit should set a Code 44. A high voltage in the circuit should set a Code 45. Codes 44 and 45 may also be set as a result of fuel system problems.

14   Refer to Section 5 for the oxygen sensor replacement procedure.

Throttle position sensor (TPS)

15   The throttle position sensor (TPS) is located on the left end of the throttle shaft on

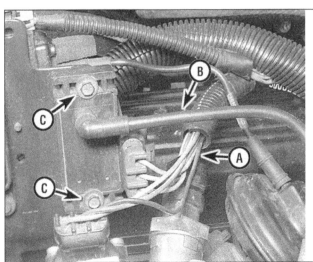

4.4 Typical mounting details for the Manifold Absolute Pressure (MAP) sensor:

A   Electrical connector
B   Vacuum hose
C   Mounting screws

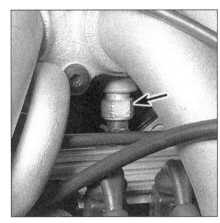

4.6 The Intake Air Temperature (IAT) sensor is located in the underside of the plenum on 1985 and later vehicles - to remove it, disconnect the wire and unscrew the sensor

the rear TBI unit on 1984 CFI-equipped engines and on the right end of the throttle shaft on the throttle body on 1985 and later TPI equipped engines **(see illustration)**.

16   By monitoring the output voltage from the TPS, the ECM can determine fuel delivery based on throttle valve angle (driver demand). A broken or loose TPS can cause intermittent bursts of fuel from the injector and an unstable idle because the ECM thinks the throttle is moving.

17   A problem in any of the TPS circuits will set either a Code 21 or 22. Once a trouble code is set, the ECM will use an artificial default value for TPS and some vehicle performance will return.

18   Should the TPS require replacement, the complete procedure is contained in Section 9 (CFI) or 10 (TPI) of Chapter 4.

Park/Neutral switch (automatic transmission-equipped models only)

19   The Park/Neutral (P/N) switch indicates to the ECM when the transmission is in Park or Neutral. This information is used for the transmission Converter Clutch (TCC) and the Idle Air Control (IAC) valve operation. **Caution:** *The vehicle should not be driven with the Park/Neutral switch disconnected because idle quality will be adversely affected and a false Code 24 (failure in the Vehicle Speed Sensor circuit) may be set. For more information regarding the P/N switch, which is part of the Neutral/start and back-up light switch assembly, see Chapters 7 and 12.*

Cold start module (1984 models only)

20   The cold start module signals the ECM when a given number of hours of engine running time have passed. These hours are about the same as a given number of miles of driving. The purpose of the cold start module is to control emissions for 10 seconds after the engine starts below 175-degrees F, until a certain mileage is reached.

21   The module has three wires - 12-volt ignition, ground and a lead to the ECM. Before the preset number of hours is reached, the voltage to the ECM is zero. When the module reaches the required number of hours, it completes the ECM circuit and sends 12-volts to the ECM, blowing the fuse that protects the ECM.

22   When this event occurs, take the vehicle to your dealer to have the cold start module replaced.

Crank signal (1984 models only)

23   The ECM receives a 12-volt signal from the starter solenoid during cranking to allow enrichment and cancel the diagnostic mode until the engine is running or 12-volts is no longer in the circuit. If the signal from the starting solenoid is not available, the vehicle can be difficult to start.

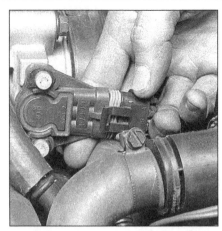

4.15 The Throttle Position Sensor (TPS) is located on the right end of the throttle shaft on 1985 and later vehicles - don't remove it until you have read the appropriate Section in Chapter 4

Air conditioning compressor clutch engagement signal (1984 models only)

24   This signal tells the ECM that the compressor clutch is engaged. The ECM uses this information to adjust the idle speed when the air conditioning system is operating.

Air conditioning On signal (1985 and later models)

25   This signal tells the ECM that the A/C selector switch is turned to the On position and that the pressure cycling switch is closed. The ECM uses this information to adjust the idle speed when the air conditioning system is working.

26   If this signal is not available to the ECM, idle may be rough, especially when the A/C compressor cycles. The voltage at ECM terminal B8 should equal battery voltage when the A/C is activated and the pressure cycling switch is closed.

Vehicle speed sensor (1985 and later models)

27   The vehicle speed sensor (VSS) sends a pulsing voltage signal to the ECM, which the ECM converts to miles per hour. This sensor controls the operation of the TCC.

Distributor reference signal (1985 and later models)

28   The distributor sends a signal to the ECM to tell it both engine rpm and crankshaft position. See Electronic Spark Timing (EST), Section 7, for further information.

---

5   Oxygen sensor

Refer to illustrations 5.1 and 5.11

General description

1   The oxygen sensor, which is located in

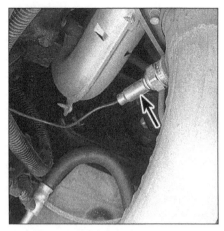

5.1 The oxygen sensor (arrow) is located on the backside of the left exhaust pipe

the backside of the left exhaust pipe just below the joint between the manifold and the pipe **(see illustration)**, monitors the oxygen content of the exhaust gas stream. The oxygen content in the exhaust reacts with the oxygen sensor to produce a voltage output, which varies from 0.1 to 0.9-volts. The ECM monitors this voltage output to determine the ratio of oxygen to fuel in the mixture. The ECM alters the air/fuel mixture ratio by controlling the pulse width (open time) of the fuel injectors. A mixture ratio of 14.7 parts air to 1 part fuel is the ideal mixture ratio for minimizing exhaust emissions, thus allowing the catalytic converter to operate at maximum efficiency. It is this ratio of 14.7 to 1 which the ECM and the oxygen sensor attempt to maintain at all times.

2   The oxygen sensor is like an open circuit and produces no voltage when it is below its normal operating temperature of about 600-degrees F (315-degrees C). During this initial period before warm-up, the ECM operates in open loop mode.

3   However, if the engine reaches normal operating temperature and/or has been running for two or more minutes, and if the oxygen sensor is producing a steady signal voltage between 0.35 and 0.55-volts, even though the TPS indicates that the engine is not at idle, the ECM will set a Code 13.

4   A delay of two minutes or more between engine start-up and normal operation of the sensor, followed by a low voltage signal or a short in the sensor circuit, will cause the ECM to set a Code 44. If a high voltage signal occurs, the ECM will set a Code 45.

5   When any of the above codes occur, the ECM operates in the open loop mode; that is it controls fuel delivery in accordance with a programmed default value instead of feedback information from the oxygen sensor.

6   The proper operation of the oxygen sensor depends on four conditions:

a)   **Electrical** - *The low voltages and low currents generated by the sensor depend upon good, clean connections, which should be checked whenever a*

5.11 To remove the oxygen sensor from the exhaust pipe, disconnect the electrical connector at the end of the pigtail and unscrew the sensor (arrow) - the sensor is easier to remove if you run the engine for about a minute to heat up (expand) the exhaust pipe

malfunction of the sensor is suspected or indicated.

b) **Outside air supply** - *The sensor is designed to allow air circulation to the internal portion of the sensor. Whenever the sensor is removed and installed or replaced, make sure the air passages are not restricted.*

c) **Proper operating temperature** - *The ECM will not react to the sensor signal until the sensor reaches approximately 600-degrees F (315-degrees C). This factor must be taken into consideration when evaluating the performance of the sensor.*

d) **Unleaded fuel** - *The use of unleaded fuel is essential for proper operation of the sensor. Make sure the fuel you are using is of this type.*

7   In addition to observing the above conditions, special care must be taken whenever the sensor is serviced.

a) *The oxygen sensor has a permanently attached pigtail and connector that should not be removed from the sensor. Damage or removal of the pigtail or connector can adversely affect operation of the sensor.*

b) *Grease, dirt and other contaminants should be kept away from the electrical connector and the louvered end of the sensor.*

c) *Do not use cleaning solvents of any kind on the oxygen sensor.*

d) *Do not drop or roughly handle the sensor.*

e) *The silicone boot must be installed in the correct position to prevent the boot from being melted and to allow the sensor to operate properly.*

Replacement

Note: *Because the oxygen sensor is located in the exhaust pipe, it may be too tight to remove when the engine is cold. If you find it*

difficult to loosen, start and run the engine for a minute or two, then shut it off. Be careful not to burn yourself during the following procedure.

8   Disconnect the cable from the negative terminal of the battery.

9   Raise the vehicle and place it securely on jackstands.

10   Carefully unsnap the pigtail lead retaining clip, if used.

11   Disconnect the electrical connector from the sensor **(see illustration)**.

12   Note the position of the silicone boot, if equipped, and carefully unscrew the sensor from the exhaust pipe. **Caution:** *Excessive force may damage the threads.*

13   Anti-seize compound must be used on the threads of the sensor to facilitate future removal. The threads of new sensors will already be coated with this compound, but if an old sensor is removed and reinstalled, recoat the threads.

14   Install the sensor and tighten it to the specified torque.

15   Reconnect the electrical connector of the pigtail lead to the main engine wiring harness.

16   Snap the pigtail retaining clip closed.

17   Lower the vehicle and reconnect the cable to the negative terminal of the battery.

6   Electronic Spark Timing (EST) system

Refer to illustration 6.4

General description

1   To provide improved engine performance, fuel economy and control of exhaust emissions, the Electronic Control Module (ECM) controls distributor spark advance (ignition timing) with the Electronic Spark Timing (EST) system.

2   The ECM receives a reference pulse from the distributor, which indicates both engine rpm and crankshaft position. The ECM then determines the proper spark advance for the engine operating conditions and sends an EST pulse to the distributor.

Check

3   The ECM will set EST at a specified value when the diagnostic test terminal in the ALCL connector is grounded. To check for EST operation, the timing should be checked at 2000 rpm with the terminal ungrounded. Then ground the test terminal. If the timing changes at 2000 rpm, the EST is operating. A fault in the EST system will usually set Trouble Code 42.

Setting timing

4   To set the initial base timing, disconnect the timing connector near the brake booster **(see illustration)**.

5   Set the timing as specified on the VECI label. This will cause a Code 42 to be stored in the ECM memory. Be sure to clear the mem-

ory after setting the timing (see Section 2).

6   For further information regarding the testing and component replacement procedures for the HEI/EST distributor, refer to Chapter 5.

7   Electronic Spark Control (ESC) system

Refer to illustrations 7.8 and 7.13

General description

1   Irregular octane levels in modern gasoline can cause detonation in a high performance engine. Detonation is sometimes referred to as "spark knock." This condition causes the pistons and rings to vibrate and rattle, producing a characteristic knocking or pinging sound.

2   The Electronic Spark Control (ESC) system is designed to retard spark timing up to 20-degrees to reduce spark knock in the engine. This allows the engine to use maximum spark advance to improve driveability and fuel economy.

3   The ESC knock sensor, which is located on the lower right side of the engine block (some models use two sensors, one on the right side of the block and one on the left side of the block, service procedures are the same for both sensors), sends a voltage signal of 8 to 10-volts to the ECM when no spark knock is occurring and the ECM provides normal advance. When the knock sensor detects abnormal vibration (spark knock), the ESC module turns off the circuit to the ECM. The ECM then retards the EST distributor until spark knock is eliminated.

4   Failure of the ESC knock sensor signal or loss of ground at the ESC module will cause the signal to the ECM to remain high. This condition will result in the ECM controlling the EST as if no spark knock is occurring. Therefore, no retard will occur and spark knock may become severe under heavy engine load conditions. At this point, the ECM will set a Code 43.

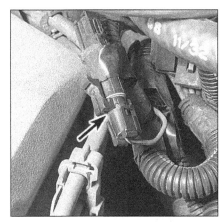

6.4 To set the initial base timing, disconnect the HEI/EST bypass connector (arrow) and set the timing as specified on the VECI label

7.8 The Electronic Spark Control (ESC) sensor is located in the lower right side of the block, just in front of the starter motor

7.13 The Electronic Spark Control (ESC) module is attached to the right side of the evaporator blower with two bolts (arrows)

8.5 On 1985 and later vehicles, the EGR solenoid is located at the right rear corner of the intake manifold next to the distributor (arrow)

5   Loss of the ESC signal to the ECM will cause the ECM to constantly retard EST. This will result in sluggish performance and cause the ECM to set a Code 43.

Component replacement

ESC sensor

6   Disconnect the cable from the negative terminal of the battery.
7   Raise the vehicle and support it on jackstands. Refer to Chapter 1 and drain the cooling system.
8   Disconnect the wiring harness connector from the ESC sensor **(see illustration)**.
9   Remove the ESC sensor from the block. Coolant will flow out of the hole, so be careful not to get any in your eyes.
10   Apply thread sealant to the ESC sensor threads.
11   Installation is the reverse of the removal procedure. Be sure to refill the cooling system.

ESC module

12   Disconnect the cable from the negative terminal of the battery.
13   Disconnect the electrical connector from the module, which is located on the right side of the evaporator blower assembly **(see illustration)**.
14   Unscrew and remove the module.
15   Installation is the reverse of removal.

8   Exhaust Gas Recirculation (EGR) system

Refer to illustrations 8.5, 8.14 and 8.16

General description

1   The Exhaust Gas Recirculation (EGR) system is used to lower NOx (oxides of nitrogen) emission levels by decreasing combustion temperature. The main element of the system is the EGR valve, mounted on the intake manifold, which feeds small amounts of exhaust gas back into the combustion chamber.

2   The EGR valve is opened by manifold vacuum to allow exhaust gases to flow into the intake manifold. The EGR valve is usually open during warm engine operation and any-time the engine is running above idle speed. The amount of gas recirculated is controlled by variations in vacuum and exhaust back-pressure.
3   The valve used on this engine is called a negative backpressure valve. It varies the amount of exhaust gas flow into the manifold depending on manifold vacuum and variations in exhaust backpressure.
4   The diaphragm on this valve has an internal vacuum bleed hole that is held closed by a small spring when there is no exhaust backpressure. Engine vacuum opens the EGR valve against the pressure of a large spring. When manifold vacuum combines with negative exhaust backpressure, the vacuum bleed hole opens and the EGR valve closes.
5   An ECM controlled solenoid **(see illustration)** is used in the vacuum line in order to maintain finer control of EGR flow. The ECM uses information from the coolant temperature, throttle position and manifold pressure sensors to regulate the vacuum solenoid.
6   During cold operation and at idle, the solenoid circuit is grounded by the ECM to block vacuum to the EGR. When the solenoid circuit is not grounded by the ECM, vacuum is allowed to the EGR.
7   With the engine stopped, turning the ignition key to On will turn on the solenoid, allowing vacuum to the EGR valve. Grounding the ALCL diagnostic test terminal will remove power to the solenoid and allow vacuum to the EGR valve.
8   Too much EGR flow weakens combustion, causing the engine to run roughly or stall. During cold operation it may cause the engine to stop after a cold start. At idle, excessive EGR flow may cause the engine to run roughly or stop after decelerating. Too much EGR flow while cruising may cause the engine to surge.
9   Too little or no EGR flow allows combustion temperatures to get too high during acceleration and load conditions. The result

can be spark knock (detonation), engine overheating and emission test failure.

1984 through 1991 models
Check
1984 models only
10   Make a physical inspection of the hoses and electrical connections to ensure that nothing is loose.
11   With the ignition off, check the EGR valve to make sure it is closed by pushing up on the underside of the diaphragm.
12   If the valve is open, disconnect the vacuum hose at the EGR valve and recheck the diaphragm. If the diaphragm is still open, clean the EGR valve and passages as described in Steps 36 on. Replace the EGR valve if necessary. If the diaphragm is closed, connect a vacuum gauge in place of the EGR valve. Proceed to Step 16.
13   If the valve is closed, install a tachometer and bring the engine to normal operating temperature. Do not ground the Test terminal. With the engine idling, the rpm should drop as the EGR valve is opened by pushing up on the underside of the diaphragm. If there is no change in rpm, clean the EGR valve and passages. Replace the EGR valve if necessary.
14   If the rpm drops, disconnect the EGR control solenoid, then check for movement of the diaphragm as the rpm is increased to 1200 rpm and returned to idle. If there is no movement, check vacuum at the engine side of the EGR solenoid **(see illustration)**. It should have at least ten inches of vacuum. If it doesn't, repair the cause of poor vacuum to the solenoid. If it has ten inches of vacuum, check to make sure that it has at least ten inches of vacuum on the EGR valve side of the solenoid. If it does, check for a restricted hose between the EGR valve and the solenoid. If the hose is okay, the EGR valve is faulty. If it does not have at least ten inches of vacuum on the EGR valve side of the solenoid, the EGR solenoid is faulty.
15   If there is movement, connect a vacuum

8.14 The engine side of the EGR solenoid should have at least ten inches of vacuum

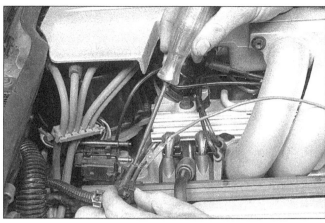

8.16 Connect a test light across the EGR solenoid wire harness terminals after checking the solenoid for vacuum - the light should come on

gauge in place of the EGR valve. There should be at least ten inches of vacuum. If not, perform the sequence of tests outlined in Step 14.

16 If there are at least ten inches of vacuum, leave the vacuum gauge connected, turn the ignition on and, with the engine stopped, connect a test light across the solenoid harness connector terminals **(see illustration)**. The light should go on.

17 If the light does not come on, probe both harness connector terminals with a test light connected to ground. If there is no voltage at either terminal, repair the open ignition circuit to the solenoid. If the light goes on at one terminal, check for an open in the circuit from the EGR solenoid to the ECM (see the wiring diagrams at the end of Chapter 12). If there is no open in the EGR solenoid circuit, either the ECM connector terminal is faulty or the ECM itself is faulty. **Note:** *Before replacing the ECM, use an ohmmeter and check the resistance between the solenoid terminals of the TCC and the EGR. Refer to the ECM wiring diagrams (at the end of Chapter 12) for coil terminal identification for both the solenoid(s) and relay(s) to be checked. Replace any solenoid where the resistance measures less than 20 ohms.*

18 If the light does come on, ground the diagnostic test terminal and note whether the light comes on. If the light comes on, check for a short to ground in the EGR solenoid circuit. If it's okay, the ECM is faulty. **Note:** *Before replacing the ECM, use an ohmmeter to check the resistance between the solenoid terminals of the TCC and the EGR. Refer to the ECM wiring diagrams (at the end of Chapter 12) for coil terminal identification for both the solenoid(s) and relay(s) to be checked. Replace any solenoid where the resistance measures less than 20 ohms. If the light does not come on, reconnect the solenoid, start the engine and note the vacuum gauge reading while the engine is idling. If there is vacuum, replace the EGR control solenoid. If there is no vacuum, the EGR control solenoid is okay. Check the valve for binding, leakage and loose mounting bolts.*

1985 through 1991 models

19 Before performing the following sequence of test procedures, check for ported vacuum to the EGR solenoid and check the hoses for leaks and restrictions. There should be at least seven inches of vacuum at 2000 rpm. **Note:** *The following test sequence assumes there is no Code 32.*

20 With the ignition on and the engine stopped, ground the diagnostic terminal, disconnect the EGR solenoid vacuum harness and apply ten inches of vacuum to the manifold side of the solenoid. It should be able to hold this vacuum. If it can, proceed to Step 24.

21 If it cannot hold vacuum, disconnect the EGR solenoid electrical connector and connect a test light between the harness terminals.

22 If the light comes on, replace the solenoid.

23 If the light does not come on, probe each harness connector terminal with a test light connected to ground. If there is no light, repair the open in the circuit from the fuse box to the EGR solenoid (see the wiring diagrams at the end of Chapter 12). If the light comes on at one terminal, check for an open in the circuit from the EGR solenoid to the ECM. If the circuit has no open, check the resistance of the solenoid - it should have more than 20 ohms. If it doesn't, replace the solenoid and the ECM. If it does, either the ECM connection or the ECM itself is faulty. **Note:** *Before replacing the ECM, check the resistance of each ECM controlled relay and solenoid coil with an ohmmeter. Refer to the ECM wiring diagrams (at the end of Chapter 12) for coil terminal identification for both the solenoid(s) and relay(s) to be checked. Replace any relay or solenoid if the coil resistance measures less than 20 ohms.*

24 If the solenoid can hold vacuum, unground the diagnostic terminal - the vacuum should drop.

25 If there is no drop in vacuum, disconnect the solenoid electrical connector and note the vacuum reading. If the vacuum reading drops, repair the short to ground in the EGR solenoid circuit. If there is no short to ground in the EGR

solenoid circuit, the ECM is faulty. **Note:** *Before replacing the ECM, check the resistance of each ECM controlled relay and solenoid coil with an ohmmeter. Refer to the ECM wiring diagrams (at the end of Chapter 12) for coil terminal identification for both the solenoid(s) and relay(s) to be checked. Replace any relay or solenoid if the coil resistance measures less than 20 ohms. If there is no drop in vacuum reading, replace the EGR solenoid.*

26 If there is a drop in vacuum, turn the ignition off, connect a vacuum pump to the EGR valve and use a mirror to watch the valve diaphragm while applying vacuum. The diaphragm should move freely and hold vacuum for at least 20 seconds. If it doesn't, replace the EGR valve.

27 If the diaphragm does hold vacuum for at least 20 seconds, apply vacuum to the EGR valve, start the engine and immediately note the vacuum reading. The valve is good if it moves to its seated position (valve closed) and if the vacuum reading drops while starting the engine. If the valve does not move to its seated position and/or vacuum doesn't drop while starting the engine, remove the EGR valve and check for plugged passages. If the passages are not plugged, replace the EGR valve. If the valve moves to its seated position and the vacuum reading drops while starting the engine, check the Park/Neutral switch on the automatic transmission (if equipped) by connecting a test light between the EGR solenoid harness terminals, and, with the engine at normal operating temperature, accelerate the engine to about 1500 rpm in Park (observe the light, it should stay on). Repeat this test in Drive - the test light should dim or go out. If both conditions are met, then the switch is okay.

Component replacement

EGR valve

28 If your vehicle is a 1984 model, remove the air cleaner. If your vehicle is a 1985 or later model, remove the plenum (Chapter 4).

29 Disconnect the EGR valve vacuum line from the valve.

30 Remove the EGR valve mounting bolts.

31   Remove the EGR valve from the manifold.

32   If the EGR passages in the manifold have deposits built-up in them they should be cleaned. Care should be taken to ensure that all loose particles are completely removed to prevent them from clogging the EGR valve or from being ingested into the engine. **Caution:** *Do not wash the EGR valve in solvents or degreaser - permanent damage to the valve diaphragm may result. Sand blasting of the valve is also not recommended since it can affect the operation of the valve.*

33   Buff the exhaust deposits from the mounting surface and around the valve with a wire brush.

34   Look for exhaust deposits in the valve outlet. Remove deposit buildup with a screwdriver.

35   Clean the mounting surfaces of the intake manifold and the valve assembly.

36   If the valve itself is operating correctly, it can be reused after it has been cleaned and checked for deposits. Hold the valve in your hand. Tap on the end of the round pintle using a light snapping action with a soft-face hammer. This will remove the exhaust deposits from the valve seat. Remove all loose particles.

37   Clean the mounting surface of the valve and the pintle with a wire brush.

38   Depress the valve diaphragm and check the seating area for cleanliness and signs of rubbing by looking through the valve outlet. If the pintle or the seat are not completely clean, repeat the procedure in Step 36.

39   Hold the bottom of the valve securely and try to rotate the top of the valve back-and-forth. Replace the valve if any looseness is felt.

40   Inspect the valve outlet for deposits. Remove any deposit buildup with a screwdriver or other suitable sharp tool.

41   Clean the manifold mounting surface.

42   Using a new gasket, install the old (cleaned) or new EGR valve on the intake manifold.

43   Install the EGR valve mounting bolts and tighten them to the specified torque.

44   Attach the vacuum hose to the valve.

45   If your vehicle is a 1984 model, install the air cleaner. If your vehicle is a 1985 or later model, install the plenum (Chapter 4).

EGR solenoid

46   Disconnect the cable from the negative terminal of the battery.

47   If your vehicle is a 1984 model, remove the air cleaner.

48   Disconnect the electrical connector at the solenoid.

49   Disconnect the vacuum hoses from the solenoid.

50   Remove the mounting nut and detach the solenoid.

51   Install the new solenoid and tighten the nut to the specified torque. The remainder of the installation procedure is the reverse of removal.

1992 and later models
Check

52   Start the engine, warm it up to normal operating temperature and allow the engine to idle. Manually lift the EGR diaphragm. The engine should stall, or at least the idle should drop considerably. **Note:** *The EGR valve is located at the rear of the intake manifold* **(see illustration 2.1c).**

53   If the EGR valve appears to be in proper operating condition, carefully check all hoses connected to the valve for breaks, leaks or kinks. Replace or repair the valve/hoses as necessary.

54   With the engine idling at normal operating temperature, disconnect the vacuum hose from the EGR valve and connect a vacuum pump. When vacuum is applied the engine should stumble or die, indicating the vacuum diaphragm is operating properly. **Note:** *Some models use a backpressure-type EGR valve. On models so equipped, backpressure must be created in the exhaust system before the vacuum pump will actuate the valve. To create backpressure, install a large socket into the tailpipe and clamp it to the pipe to prevent it from dropping out during the test. The 1/2 inch hole in the socket will allow the engine to idle but still create backpressure on the EGR valve. Don't restrict the exhaust system any longer than necessary to perform this test. Replace the EGR valve with a new one if the test does not affect the idle.*

55   Check the operation of the EGR control solenoid. Check for vacuum to the solenoid. If vacuum exists, check for battery voltage to the solenoid.

56   Further testing of the EGR system, vacuum solenoid and the ECM will require a SCAN tool to access computer information that directly controls the EGR system. Have the vehicle checked by a dealer service department or other qualified repair facility.

Component replacement
EGR valve

57   Disconnect the vacuum hose at the EGR valve.

58   Remove the nuts or bolts that secure the valve to the intake manifold or adapter.

59   Lift the EGR valve from the engine.

60   Clean the mounting surfaces of the EGR valve. Remove all traces of gasket material.

61   Place the new EGR valve, with a new gasket, on the intake manifold or adapter and tighten the attaching nuts or bolts.

62   Connect the vacuum signal hose.

EGR vacuum solenoid

63   Remove the hoses from the EGR vacuum control solenoid, labeling them to ensure proper installation.

64   Remove the solenoid and replace it with the new one.

EGR valve cleaning

65   With the EGR valve removed, inspect the passages for excessive deposits.

66   It is a good idea to place a rag securely

in the passage opening to keep debris from entering. Clean the passages by hand, using a drill bit.

9   Air Injection Reaction (AIR) system

Refer to illustrations 9.2, 9.22, 9.23, 9.32, 9.41, 9.47, 9.52 and 9.56

General description

1   The Air Injection Reaction (AIR) system reduces hydrocarbon (HC), carbon monoxide (CO) and nitrous oxide (NOx) emissions in the exhaust and enables the catalytic converter to heat up quickly after engine start-up so that conversion of the exhaust gases can occur as soon as possible.

2   The AIR system includes an air pump, a control valve (which is really two - switching and divert - valves), a check valve, the catalytic converter and the plumbing which connects these components **(see illustration)**. A positive displacement, vane-type pump, driven by a belt on the front of the engine, supplies air to the system. The pump is permanently lubricated and requires no maintenance.

3   Intake air passes through a centrifugal filter fan at the front of the pump where foreign materials are separated from the air by centrifugal force. The fan should not be cleaned nor should it ever be removed from the pump unless it is damaged, as removal will destroy the fan.

4   On 1984 vehicles, air flows from the pump through an ECM controlled valve called an AIR control valve and through check valves to either the exhaust manifolds or the catalytic converter. There are really two valves inside the AIR control valve housing. The diverter valve sends air to the air cleaner, when necessary, to protect the catalytic converter, or to the other valve, which is called a switching valve. The switching valve directs air either to the exhaust ports or to the catalytic converter, depending on operating conditions. The one piece control valve housing is located at the front of the engine in

9.2 Air Injection Reaction (AIR) pump location (arrow)

close proximity to the AIR pump and is connected to it by a short metal pipe called an adapter. The check valves prevent exhaust gas backflow into the pump in the event of a backfire in the exhaust system or AIR pump drivebelt failure.

5   On 1985 and later models, a basically similar device combines the diverter valve with a solenoid controlled vent. In its energized state, the air is diverted to the switching valve; when de-energized, air is diverted to the silencer, a short metal pipe with a small muffler attached to its end. The silencer assembly is bolted to the right side of the water pump. Aside from where diverted air is dumped, the differences between 1984 and later AIR control valves are minor. Later units are actually two valves connected by a rubber hose instead of a one-piece housing, are located a little farther from the pump and are connected to the pump with a rubber hose instead of a metal pipe.

6   The air switching valve directs air to the exhaust ports during cold engine operation or whenever the system is operating in open loop mode. It sends air to the catalytic converter when the system is in closed loop.

7   When it's necessary to divert air, the ECM turns off the solenoid in the diverter valve, routing air to the air cleaner in 1984 vehicles and to the silencer in 1985 and later vehicles.

8   Air is diverted to the air cleaner or silencer under all of the following conditions:

If the engine is running too rich.
Any time the ECM recognizes a problem and sets the Check Engine light.
During deceleration.
During high rpm operation when air pressure is greater than the setting for the internal relief valve.

9   If no air flow is routed to the exhaust stream at the exhaust ports or the catalytic converter pipe, the HC and CO emission levels will be too high. But if air is flowing to the exhaust ports all the time, it increases the temperature of the converter to unacceptable levels. Similarly, air flowing constantly to the catalytic converter causes overheating of the converter if the engine is running too rich. Therefore, both the diverter valve and the switching valve are operated by ECM-controlled solenoids to ensure that they respond quickly to the engine's changing needs.

10   An open in the diverter valve electrical circuit will divert all air flow to the air cleaner (1984) or silencer (1985 on). If an open occurs in the switching valve electrical circuit, air will be routed to the converter at all times. Mechanical failures of the valves themselves will result in air flowing incorrectly to the exhaust ports or the converter.

1984 through 1991 models

Check

Air pump

11   Disconnect the adapter pipe (1984) or rubber hose (1985 on) from the AIR pump.

Accelerate the engine to about 1500 rpm and feel the air flow from the pump. If air flow increases as the engine is accelerated, the pump is operating properly.

12   If air flow does not increase, or is not present, check the drivebelt for proper tension. If the belt is tensioned properly, listen for a leaking pressure relief valve (an air leak is audible even while the pump is operating).

13   The AIR pump is not completely silent. Under normal operating conditions, its noise rises in pitch as the engine speed increases. But a harsh, whining sound is a symptom of trouble. To determine if the AIR pump is emitting excessive noise, operate the engine with the pump drivebelt removed.

14   If the noise vanishes when the belt is removed, rotate the pump by hand to see if it has seized. Inspect the hoses, tubes and all connections for leaks and proper routing. Check for air flow from the diverter and/or switching valve(s). Check the AIR pump mounting bolts to make sure they are tight. If no irregularities exist and the AIR injection pump noise is still excessive, replace the pump.

Hoses and pipes

15   Inspect the hoses and pipes for deterioration and holes. Check the routing of all hoses and pipes - close proximity to hot or sharp engine parts may cause wear. Check all hose and pipe connections and clamps for tightness.

16   If a leak is suspected on the pressure side of the system, or if a hose or pipe has been disconnected on the pressure side, the connections should be checked for leaks with a soapy water solution. With the pump running, bubbles will form if a leak exists.

Check valve

17   Inspect the check valve whenever the hose is disconnected from it or whenever check valve failure is suspected. If the pump is inoperative and an inspection reveals evidence of exhaust gases in the pump, check valve failure has occurred.

18   Blow through the check valve (toward the cylinder head), then attempt to suck back through the check valve. Flow should only be in one direction (toward the exhaust valve). Replace a valve that does not operate properly.

Air management system

Note: *If the above quick checks can't pinpoint the cause of a malfunction in the air management system, there may be an electrical problem. The following sequence of tests will help you find it.*

19   The air management system is controlled by diverter and switching valves, each with its own ECM-controlled vacuum solenoid. When a solenoid is grounded by the ECM, manifold vacuum will activate a valve and direct pump air as follows:

a) *While the diverter solenoid is ungrounded by the ECM, air is diverted to the air cleaner (1984) or silencer (1985 on).*

b) *As soon as the diverter solenoid is grounded by the ECM, air is routed to the air switching valve.*

c) *When the switching solenoid is not grounded by the ECM, air is sent to the catalytic converter.*

d) *Should the switching solenoid be grounded by the ECM, air is sent to the exhaust port.*

20   To check the air management system, start the engine. With the diagnostic terminal ungrounded and the engine running at part throttle below 2000 rpm, air should be felt at the outlet to the exhaust ports during open loop operation (which takes from 6 to 120 seconds on a warm engine) and should be switched to the converter as the system goes closed loop.

21   The air management system will do one of four things when subjected to the above conditions:

It will function correctly (proceed to Step 22).
It will constantly send air to the exhaust ports (proceed to Step 24).
It will constantly send air to the divert valve (proceed to Step 25).
It will constantly send air to the catalytic converter (proceed to Step 25).

22   If the system functions correctly, disconnect the HEI bypass timing connector **(see illustration)**. This will set a Code 42. When a code is set, the ECM opens the ground to the AIR control valve and allows air to divert. Air should be diverted as soon as the Check Engine light comes on. If it does divert, the system is functioning properly. Reconnect the HEI timing connector and clear the ECM memory.

23   If the AIR control valve does not divert as soon as the Check Engine light comes on, turn the ignition to On and, with the engine stopped, disconnect the diverter valve solenoid electrical connector and connect a

9.22 Disconnecting the HEI bypass timing connector (arrow) will set a Code 42, which will cause the ECM to open the ground circuit from the AIR control valve and to flash the Check Engine light - air should divert as soon as the Check Engine light comes on

test light between the harness connector terminals **(see illustration)**. If the light doesn't come on, replace the control valve assembly (1984) or the diverter valve (1985 on). If it does come on, check for a short to ground in the circuit from the solenoid to the ECM (refer to the air management system wiring diagrams at the end of Chapter 12). If the circuit from the solenoid to the ECM is okay, check the resistance between the solenoid terminals. It should be more than 20 ohms. If it isn't, replace the control solenoid and the ECM. If it is more than 20 ohms, the ECM may be faulty. **Note:** *Before replacing the ECM, check the resistance of each ECM controlled relay and solenoid coil with an ohmmeter. Refer to the ECM wiring diagrams (at the end of Chapter 12) for coil terminal identification for both the solenoid(s) and relay(s) to be checked. Solenoids are turned on and off by ECM internal electronic switches called "drivers." Each driver is part of a group of four called "quad-drivers." Failure of one driver can damage any other driver within the set. Solenoid coil resistance must measure more than 20 ohms. Less resistance will cause early failure of the ECM driver. Using an ohmmeter, check the resistance of the canister purge, diverter and switching solenoid coils before installing a replacement ECM. Replace any relay or solenoid if the coil resistance measures less than 20 ohms.*

24   If the system constantly sends air to the exhaust ports, turn the ignition to On and, with the engine stopped and the test terminal not grounded, disconnect the switching valve solenoid electrical connector and connect the test light between the harness connector terminals. Note whether the test light comes on. If the test light doesn't come on, replace the air switching valve. If the light comes on, check for a short to ground in the circuit from the air switching valve to the ECM (refer to the air management system wiring diagrams at the end of Chapter 12). If circuit 436 is okay, replace the ECM. **Note:** *Before replacing the ECM, check the resistance of each ECM controlled relay and solenoid coil with an ohmmeter. Refer to the ECM wiring diagrams (at the end of Chapter 12) for coil terminal identification for both the solenoid(s) and relay(s) to be checked. Solenoids are turned on and off by ECM internal electronic switches called "drivers." Each driver is part of a group of four called "Quad-Drivers." Failure of one can damage any other driver within the set. Solenoid coil resistance must measure more than 20 ohms. Less resistance will cause early failure of the ECM driver. Using an ohmmeter, check the resistance of the canister purge, switching and diverter solenoid coils before installing a replacement ECM. Replace any relay or solenoid if the coil resistance measures less than 20 ohms.*

25   If the system constantly sends air to the diverter valve or to the catalytic converter, check the circuits, as follows:

26   With the ignition on and the engine stopped, ground the test terminal. Disconnect the applicable solenoid wiring harness

9.23 If the air management system does not divert air as soon as the Check Engine light comes on, turn the ignition to On and, with the engine stopped, disconnect the diverter valve solenoid electrical connector (the lower connector) and hook up a test light between the harness connector terminals

and connect the test light between the harness connector terminals. If the test light comes on, check vacuum at the valve to make sure that it is at least 10-inches while the engine is idling. If it is, either the valve connection or the valve is faulty.

27   If the test light does not come on, connect it between each connector terminal and ground. If the light comes on at one terminal, check for an open in the wire between the applicable solenoid and its ECM terminal. If there is no open, check the resistance of the applicable solenoid. If it is under 20 ohms, replace the solenoid and the ECM. If it isn't, either the ECM connector or the ECM itself is faulty. **Note:** *Before replacing the ECM, check the resistance of each ECM controlled relay and solenoid coil with an ohmmeter. Refer to the ECM wiring diagrams (at the end of Chapter 12) for coil terminal identification for both the solenoid(s) and relay(s) to be checked. Solenoids are turned on and off by ECM internal electronic switches called "drivers." Each driver is part of a group of four called "quad-drivers." Failure of one can damage any other driver within the set. Solenoid coil resistance must measure more than 20 ohms. Less resistance will cause early failure of the ECM driver. Using an ohmmeter, check the resistance of the canister purge, switching and diverter solenoid coils before installing a replacement ECM. Replace any relay or solenoid if the coil resistance measures less than 20 ohms.*

28   If the light does not come on at all, repair the open in the circuit from the fusebox to the ignition switch. If the light comes on at both terminals, repair the short in the circuit to the ECM. Then test it again.

Component replacement

Pump centrifugal filter fan

Caution: *The centrifugal filter fan should not*

9.32 To replace the AIR pump filter fan, remove the pulley bolts and the pulley, then insert a pair of needle nose pliers into the vanes and pull the old filter off

be cleaned with either compressed air or solvents. Nor should it be removed from the pump unless it is damaged, as removal usually destroys it.

29   Disconnect the cable from the negative terminal of the battery. Loosen the AIR pump drive pulley bolts.

30   If your vehicle is a 1984 model, lift the drive tensioner to the raised position. If you have a 1985 or later model vehicle, push the tensioner to the left. Remove the drivebelt (Chapter 1).

31   Remove the drive pulley bolts and the pulley from the AIR pump.

32   Carefully pull the pump centrifugal filter fan from the AIR pump by inserting a pair of needle-nose pliers into the edge **(see illustration)**. **Caution:** *Do not allow any filter fragments to enter the AIR pump intake hole. Do not use a screwdriver to pry off the filter fan - it will damage the pump sealing lip. Do not drive the metal hub from the filter fan.*

33   Install the new filter fan, place the pump pulley against it and tighten the pump pulley bolts a little at a time. This will press the centrifugal filter fan into its cavity in the pump housing. **Caution:** *Do not drive the filter fan on with a hammer. A slight amount of interference with the housing bore is normal. After a new filter fan has been installed, it may squeal upon initial operation or until the outer sealing lip has been worn in. This may require a short period of pump operation at various engine speeds.*

34   The remainder of installation is the reverse of the removal procedure.

AIR injection pump

35   Hold the pump pulley from turning by compressing the drivebelt, then loosen the pump pulley bolts.

36   If your vehicle is a 1984 model, lift the drive tensioner to the raised position. If it's a 1985 or later model vehicle, push the drive tensioner to the left.

37   Remove the drivebelt (Chapter 1).

38   Remove the pulley bolts and the pulley.

9.41 Components of the AIR pump mounting

1   AIR pump
2   AIR pump support bracket
3   Alternator bracket

9.47 To remove the control valve, unplug the electrical
connectors (1) and the vacuum lines (2) from the diverter and
switching valves

39   If you are replacing the pump centrifugal filter fan, insert needlenose pliers and pull the filter fan from the hub (see Steps 32 and 33 above).

40   If your vehicle is a 1984 model, disconnect the adapter pipe from the AIR pump. If it's a 1985 or later model vehicle, loosen the hose clamp and detach the hose from the AIR pump.

41   Remove the five bracket bolts (two in front and three in the rear) that mount the AIR pump (see illustration).

42   Loosen the AIR pump-to-alternator bracket bolt at the alternator and swing the bracket up and out of the way. Remove the AIR pump.

43   Installation is the reverse of removal.

AIR control valve (diverter and switching valves)

Note: On 1984 vehicles, failure of either valve necessitates the replacement of the entire (one-piece) AIR control valve assembly. On newer vehicles, even though the diverter and switching valves are referred to jointly as the AIR control valve, either one may be purchased individually.

44   Disconnect the cable from the negative terminal of the battery.

45   If your vehicle is a 1984 model, remove the air cleaner.

46   If your vehicle is a 1984 model, remove the adapter bolts and the adapter (the pipe between the AIR pump and the control valve). If your vehicle is a 1985 or later model, disconnect the rubber hose that connects the AIR pump and the control valve.

47   Disconnect the electrical connectors and vacuum lines from the switching valve and the diverter valve (see illustration).

48   Loosen the hose clamps and detach all hoses from both valves.

49   Remove the bracket bolts, the brackets and the AIR control valve assembly. If you are only replacing a single valve on a vehicle manufactured after 1984, disconnect the two components and discard the faulty part. Con-

9.52 Use a backup wrench on the air
injection pipe to prevent it from being
twisted when the check valve is removed

nect the new component.

50   Installation is the reverse of the removal procedure.

Check valve

Note: There is a check valve for each cylinder bank. The following removal procedure applies to either valve.

51   Loosen the hose clamp and disconnect the hose from the check valve.

52   Using a backup wrench to prevent the air injection pipe from turning, loosen and unscrew the check valve from the pipe (see illustration).

53   Installation is the reverse of the removal procedure. Be sure to coat the threads with anti-seize compound.

1992 and later models

Check

54   Because of the complexity of this system it is difficult for the home mechanic to make a proper diagnosis. If the system is suspected of not operating properly, individual components can be checked.

55   Begin any inspection by carefully check-

9.56 On 1992 and later models, the
electric AIR pump (arrow) is located below
the left headlight

ing all hoses, vacuum lines and wires. Be sure they are in good condition and that all connections are tight and clean.

56   To check the pump, allow the engine to reach normal operating temperature and then idle speed. Locate the hose running from the air pump and squeeze it to feel the pulsation (see illustration). Have an assistant increase the engine speed and check for a parallel increase in airflow. If this is observed as described, the pump is functioning properly. If it is not operating in this manner, a faulty pump is indicated. If OK, raise the rpm above 2,850 rpm and check to make sure the ECM turns OFF the AIR system,

57   The check valve can be inspected by first removing it from the air line. Attempt to blow through it from both directions. Air should only pass through it in the direction of normal airflow. If it is either stuck open or stuck closed the valve should be replaced.

58   Remove the AIR system relay (located on the AIR pump motor) and with the ignition key ON (engine not running) check for power to the relay. Battery voltage should be available.

10.17 To remove the EECS solenoid, remove the bolts (1) and the cover (2), then disconnect the wiring harness and both hoses - to remove the canister itself, disconnect the hoses (3) and loosen the pinch bolt (4)

10.25 The canister purge solenoid on 1992 and later models is located on the right side of the intake manifold (arrow)

59   Place a jumper wire between A and E on the relay electrical connector . This should activate the AIR pump. If the pump activates with the jumper wire and not with the relay in place, replace the relay with a new part.
60   If the jumper wire does not activate the AIR pump, check the wiring harness and the 25 amp inline AIR pump fuse (located on the AIR pump motor - 1992 through 1994 models only) and the 20 amp maxi-fuse located in the underhood fuse box.

AIR pump replacement

61   Disconnect the electrical connectors from the AIR pump and relay.
62   Remove the AIR pump mounting bolts and the silencer clamp.
64   Remove the AIR pump from the engine compartment.
65   Installation is the reverse of removal.

---

10   Evaporative Emission Control System (EECS)

Refer to illustrations 10.17 and 10.25

General description

1   This system is designed to trap and store fuel vapors that evaporate from the fuel tank, throttle body and intake manifold.
2   The Evaporative Emission Control System (EECS) consists of a charcoal-filled canister and the lines connecting the canister to the fuel tank, ported vacuum and intake manifold vacuum.
3   Fuel vapors are transferred from the fuel tank, throttle body and intake manifold to a canister where they are stored when the engine is not operating. When the engine is running, the fuel vapors are purged from the canister by intake air flow and consumed in the normal combustion process.
4   The ECM operates a solenoid valve

which controls vacuum to the purge valve in the charcoal canister. Under cold engine or idle conditions, the solenoid is turned on by the ECM, which closes the valve and blocks vacuum to the canister purge valve. The ECM turns off the solenoid valve and allows purge when the engine is warm, after the engine has been running a specified time, above a specified vehicle speed and above a specified throttle opening.
5   The control valve opens with vacuum applied and closes with no vacuum. This prevents purge to the intake manifold under conditions of low ported vacuum, such as deceleration.
6   If the solenoid is open, or is not receiving power, the canister can purge to the intake manifold at all times. This can allow extra fuel at idle or during warm-up, which can cause rough or unstable idle, or rich operation during warm-up.
7   Poor idle, stalling and poor driveability can be caused by an inoperative purge valve, a damaged canister, split or cracked hoses or hoses connected to the wrong tubes.
8   Evidence of fuel loss or fuel odor can be caused by liquid fuel leaking from fuel lines or the TBI, a cracked or damaged canister, an inoperative bowl vent valve, an inoperative purge valve, disconnected, misrouted, kinked, deteriorated or damaged vapor or control hoses or an improperly seated air cleaner or air cleaner gasket.

Check

9   Inspect each hose attached to the canister for kinks, leaks and breaks along its entire length. Repair or replace as necessary.
10   Inspect the canister. It it's cracked or damaged, replace it.
11   Look for fuel leaking from the bottom of the canister. If fuel is leaking, replace canister. Check the hoses and hose routing.
12   Check the filter at the bottom of the canister. If it's dirty, plugged or damaged,

replace the canister.
13   Apply a short length of hose to the lower tube of the purge valve assembly and attempt to blow through it. Little or no air should pass into the canister (a small amount of air will pass because the canister has a constant purge hole).
14   With a hand vacuum pump, apply vacuum through the control vacuum signal tube to the purge valve diaphragm. If the diaphragm does not hold vacuum for at least 20 seconds, the diaphragm is leaking and the canister must be replaced.
15   If the diaphragm holds vacuum, again try to blow through the hose while vacuum is still being applied. An increased flow of air should be noted. If it isn't, replace the canister.

Component replacement

1991 and earlier models

Fuel vapor canister solenoid

16   Disconnect the cable from the negative terminal of the battery.
17   Remove the solenoid cover bolts and the cover **(see illustration)**.
18   Disconnect the wires and vacuum hoses from the solenoid.
19   Remove the solenoid.
20   Installation is the reverse of removal.

Fuel vapor canister

21   Disconnect the hoses from the canister.
22   Remove the pinch bolt and loosen the clamp.
23   Remove the canister.
24   Installation is the reverse of removal.

1992 and later models

Fuel vapor canister solenoid

25   Disconnect the electrical connector and vacuum hoses from the solenoid **(see illustration)**.
26   Remove the nut from the mounting stud and remove the solenoid.

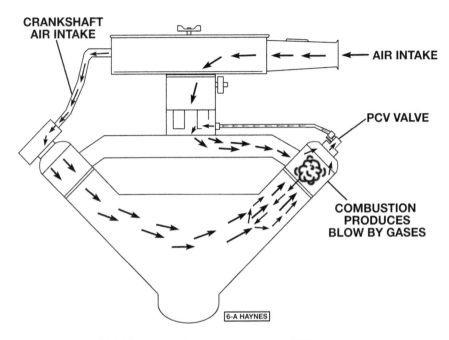

11.1 The route of vapor flow in a typical PCV system

Fuel vapor canister

27   Raise the vehicle and support on jackstands.
28   Remove the right rear tire.
29   Remove the canister splash shield and disconnect the hoses from the canister.
30   Release the canister retaining bracket latch and remove the canister.
31   Installation is the reverse of removal.

11   Positive Crankcase Ventilation (PCV) system

Refer to illustration 11.1

General description

1   The Positive Crankcase Ventilation (PCV) system reduces hydrocarbon emissions by scavenging crankcase vapors. It does this by circulating fresh air from the air cleaner through the crankcase, where it mixes with blow-by gases and is then rerouted through a PCV valve to the intake manifold **(see illustration)**.
2   The main components of the PCV system are the PCV valve, a fresh air filtered inlet and the vacuum hoses connecting these two components with the engine and the EECS system. The valve itself, which regulates the flow of gases in accordance with engine speed and manifold vacuum, is located in the left rocker arm cover. The fresh air filtered inlet is located in the air cleaner on 1984 vehicles and in the right rocker arm cover on later models (see Chapter 1).
3   To maintain idle quality, the PCV valve restricts the flow when the intake manifold vacuum is high. If abnormal operating condi-

tions arise, the system is designed to allow excessive amounts of blow-by gases to backflow through the crankcase vent tube into the air cleaner to be consumed by normal combustion.

Check

4   The PCV system can be checked quickly and easily for proper operation. It should be checked regularly because carbon and gunk deposited by blow-by gases will eventually clog the PCV valve and system hoses. The common symptoms of a plugged or clogged PCV valve include a rough idle, stalling or a slow idle speed, oil leaks, oil in the air cleaner or sludge in the engine .
5   To check for proper vacuum in the system, locate the PCV valve fresh air inlet filter (in the air cleaner on 1984 models and in the right rocker arm cover on later models).
6   Disconnect the hose leading to this filter. Be careful not to break the molded fitting on the filter.
7   With the engine idling, place your thumb lightly over the end of the hose. You should feel a slight vacuum. The suction creates a slight hissing sound, which can be heard as your thumb is released. This sound indicates that air is being drawn through the system. If a vacuum is felt, the system is functioning properly. Check that the filter is not clogged or dirty. If in doubt, replace it.
8   If there is very little or no vacuum at the end of the hose, the system is clogged and must be inspected further. A leaking valve or hose will cause a rough idle, stalling or a high idle speed.
9   Shut off the engine and locate the PCV valve (right rocker arm cover on all models). Carefully pull it from its rubber grommet.

Shake it and listen for a clicking sound. If the valve does not click freely, replace it with a new one.
10   Start the engine and run it at idle speed with the PCV valve removed. Place your thumb over the end of the valve and feel for suction. There should be a relatively strong vacuum.
11   If little or no vacuum is felt at the PCV valve, turn off the engine and disconnect the vacuum hose from the other end of the valve. Run the engine at idle speed and check for vacuum at the end of the hose just disconnected. No vacuum at this point indicates that the vacuum hose or fresh air inlet fitting on the other rocker arm cover is plugged. If it is the hose that is blocked, replace it with a new one or remove it from the engine and blow it out sufficiently with compressed air. A clogged passage at the manifold requires that the component be removed and thoroughly cleaned to remove carbon build-up. A strong vacuum felt going into the PCV valve, but little or no vacuum coming out of the valve, indicates a failure of the PCV valve itself. Replace it.
12   When purchasing a new PCV valve, make sure it is the correct one for your engine. An incorrect PCV valve may pull too little or too much vacuum, possibly leading to engine damage.

Component replacement

13   The component replacement procedures for the PCV valve and filter are in Chapter 1.

12   Thermostatic Air Cleaner (Thermac)

Note: *The Thermac system is employed only on 1984 models.*

General description

1   The thermostatic air cleaner (Thermac) system improves engine efficiency and driveability under varying climactic conditions by controlling the temperature of the air coming into the air cleaner. A uniform incoming air temperature allows leaner air/fuel ratios during warm-up, which reduces hydrocarbon emissions.
2   The Thermac system uses a damper assembly, located in the snorkel of the air cleaner housing, to control the ratio of cold and warm air directed into the throttle body. This damper is controlled by a vacuum motor that is, in turn, modulated by a temperature sensor in the air cleaner. On some engines a check valve is used in the sensor, which delays the opening of the damper flap when the engine is cold and the vacuum signal is low.
3   When the engine is cold, the damper flap blocks off the air cleaner inlet snorkel, allowing only warm air from around the exhaust manifold to enter the engine. As the engine warms up, the flap gradually opens

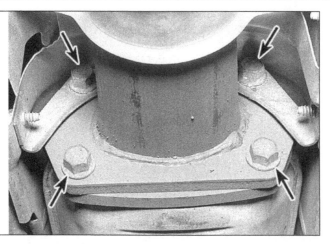

13.6 Remove these four bolts from the flange between the front of the catalytic converter and the exhaust pipe - there are four more bolts in exactly the same positions at the other end of the converter that must also be removed

the snorkel passage, increasing the amount of cold air allowed into the air cleaner. By the time the engine has reached its normal operating temperature, the flap opens completely, allowing only cold, fresh air to enter.

4   Hesitation during warm-up can be caused by the heat stove tube becoming disconnected, the vacuum diaphragm motor becoming inoperative (open to the snorkel), loss of manifold vacuum, a sticking damper door, a missing or leaking seal between the TBI and the air cleaner, a missing air cleaner cover seal, a loose air cleaner cover or a loose air cleaner.

5   Lack of power, pinging or spongy throttle response can be caused by a damper door that does not open to the outside air or a temperature sensor that doesn't bleed off excess vacuum.

Check

6   Refer to Chapter 1 for maintenance and checking procedures for the Thermac system. If any of the problems mentioned above are discovered while performing routine maintenance checks, refer to the following procedures.

7   If the damper door does not close off snorkel air when the engine is started cold, disconnect the vacuum hose at the snorkel vacuum motor and place your thumb over the hose end, checking for vacuum. If there is vacuum going to the motor, check that the damper door and link rod are not frozen or binding inside the air cleaner snorkel.

8   If a vacuum pump is available, disconnect the vacuum hose and apply vacuum to the motor to make sure the damper door moves. Replace the vacuum motor if the application of vacuum does not open the door.

9   If the above test indicates that there is no vacuum going to the motor, check the

hoses for cracks, crimps and loose fitting connections. If the hoses are clear and in good condition, replace the temperature sensor inside the air cleaner housing.

Component replacement

Note: *Check on parts availability before attempting component replacement. Some components may no longer be available for all models, requiring replacement of the complete air cleaner assembly.*

Air cleaner vacuum motor

10   Remove the air cleaner assembly from the engine and disconnect the vacuum hose from the motor.

11   Using a 1/16-inch drill bit, drill out the two spot welds which secure the vacuum motor retaining strap to the snorkel tube. Enlarge the hole with bigger drill bits as required to remove the retaining strap. Do not damage the snorkel tube.

12   Remove the motor retaining strap.

13   Lift up the motor, cocking it to one side to unhook the motor linkage at the control damper assembly.

14   To install, drill a 7/64-inch hole in the snorkel tube at the center of the vacuum motor retaining strap.

15   Insert the vacuum motor linkage into the control damper assembly.

16   Using the sheet metal screw supplied with the motor service kit, attach the motor and retaining strap to the snorkel. Make sure the sheet metal screw does not interfere with the operation of the damper door. Shorten the screw if necessary.

17   Connect the vacuum hose to the motor and install the air cleaner assembly.

Air cleaner temperature sensor

18   Remove the air cleaner from the engine

and disconnect the vacuum hoses at the sensor.

19   Carefully note the position of the sensor. The new sensor must be installed in exactly the same position.

20   Pry up the tabs on the sensor retaining clip and remove the sensor and clip from the air cleaner.

21   Install the new sensor with a new gasket in the same position as the old one.

22   Press the retaining clip onto the sensor. Do not damage the control mechanism in the center of the sensor.

23   Connect the vacuum hoses and attach the air cleaner to the engine.

13   Catalytic converter

Refer to illustration 13.6

General description

1   The catalytic converter is an emission control device added to the exhaust system to reduce pollutants from the exhaust gas stream. A single-bed converter design is used in combination with a three-way (reduction) catalyst. The catalytic coating on the three-way catalyst contains platinum and rhodium, which lowers the levels of oxides of nitrogen (NOx) as well as hydrocarbons (HC) and carbon monoxide (CO).

Check

2   The test equipment for a catalytic converter is expensive and highly sophisticated. If you suspect that the converter on your vehicle is malfunctioning, take it to a dealer or authorized emissions inspection facility for diagnosis and repair.

3   The converter is located underneath the passenger compartment, just behind the junction of the left and right side exhaust pipes. Therefore, whenever the vehicle is raised for servicing of underbody components, check the converter for leaks, corrosion and other damage. If damage is discovered, the converter can simply be unbolted from the exhaust system and replaced.

Component replacement

4   Raise the vehicle and place it securely on jackstands.

5   Disconnect the converter air pipe fitting on the right side of the converter.

6   Remove the four converter-to-exhaust pipe flange bolts from each end of the converter **(see illustration)**.

7   Remove the converter.

8   Installation is the reverse of removal.

Chapter 7 Part A
Manual transmission and overdrive unit

Contents

Specifications

Torque specifications

Ft-lbs (unless otherwise indicated)

Four-speed/with overdrive

Shifter rod adjusting nuts	20 to 30
Shift lever-to-transmission nuts	15 to 24
Shifter cross bolt	
1984 and 1985	40 to 51
1986 through 1988	15 to 23
Shifter bracket bolt	15 to 22
Shifter-to-underbody bolt	22 to 30
Transmission-to-clutch housing bolts	45 to 60
Transmission drain plug	15 to 25
Transmission fill plug	25 to 35
Overdrive unit-to-transmission	34 to 36
Overdrive unit oil pan bolts	6 to 8
Overdrive unit-to-oil cooler lines	8 to 12

ZF six-speed

Transmission-to-clutch housing bolts	37
Transmission drain and fill plugs	26
Shift lever cover nuts	89 in-lbs

1.2 Underside view of the transmission

1	Engine (oil pan)	4	Shift linkage	7	Cooler lines
2	Clutch housing (bellhousing)	5	Overdrive unit fluid level check plug	8	Transmission oil check and fill plugs
3	Transmission	6	Overdrive unit (oil pan)		

1   General information

Refer to illustration 1.2

The vehicles covered by this manual are equipped with a 4-speed automatic transmission, a 4-speed manual transmission with automatic overdrive or a 6-speed manual transmission. All information on the manual transmission, along with the overdrive unit attached to it, is included in this Part of Chapter 7. Information for the automatic transmission can be found in Part B of this Chapter.

The 4-speed manual transmission with automatic overdrive, used in models manufactured from 1984 through 1988, is essentially two separate units coupled together **(see illustration)**. The transmission is a conventional 4-speed, commonly designated as an 83 mm type. The overdrive unit offers a two-speed operation with either a 1:1 or a 0.68:1 ratio, controlled by the ECM computer or a driver-operated switch.

Beginning in 1989, the 4-speed overdrive transmission was replaced by a ZF 6-speed unit, which was adopted primarily to increase fuel economy. To ensure compliance with federal fuel economy standards, the 6-speed transmission inhibits the use of 2nd and 3rd gears when shifting out of 1st gear under the following conditions; coolant temperature is above 122-degrees F), throttle is open 35-percent or less and the vehicle's speed is between 12 and 19 mph.

Due to the complexity, unavailability of replacement parts and the special tools necessary, internal repair procedures for these two units are not recommended for the home mechanic. The information contained within this manual will be limited to general diagnosis, external adjustments and removal and installation.

Depending on the expense involved in having a faulty transmission overhauled, it may be an advantage to consider replacing the unit with either a new or rebuilt one. Your local dealer or transmission shop should be able to supply you with information concerning cost, availability and exchange policy. Regardless of how you decide to remedy a transmission problem, you can still save considerable expense by removing and installing the unit yourself.

2   Transmission shift linkage - adjustment

Note: *This procedure applies to 1988 and earlier models only. The shift linkage is not adjustable on 1989 and later models.*

1   Disconnect the negative cable at the battery. Place the cable out of the way so it cannot accidentally come in contact with the negative terminal of the battery, as this would once again allow power into the electrical system of the vehicle.

2   Remove the left (driver's) seat, disconnecting any electrical wiring. Refer to Chapter 11 if necessary.

3   Remove the shift knob at the top of the shifter. The method for doing this depends on the date of manufacture of your vehicle.

  a) *For early models, where the overdrive switch is mounted on the console, carefully pry up the button on the end of the knob and then remove the T-pin. Unscrew the knob, being careful not to lose the spring that is just beneath the knob.*

b) *For later models, where the overdrive switch is actuated by the shifter button, carefully pry the button out of the shift knob and then unscrew the T-rod. Count the number of turns while removing this rod so that it can be reinstalled at the same height. Unscrew the shift knob, being careful not to lose the spring that is just beneath the knob.*

4   Remove the console cover and the glove box lock.
5   Remove the left side panel of the center console.
6   Remove the shift cover, exposing the shifter components.
7   Loosen the adjusting nuts on the shifter rods.
8   With the shifter and the transmission in Neutral, install the gage pin. If a gage pin is not available, a drill bit large enough to fit into the gage pin hole with minimal clearance can be used. The bit must fit tightly to ensure that all three levers are held in precise alignment.
9   Equalize the swivels on all three shift rods, then finger tighten each of the adjusting nuts against the swivels.
10   Remove the gage pin, tighten the adjusting nuts and check the operation of the shifter. If acceptable, reinstall the various components removed previously. On late-model vehicles, the overdrive button must be flush with the top of the knob. If not, adjust the T-rod up or down as necessary.

3   Steering column lock cable - adjustment

Note: *This procedure applies to 1988 and earlier models only.*
1   Expose the lock cable. It is located the front of the gear shifter under the console.
2   Locate the steering column lock cable at the base of the shifter. There is a bracket which secures the cable to the front floor - the adjusting key is located at this bracket.
3   Lift up on the adjusting key. This will release the cable.
4   Place the lock lever on the upper steering column to the locked (Park) position (toward the rear of the car).
5   Shift the transmission into Reverse.
6   Insert the gage (or a 1/16-inch drill bit) against the reverse stop and pull the reverse lever until the reverse pawl just touches the gage.
7   Push down on the adjusting key to lock the cable in this position.
8   Remove the gage and pull on the shifter to ensure the pawl hits the stop and locks the shifter in Reverse.

4   Transmission shift effort - diagnosis

Note: *This procedure applies to 1988 and earlier models only.*

5.7 The shift rod levers are disconnected at the transmission side cover (arrows). If the adjusting nuts on the rods are not disturbed, the shifter should not require adjustment after reinstallation

1   If excessive effort is required to shift the transmission into a particular gear, a relatively simple diagnosis procedure can be used to determine if the shifter mechanism or the transmission is at fault.
2   Determine which shift rod is suspect and then disconnect it from the side of the transmission.
3   Thread two nuts onto the shift shaft now extending from the transmission, lock the nuts together with a wrench and manually shift the transmission in and out of gear.
4   Using an inch-pound torque wrench, measure the torque necessary to shift the transmission into gear. It should not exceed 72 inch-pounds.
5   Torque in excess of this amount would indicate an internal problem in the transmission. An anti-chatter lubricant (Positraction additive) may help the situation. You can obtain this special lubricant, along with any further information regarding its use, at a Chevrolet dealer.
6   If the shift effort at this location is not excessive, the problem probably lies in the shift lever or shifter. Check for binding, rust, dirt, etc.

5   Transmission shifter - removal and installation

1988 and earlier models

Refer to illustration 5.7

Removal

1   Disconnect the negative cable at the battery. Place the cable out of the way so it cannot accidentally come in contact with the negative terminal of the battery, as this would once again allow power into the electrical system of the vehicle.
2   Remove the left (driver's) seat, disconnecting any necessary electrical wires. Refer to Chapter 11 if necessary.
3   Remove the shift knob at the top of the shifter. The method for doing this depends on

the date of manufacture of your vehicle. Refer to Section 2 for this procedure.
4   Remove the console cover and the glove box lock.
5   Remove the left side panel of the center console.
6   Remove the shift cover, exposing the shifter components.
7   Disconnect the three shift rod levers at the shifter **(see illustration).**
8   Disconnect the Park lock cable at the shifter and, on late models, disconnect the electrical wiring for the overdrive switch.
9   Remove the shifter cross bolt and the shifter mounting bracket and the mounting bolt at the body panel.
10   Remove the shifter.

Installation

11   Place the shifter into position and install the body mounting bolt. Tighten it securely.
12   Position the mounting bracket and install the bracket bolts. Tighten it securely.
13   Install the shifter cross bolt and tighten it securely.
14   Connect the three shift rods and then adjust the linkage as outlined in Section 2.
15   Connect the Park lock cable and adjust it as outlined in Section 3.
16   Connect the overdrive electrical switch wiring (if disconnected).
17   Reinstall the remaining components.

1989 and later models

18   Use a small screwdriver to carefully pry off the shift knob release button from the top of the shift knob. Remove the rivet pin inside the knob.
19   Unscrew the knob counterclockwise.
20   Pry out the small retainer from above the circular collar on the lever. Lift off the collar.
21   Remove the upper center console. The console is secured by several screws, some of which are hidden beneath removable plastic covers.
22   Remove the shifter boot bolts and the shifter boot.
23   Installation is the reverse of removal.

6.11 Disconnect the cooler lines on the right side of the overdrive unit

6.16 The transmission must be slid to the rear about 8 to 10 inches to clear the input shaft at the front - note the use of a special transmission jack which bolts to the transmission and incorporates safety chains

6   Transmission and overdrive unit - removal and installation

Refer to illustrations 6.11 and 6.16
Note: *On 1988 and earlier models, remove both the transmission and the overdrive unit together as a single unit and then separate the two components once removed.*

Removal

1   Disconnect the negative cable at the battery. Place the cable out of the way so it cannot accidentally come in contact with the negative terminal of the battery, as this would once again allow power into the electrical system of the vehicle.
2   On models with Crossfire Throttle Body Injection, remove the air cleaner assembly from the top of the engine.
3   Remove the distributor cap (see Chapter 1 or Chapter 5 for details).
4   On those models equipped with a TV cable, disconnect the cable at the throttle lever.
5   Raise the vehicle and support it securely on jackstands.
6   Remove the upper and lower body braces on convertible models.
7   Remove the complete exhaust system as an assembly. Refer to Chapter 4 for additional information.
8   Remove the exhaust hanger at the transmission.
9   Support the transmission with a jack, preferably a special transmission jack made for this purpose.
10   Remove the driveline beam and the driveshaft from under the vehicle (refer to Chapter 8 for additional information).
11   Disconnect the cooler lines at the overdrive unit **(see illustration)**. Place plastic bags over the ends of the lines to prevent fluid loss and contamination.
12   On 1988 and earlier models remove the three shift rod levers at the transmission side cover. On 1989 and later models remove the

shifter boot and console cover (section 5).
13   Disconnect the Throttle Valve cable (if equipped) at the overdrive unit.
14   Disconnect the various electrical connectors attached to the transmission and overdrive unit.
15   Check that all wiring, cables, brackets, etc. are disconnected from the transmission and out of the way.
16   Support the engine and remove the four bolts that attach the transmission to the clutch housing. Slide the transmission to the rear to disengage it from the clutch. Lower the transmission **(see illustration)**.
17   At this stage, the overdrive unit can be separated from the transmission by removing the bolts.
18   The clutch components can be inspected by removing the clutch cover from the engine (refer to Chapter 8). In most cases, new clutch components should be installed as a matter of course if the transmission is removed.

Installation

19   With the clutch components installed and properly aligned (see Chapter 8), bolt the overdrive unit to the transmission and raise the assembly into position. Carefully slide the input shaft into place in the clutch splines.
20   Install the four transmission-to-clutch housing bolts and tighten them securely.
21   Connect the oil cooler lines to the overdrive unit.
22   Refer to Chapter 8 to install the driveshaft and the driveline beam. **Caution:** *Do not overtighten the driveline beam as damage to the overdrive unit can result.*
23   Connect and adjust the shift linkage (see Section 2).
24   Connect the various electrical connectors and the TV cable (if equipped).
25   Refill the transmission and the overdrive unit with fluid (each unit requires different types of fluid - see Chapter 1).
26   Install the exhaust system and the braces (convertible models only).

27   Make a final check that all components are correctly installed, then lower the vehicle.
28   In the engine compartment, install the distributor cap, TV cable and air cleaner (as necessary). Reconnect the negative battery cable and test drive the vehicle.

7   Overdrive switch - removal and installation

Note: *The overdrive switch on early models, which is located on the console, is straightforward in its removal and installation. On later models, described here, the switch is attached to the front of the shift lever, just under the boot.*

Removal

1   Disconnect the negative cable at the battery. Place the cable out of the way so it cannot accidentally come in contact with the negative terminal of the battery, as this would once again allow power into the electrical system of the vehicle.
2   To gain access to the switch, each of the trim plates will have to be removed and the shifter boot lifted up.
3   Begin by removing the instrument panel trim plate. The headlight switch knob will have to be removed and possibly the lever for the tilt steering column.
4   Remove the screws on either side of the center accessory trim plate and remove the plate.
5   Remove the screws retaining the center console trim plate.
6   Disconnect the various electrical connectors attached to the console trim plate, as well as the screws securing the shift boot to the trim plate. Remove the trim plate.
7   Carefully pry the overdrive button from the top of the shift lever.
8   Unscrew the T-rod from the shift knob, counting the number of turns so it can be returned to its original location.

9   Loosen the nut that secures the switch to the shifter and then remove the switch and actuator block from the shifter. **Note:** *Be careful not to lose the return spring, which is located inside the actuator block.*

10   Disconnect the electrical lead at the switch and remove the switch and actuator assembly. The two pieces can then be separated by removing the pin that holds them together.

Installation

11   Assemble the switch and actuator block with the retaining pin. Place a dab of grease on the sliding surface of the actuator block.

12   Using a small screwdriver, compress the return spring as the switch assembly is placed in position on the shifter.

13   Secure the switch assembly with the retaining nut, using a thread-locking agent to ensure the assembly does not come loose at a later date.

14   Install the T-rod into the shifter knob, turning it the same number of turns as on disassembly.

15   Install the button on the end of the shifter. If the button is not flush with the top of the knob, remove it and screw the T-rod in or out as necessary.

16   Connect the electrical lead for the switch and install the remaining components in the reverse order of removal.

8   Transmission output seal - replacement

Refer to illustration 8.6

1   Raise the vehicle and support it securely on jackstands.

8.6 The output shaft seal is pried out of the overdrive unit using a seal remover or a long screwdriver. Work around the seal, prying it out a little at a time, being careful not to damage the output shaft

2   If equipped with a convertible top, remove the underbody braces.

3   Remove the complete exhaust assembly as a single unit (refer to Chapter 4).

4   Support the transmission with a jack.

5   Remove the driveline beam (if equipped) and the driveshaft (refer to Chapter 8).

6   Using a seal remover or a long screwdriver, pry the seal out of the end of the transmission or overdrive unit **(see illustration)**. Be careful not to damage the output shaft.

7   Coat the lip of the new seal with transmission fluid.

8   Using a large socket or piece of pipe that is the same diameter as the seal, drive the new seal into position.

9   Install the remaining components, referring to the appropriate Chapters for information. **Caution:** *Do not overtighten the driveline beam to the overdrive unit, as damage can result.*

9   Overdrive unit - inspection and overhaul

1   The overdrive unit bears many similarities to an automatic transmission, relying on many of the same components and hydraulic pressure.

2   No external adjustments are available to the home mechanic.

3   Perhaps the best maintenance for the overdrive unit is periodic fluid and filter replacement (see Chapter 1).

4   Other than seal replacement and the overdrive switch replacement (covered in previous Sections), no repair procedures are available to the home mechanic.

5   If a complete overhaul of the unit is necessary, considerable money can be saved by removing and installing the unit at home (refer to Section 6).

Notes

Chapter 7 Part B
Automatic transmission

Contents

Specifications

Torque specifications

Ft-lbs (unless otherwise indicated)

Transmission-to-engine bolts	35
Torque converter-to-driveplate bolts	
1984 through 1987	35
1988 through 1996	46
Shift lever (at transmission) bolts	24
TV cable securing screw	84 in-lbs
Inspection cover	84 in-lbs
Cooler lines (at radiator)	20
Cooler lines (at transmission)	120 in-lbs

1 General information

The automatic transmission installed in this vehicle is essentially a 4-speed overdrive unit with a clutch-type torque converter. The clutch is designed to engage at speeds above 25 mph and provides a direct connection between the engine and the drive wheels for better efficiency and fuel economy. When the shift lever is in the Overdrive D position, the overdrive gear range is automatically selected when the vehicle reaches a steady speed above 40 mph.

Due to the complexity of the clutches and the hydraulic control system, and because of the special tools and expertise required to perform an automatic transmission overhaul, it should not be undertaken by the home mechanic. Therefore, the procedures in this Chapter are limited to general diagnosis, routine maintenance, adjustment and transmission removal and installation.

If the transmission requires major repair work, it should be left to a dealer service department or an automotive or transmission repair shop. You can, however, remove and install the transmission yourself and save the expense, even if the repair work is done by a transmission specialist.

Adjustments that the home mechanic can perform include those involving the throttle valve (TV) cable and the shift linkage. **Caution:** *Never tow a disabled vehicle with an automatic transmission at speeds greater than 30 mph or distances over 50 miles.*

3.4 To check for free operation, pull forward on the inner cable and then let go. The end of the cable should then slide back until it touches the cable terminal (Tuned Port Injection shown - Crossfire throttle body injection similar)

1	Cable end	3	Inner cable
2	Cable terminal		

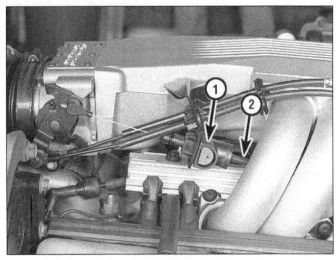

3.8 Throttle Valve (TV) cable adjustment (Tuned Port Injection shown - Crossfire throttle body injection similar)

1	Re-adjust tab	2	Slider

2   Diagnosis - general

1   Automatic transmission malfunctions may be caused by a number of conditions, such as poor engine performance, improper adjustments, hydraulic malfunctions and mechanical problems.
2   The first check should be of the transmission fluid level and condition. Refer to Chapter 1 for more information. Unless the fluid and filter have been recently changed, drain the fluid and replace the filter (also in Chapter 1).
3   Road test the vehicle and drive in all the various selective ranges, noting discrepancies in operation. Check as follows:
   Overdrive range: While stopped, position the lever in the Overdrive range and accelerate. Check for a 1-2 shift, 2-3 shift and 3-4 shift. Also, the converter clutch should apply in 2nd or 3rd gear, depending on calibration. Check for part-throttle downshift by depressing the accelerator 3/4 of the way - the transmission should downshift. Check for a full detent downshift by depressing the throttle all the way - the transmission should immediately downshift.
   Drive range: At speed in Fourth gear (Overdrive range), manually shift the transmission to Drive range. The transmission should shift to 3rd gear immediately. It should not shift back to Overdrive. Also check for part throttle and full throttle downshifts in this range.
   Drive 2 range: While in the 3rd gear range, shift to 2nd gear. The transmission should downshift immediately. While in the 2nd gear range, check for downshift at part and full throttle.
   Low range: Position the lever in Low (1) position and check the operation.
   Overrun braking: This can be checked by manually shifting to a lower range. Engine

rpm should increase and a braking effect should be noticed.
   Reverse: Position the shifter in Reverse and check operation.
4   Verify that the engine is not at fault. If the engine has not received a tune-up recently, refer to Chapter 1 and make sure all engine components are functioning properly.
5   Check the adjustment of the throttle valve (TV) cable (see Section 3). **Note:** *1994 and later models no longer use a TV cable, all shifting is performed by solenoids and valves controlled by the ECM.*
6   Check the condition of all vacuum and electrical lines and fittings at the transmission, or leading to it.
7   Check for proper adjustment of the shifter control cable (Section 5).
8   If at this point a problem remains, there is one final check before the transmission is removed for overhaul. The vehicle should be taken to a transmission specialist who will connect a special oil pressure gauge and check the line pressure in the transmission.

3   Throttle valve (TV) cable - description, inspection and adjustment

Refer to illustrations 3.4 and 3.8
Note: *This Section applies to 1993 and earlier models only. 1994 and later models no longer use a TV cable, all shifting is performed by solenoids and valves controlled by the ECM.*

Description

1   The throttle valve cable used on these transmissions should not be thought of as merely a "downshift" cable, as in earlier years. The TV cable controls line pressure, shift points, shift feel, part throttle downshifts and detent downshifts.

2   If the TV cable is broken, sticky, misadjusted or is the incorrect part for the model, the vehicle will experience a number of problems.

Inspection

3   Inspection should be made with the engine running at idle speed with the selector lever in Neutral. Set the parking brake firmly and block the wheels to prevent any vehicle movement. As an added precaution, have an assistant in the driver's seat applying the brake.
4   Grab the inner cable a few inches behind where it attaches to the throttle linkage and pull the cable forward. It should easily slide through the cable terminal that connects to the throttle linkage **(see illustration)**.
5   Release the cable and it should return to its original location with the cable stop against the cable terminal.
6   If the TV cable does not operate as above, the cause is a defective or misadjusted cable or damaged components at either end of the cable.

Adjustment

7   The engine should not be running during this adjustment.
8   Depress the re-adjust tab and move the slider back through the fitting away from the throttle linkage until the slider stops against the fitting **(see illustration)**.
9   Release the re-adjust tab.
10   Manually turn the throttle lever to the "wide open throttle" position, which will automatically adjust the cable. Release the throttle lever. **Note:** *Do not use excessive force at the throttle lever to adjust the TV cable. If great effort is required to adjust the cable, disconnect the cable at the transmission end and check for free operation. If it is still difficult, replace the cable. If it is now free, suspect a bent TV link in the transmission or a*

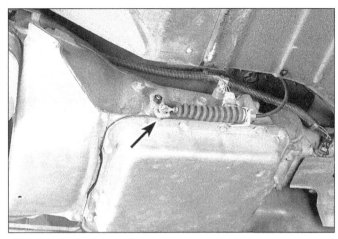

5.3 The shifter is adjusted by loosening the cable attaching nut (arrow) where it connects to the transmission, then placing both the lever and shifter in Park and tightening the nut

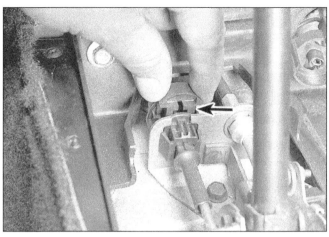

6.2 The Park lock cable is adjusted by pulling up on the adjusting key (arrow), putting the column and cable in the locked position, then pushing down on the adjusting key

problem with the throttle lever.

11   After adjustment, check for proper operation as described in Steps 3 through 6 above.

4   Throttle valve (TV) cable - replacement

Note: *This Section applies to 1993 and earlier models only. 1994 and later models no longer use a TV cable; all shifting is performed by solenoids and valves controlled by the ECM.*

1   Disconnect the negative cable at the battery. Place the cable out of the way so that it cannot accidentally come into contact with the terminal, which would once again allow current flow.
2   Push in on the re-adjust tab and move the slider back through the fitting in the direction away from the throttle lever (towards the rear of the vehicle).
3   Disconnect the cable terminal from the throttle lever by pushing forward and then off.
4   Compress the locking tabs and disconnect the cable assembly from the bracket.
5   Raise the vehicle and place it firmly on jackstands.
6   If equipped with a convertible top, remove the underbody braces that run under the transmission.
7   Remove the complete exhaust system (refer to Chapter 4).
8   Release the cable from its mounting near the top of the transmission.
9   Remove the screw and washer securing the cable to the transmission. Pull the cable up and disconnect it from the link at the end of the cable.
10   Place a new seal in the transmission case hole.
11   Connect the transmission end of the cable to the link and secure it to the transmission case with the screw.
12   Route the new cable up and secure it to the clips at the top of the transmission.
13   Install the exhaust system and the

underbody braces (if equipped) and lower the vehicle.
14   Pass the cable through the bracket and secure it at the bracket with the locking tabs.
15   Connect the cable terminal to the throttle lever and adjust the cable as described in the previous Section.
16   Connect the negative battery cable and check for proper operation.

5   Shifter control cable - adjustment

Refer to illustration 5.3

1   Place the shifter in the Park position.
2   Raise the vehicle for access underneath and support it securely on jackstands. Block the wheels and set the parking brake to prevent any movement.
3   Loosen the nut that attaches the cable to the transmission shift lever **(see illustration)**.
4   Manually place the transmission shift lever in the Park position by turning it clockwise to the last stop.
5   Tighten the cable attaching nut at this position, lower the vehicle and check for proper operation.

6   Steering column lock cable - adjustment

Refer to illustration 6.2

1   Place the shifter in the Park position.
2   The lock cable runs from the front of the gear shifter to the middle of the steering column. Expose the shifter mechanism and pull up on the adjusting key. It is located at the point where the cable is attached to the body bracket **(see illustration)**.
3   Reaching up under the dash, place the column and cable in the locked Park position and then push down on the adjusting key to lock the cable in place.

7   Transmission output seal - replacement

Refer to illustration 7.6

1   Raise the vehicle and support it firmly on jackstands.
2   If equipped with a convertible top, remove the upper and lower underbody braces.
3   Remove the complete exhaust system as described in Chapter 4.
4   Support the transmission with a jack.
5   Remove the driveline support beam and the driveshaft (see Chapter 8).
6   Using a seal remover or a long screwdriver, pry out the old seal from the end of the transmission **(see illustration)**.
7   Compare the new seal with the old to make sure they are the same.
8   Coat the lips of the new seal with automatic transmission fluid.
9   Drive the new seal into position using a large socket or a piece of pipe that is the same diameter as the seal.

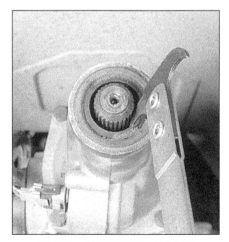

7.6 Use a seal remover or a long screwdriver to carefully pry the seal out of the end of the transmission

10   Reinstall the various components in the reverse order of removal, referring to the necessary Chapters as needed.

8   Transmission - removal and installation

Refer to illustration 8.16

Removal

1   Disconnect the negative cable at the battery. Place the cable out of the way so that it cannot accidentally come into contact with the terminal, which would again allow current flow.
2   Disconnect the TV cable at the throttle lever, if equipped.
3   Remove the transmission fluid dipstick.
4   On 1991 and earlier models, remove the distributor cap to prevent damage, as the engine may move slightly as the transmission is removed.
5   Raise the vehicle and support it firmly on jackstands.
6   If equipped with a convertible top, remove the upper and lower body braces.
7   Remove the complete exhaust system as a unit (see Chapter 4).
8   Disconnect and label the electrical connections at the transmission.
9   Disconnect the shift control cable at the transmission.
10   Remove the cover and carefully scribe a mark on the driveplate and torque converter so that these components can be returned to their original positions.
11   Remove the torque converter-to-driveplate nuts and/or bolts, turning the engine over by hand at the crankshaft as necessary.
12   Position a transmission jack under the transmission and secure the transmission to the jack. Raise the transmission slightly.
13   Remove the driveline support beam and the driveshaft as outlined in Chapter 8.
14   Lower the transmission slightly until proper access can be obtained to disconnect the oil cooler lines and the TV cable (if equipped).

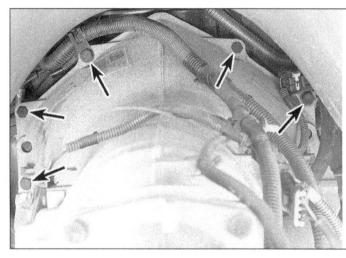

8.16 To detach the transmission from the engine, remove these six bolts (arrows, lower right bolt not visible in this photo)

15   Support the engine with an engine hoist so that movement will be kept to a minimum.
16   Remove the transmission-to-engine bolts **(see illustration)**.
17   Make a final check that all wiring, cables, etc. are disconnected from the transmission and then separate the transmission from the engine. As this is done, the torque converter must be held back against the transmission. A special holding fixture is available or you can use a piece of bar stock across the front of the transmission for this purpose.

Installation

18   Raise the transmission into position. Before installing the driveplate-to-converter bolts, make sure the weld nuts on the converter are flush with the driveplate and the converter rotates freely. Align the marks you made on removal and tighten the converter-to-driveplate bolts finger tight. Once all bolts are tightened in this manner, tighten to the specified torque.
19   If the oil filler tube was removed, install a new seal where the tube enters the transmission.
20   Install the various components in the reverse order of removal, referring to the proper Chapters where necessary for addi-

tional information. **Caution:** *Do not over torque the bolts attaching the driveline support beam to the transmission as damage to the transmission may result.*
21   After all components are installed, adjust the shifter control cable and the TV cable referring to Sections 3 and 5 of this Chapter.
22   If the transmission has been drained and a new torque converter installed or if the fluid has been drained from the existing one, a slightly different fluid filling procedure is required:
  a) *Add approximately 23.0 pints of fluid through the filler tube.*
  b) *With the shifter in Park, depress the accelerator enough to get the engine operating at a fast idle speed - do not race the engine.*
  c) *With your foot firmly on the brake pedal, move the shifter through each gear.*
  d) *After 1 to 3 minutes of running at idle, check the fluid level with the shifter in Park (engine still running).*
  e) *Add enough fluid to bring the level to a point between the two dimples on the dipstick. Add fluid only a little at a time, waiting 1 to 3 minutes, to prevent overfilling the transmission.*

Chapter 8
Clutch and drivetrain

Contents

Specifications

Clutch

Fluid type	Brake fluid conforming to DOT 3 specifications
Disc runout	0.020 in maximum

Rear axle

Lubricant	SAE 80W or SAE 80W-90 GL-5 gear lubricant with special limited-slip additive
Ring gear diameter (Model 36)	7.875 in
Ring gear diameter (Model 44)	8.5 in

Torque specifications

	Ft-lbs
Clutch housing to engine	
4-speed transmission	25 to 35
6-speed transmission	37
Manual transmission to clutch housing	
4-speed transmission	45 to 60
6-speed transmission	36
Pressure plate to flywheel	25 to 35
Flywheel/driveplate to engine	63 to 85
Clutch pedal pivot bolt	22 to 30
Clutch fork ballstud nut	37
Clutch master cylinder	
4-speed transmission	15 to 22
6-speed transmission	25
Clutch slave cylinder to clutch housing	
4-speed transmission	30 to 44
6-speed transmission	19
Universal joint strap bolts	150
Cover beam-to-carrier	
Model 36	21 to 24
Model 44	32 to 38
Pinion nut	190 to 210

Torque specifications

	Ft-lbs
Differential fill plug	
1984 through 1987	8 to 11
1988 through 1996	30
Trunnion (spider) strap at pinion..	15 to 20
Trunnion (spider) strap at yoke..	22 to 30
Cover beam-to-body	81 to 96
Driveline support beam to transmission bolt	
1984 through 1987	33 to 40
1988 through 1996	37
Driveline support beam to differential bolt	
1984 through 1987	51 to 67
1988 through 1996	60

1   General information

The information in this Chapter deals with the components from the rear of the engine to the rear wheels, except for the transmission, which is dealt with in the previous Chapter. For the purposes of this Chapter, these components are grouped into three categories; clutch, driveshaft and rear axle. Separate Sections within this Chapter offer general descriptions and checking procedures for each of these three groups.

Although the Corvette does use some standard rear-wheel drive pieces, many of the items covered in this Chapter will be unique to this vehicle. Of particular interest is the driveline support beam that runs from the transmission to the rear axle assembly. This piece acts as a mount for the transmission/engine, as well as serving as a main structural member to the driveline. The materials used in the manufacture of some of the components also differ, with aluminum used extensively in place of steel.

As nearly all the procedures covered in this Chapter involve working under the vehicle, make sure it is firmly supported on sturdy jackstands or on a hoist where the vehicle can be easily raised and lowered.

2   Clutch - description and check

Refer to illustration 2.1

1   All models equipped with a manual transmission feature a single dry plate, diaphragm spring-type clutch **(see illustration)**. 1984 through 1988 models are equipped with a single piece flywheel and 1989 through 1996 models use a dual mass type flywheel which has a 2-piece unit connected by two rows of primary and secondary torsional damper springs. It has an interface that rides on the input shaft splines on a ball bearing arrangement to minimize driveline vibration. The clutch for the dual mass flywheel is conventional except that it doesn't contain torsional damper springs to control vibration, since the flywheel springs perform this function. The clutch actuation is through a hydraulic system.

2   When the clutch pedal is depressed,

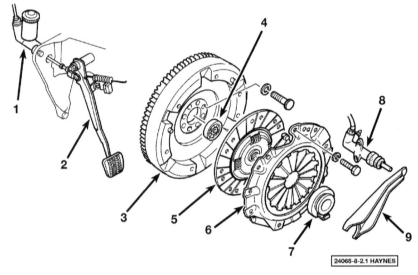

2.1 Exploded view of a typical clutch assembly

1	Clutch master cylinder	6	Pressure plate and cover assembly
2	Clutch pedal	7	Release bearing
3	Flywheel	8	Clutch release/slave cylinder
4	Pilot bearing	9	Clutch release lever (or "fork")
5	Clutch plate		

hydraulic fluid (under pressure from the clutch master cylinder) flows into the slave cylinder. Because the slave cylinder is connected to the clutch fork, the fork moves the release bearing into contact with the pressure plate release fingers, disengaging the clutch plate.

3   The hydraulic system locates the clutch pedal and provides clutch adjustment automatically, so no adjustment of the linkage or pedal is required.

4   Terminology can be a problem regarding the clutch components because common names have in some cases changed from that used by the manufacturer. For example, the driven plate is also called the clutch plate or disc, the clutch release bearing is sometimes called a throwout bearing, the slave cylinder is sometimes called the operating cylinder.

5   Other than to replace components with obvious damage, some preliminary checks

should be performed to diagnose a clutch system failure.

a) *The first check should be of the fluid level in the clutch master cylinder. If the fluid level is low, add fluid as necessary and re-test. If the master cylinder runs dry, or if any of the hydraulic components are serviced, bleed the hydraulic system as described in Section 10.*

b) *To check "clutch spin down time", run the engine at normal idle speed with the transmission in Neutral (clutch pedal up - engaged). Disengage the clutch (pedal down), wait nine seconds and shift the transmission into Reverse. No grinding noise should be heard. A grinding noise would indicate component failure in the pressure plate assembly or the clutch disc.*

c) *To check for complete clutch release, run the engine (with the brake on to prevent movement) and hold the clutch*

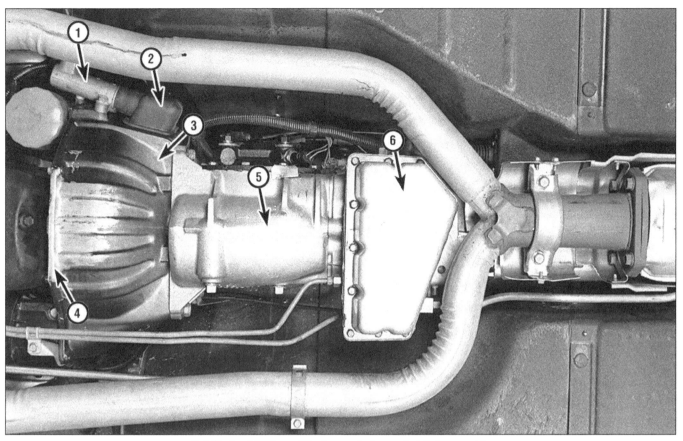

3.4 Underside view of the transmission and clutch assembly

1	Clutch slave cylinder	3	Clutch housing	5	Transmission
2	Slave cylinder dust boot	4	Clutch housing cover	6	Overdrive unit oil pan

pedal approximately 1/2-inch from the floor mat. Shift the transmission between 1st gear and Reverse several times. If the shift is not smooth, component failure is indicated.

d) *Visually inspect the clutch pedal bushing at the top of the clutch pedal to make sure there is no sticking or excessive wear.*

e) *Under the vehicle, check that the clutch fork is solidly mounted on the ball stud.*

---

3 Clutch components - removal, inspection and installation

Refer to illustrations 3.4, 3.7, 3.12, 3.14 and 3.21

Removal

1   Access to the clutch components is normally accomplished by removing the transmission, leaving the engine in the vehicle. If, of course, the engine is being removed for major overhaul, then the opportunity should always be taken to check the clutch for wear and replace worn components as necessary. The following procedures will assume that the engine will stay in place.

2   Referring to Chapter 7 Part A, remove the transmission from the vehicle. Support

the engine while the transmission is out. Preferably, an engine hoist from above should be used. However, if a jack is used underneath the engine, make sure a piece of wood is used between the jack and oil pan to spread the load. **Caution:** *The pickup for the oil pump is very close to the bottom of the oil pan. If the pan is bent or distorted in any way, engine oil starvation could occur.*

3   Remove the slave cylinder (see Section 6).

4   Remove the clutch housing cover (inspection cover) bolts and then the clutch housing-to-engine bolts. Remove the housing **(see illustration)**. It may have to be gently pried off the alignment dowels with a screwdriver or pry bar.

5   The clutch fork and release bearing can remain attached to the housing for the time being.

6   To support the clutch disc during removal, install a clutch alignment tool through the middle of the clutch.

7   Carefully inspect the flywheel and pressure plate for indexing marks. These can be an X, an O, or a white-painted letter. If these cannot be found, scribe your own marks so that the pressure plate and the flywheel will be in the same alignment upon installation **(see illustration)**.

8   Turning each bolt only one turn at a

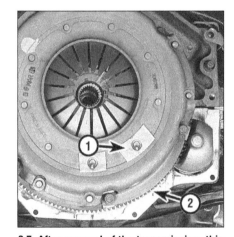

3.7 After removal of the transmission, this will be the view of the clutch components

1   *Pressure plate assembly (clutch disc inside)*
2   *Flywheel*

time, slowly loosen the pressure plate-to-flywheel bolts. Work in a diagonal pattern, again, loosening each bolt a little at a time until all spring pressure is relieved. Then grasp the pressure plate firmly and completely remove the bolts, followed by the pressure plate and clutch disc.

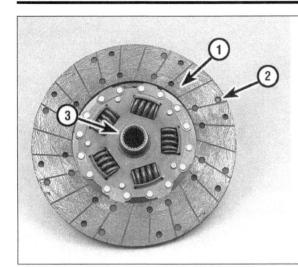

3.12 The clutch disc
1   **Lining** - this will wear down in use
2   **Rivets** - these secure the lining and will damage the flywheel or pressure plate if allowed to contact the surfaces
3   **Markings** - "Flywheel side" or similar

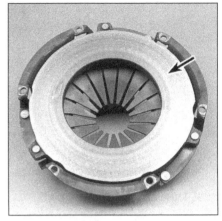

3.14 The machined face of the pressure plate must be inspected for scoring and any other damage

Inspection

9   Ordinarily, when a problem occurs in the clutch, it can be attributed to wear of the clutch driven plate assembly (clutch disc). However, all components should be inspected at this time.
10   Inspect the flywheel as discussed in the next Section.
11   Inspect the pilot bearing (see Section 5).
12   Inspect the linings on the clutch disc. There should be at least 1/16-inch of lining above the rivet heads. Check for loose rivets, distortion, cracks, broken springs and any other obvious damage **(see illustration)**. As mentioned above, ordinarily the disc is replaced as a matter of course, so if in doubt about the quality, replace it with a new one.
13   Ordinarily, the release bearing is also replaced along with the clutch disc.
14   Check the machined surfaces of the pressure plate **(see illustration)**. If the surface is grooved or otherwise damaged, replace the pressure plate assembly. Also check for obvious damage, distortion, cracking, etc. If a new pressure plate is indicated, new or factory-rebuilt units are available.

Installation

15   Before installation, carefully wipe clean the flywheel and pressure plate machined surfaces. It is important that no oil or grease is on these surfaces or the lining of the clutch disc. Handle these parts only with clean hands.
16   Position the clutch disc and pressure plate in position with the clutch disc held in place with the alignment tool. Make sure the disc is installed properly (most replacement discs will be marked "flywheel side" or similar).
17   Tighten the pressure plate-to-flywheel bolts only finger tight, working around the pressure plate.
18   Position the disc until it is exactly in the center and the splines of the transmission input shaft will pass easily through the disc and into the pilot bearing.
19   Slowly, one turn at a time, tighten the pressure plate-to-flywheel bolts. Work in a diagonal pattern to prevent distorting the cover as the bolts are tightened to the proper torque specification.
20   Using a high melting point grease, lubricate the entire inner surface of the release

bearing. Make sure the groove inside is completely filled. Also place grease on the ball socket and the fork fingers.
21   Attach the clutch fork to the clutch housing and install the release bearing on the fork. Make sure the fork fingers and the spring tabs are seated in the bearing groove **(see illustration)**.
22   Install the clutch housing and tighten the bolts to the proper torque specification.
23   Install the transmission, slave cylinder and all components removed previously, tightening all fasteners to the proper torque specifications.
24   Adjust the shift linkage as outlined in Chapter 7 Part A.

---

4   Flywheel/driveplate - inspection, removal and installation

---

Refer to illustration 4.3
1   The flywheel as used on manual transmission vehicles, or the driveplate (flexplate)

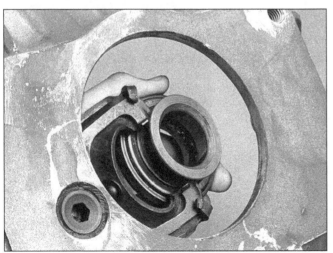

3.21 When installing the release bearing, make sure the fingers and the tabs fit into the bearing recess

4.3 The flywheel is held to the engine with six bolts. Sometimes it is necessary to place a screwdriver in the flywheel teeth to keep it from moving while loosening the bolts

5.5 A special puller is used here to pull the pilot bearing out of the crankshaft

5.9 By packing the opening behind the bearing with grease, a proper-size rod will then force the grease against the backside of the bearing, pushing it out of the bore

used on vehicles with an automatic transmission, is accessible only after removing the engine or the transmission.

2   Visually inspect the flywheel/driveplate for cracks, heat checking or other obvious defects. If the imperfections are slight, a machine shop can machine the surface flat and smooth.

3   To remove the flywheel/driveplate, remove the bolts which attach it to the engine **(see illustration)**. Be aware that it is fairly heavy and should be well supported as the final bolt is removed - do not drop it.

4   Upon installation, note that the flywheel/driveplate fits over an alignment dowel so that it will go on only one way. Tighten the attaching bolts, in several steps, to the proper torque specification following a crisscross tightening sequence.

5   Pilot bearing - removal, inspection and installation

Refer to illustrations 5.5 and 5.9

1   The clutch pilot bearing is an oil impregnated type bearing which is pressed into the rear of the crankshaft. Its primary purpose is to support the front of the transmission input shaft. The pilot bearing should be inspected whenever the clutch components are removed from the engine. Due to its inaccessibility, if you are in doubt as to its condition, replace it with a new one. **Note:** *If the engine has been removed from the vehicle, disregard the following steps that do not apply.*

2   Remove the transmission (refer to Chapter 7 Part A).

3   Remove the clutch components (Section 3).

4   Using a clean rag, wipe the bearing clean and inspect for any excessive wear, scoring or obvious damage. A flashlight will be helpful to direct light into the recess.

5   Removal can be accomplished with a

special puller **(see illustration)**, but an alternative method also works very well.

6   Find a solid steel bar which is slightly smaller in diameter than the bearing (19/32-inch should be very close). Alternatives to a solid bar would be a wood dowel or a socket with a bolt fixed in place to make it solid.

7   Check the bar for fit - it should just slip into the bearing with very little clearance.

8   Pack the bearing and the area behind it (in the crankshaft recess) with heavy grease. Pack it tightly to eliminate as much air as possible.

9   Insert the bar into the bearing bore and lightly hammer on the bar, which will force the grease to the backside of the bearing and push it out **(see illustration)**. Remove the bearing and clean all grease from the crankshaft recess.

10   To install the new bearing, lubricate the outside surface with oil then drive it into the recess with a soft-face hammer.

11   Install the clutch components, transmission and all other components removed previously, tightening all fasteners properly.

6   Clutch slave cylinder - removal, overhaul and installation

Refer to illustration 6.3
Note: *Before beginning this procedure, contact local parts stores or dealers concerning the purchase of a rebuild kit or a new slave cylinder. Availability and cost of the necessary parts may dictate whether the cylinder is rebuilt or replaced with a new one. If it is decided to rebuild the cylinder, inspect the inside bore as described in Step 9 before purchasing parts.*

Removal

1   Disconnect the negative cable at the battery. Place the cable out of the way so that it cannot accidentally come into contact with the terminal, which would again allow current to flow.

2   Raise the vehicle and support it securely on jackstands.

3   Disconnect the hydraulic line at the

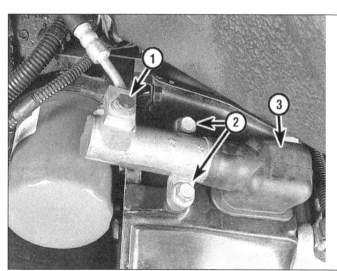

6.3 The clutch slave cylinder is attached to the clutch housing

1 *Hydraulic line fitting bolt*
2 *Mounting bolts*
3 *Rubber dust boot*

slave cylinder **(see illustration)**. Have a small can and rags handy, as some fluid will be spilled as the line is removed.
4   With your fingers, work the rubber dust boot out of the clutch housing, then remove the two mounting bolts.
5   Remove the slave cylinder.

Overhaul

6   Remove the pushrod and the rubber dust cover from the cylinder.
7   Remove the retaining clip with a small screwdriver.
8   Tap the cylinder on a block of wood to eject the plunger and seal. Also remove the spring from inside the cylinder.
9   Carefully inspect the inside bore of the cylinder. Check for deep scratches, scores or ridges. The bore must be smooth to the touch. If any imperfections are found, the slave cylinder must be replaced with a new one.
10   Using the new parts in the rebuild kit, assemble the components using plenty of fresh brake fluid for lubrication. Note the proper direction of the spring and the seal.

Installation

11   Previous to installation, the system must be bled of all air.
12   Connect the hydraulic line to the slave cylinder. Tighten the connection fully.
13   Fill the clutch master cylinder with brake fluid (conforming to DOT 3 specifications).
14   Bleed the system as described in Section 10.
15   Install the slave cylinder on the clutch housing, working the rubber dust cover into place and tightening the bolts to the proper torque.
16   Lower the vehicle and connect the negative battery cable.

---

7   Clutch master cylinder - removal, overhaul and installation

Note: *Before beginning this procedure, contact local parts stores or dealers concerning the purchase of a rebuild kit or a new master cylinder. Availability and cost of the necessary parts may dictate whether the cylinder is rebuilt or replaced with a new one. If it is decided to rebuild the cylinder, inspect the inside bore as described in Step 15 before purchasing parts.*

Removal

1   Disconnect the negative cable at the battery. Place the cable out of the way so it cannot accidentally come in contact with the negative terminal of the battery, as this would once again allow power into the electrical system of the vehicle.
2   Remove the hush panel from under the dashboard. This is required to gain access to the top of the clutch pedal.
3   Under the dashboard, disconnect the

pushrod from the top of the clutch pedal. It is held in place with a spring clip.
4   Disconnect the hydraulic line at the clutch master cylinder. The master cylinder is located on the firewall, just to the right of the battery. If available, use a flare nut wrench on the fitting, which will prevent the fitting high points from being rounded off. Have rags handy as some fluid will be lost as the line is removed. **Caution:** *Do not allow brake fluid to come into contact with paint as it will damage the finish.*
5   Remove the two bolts that secure the master cylinder to the engine. Remove the master cylinder, again being careful not to spill any of the fluid.

Overhaul

6   Remove the reservoir cap and drain all fluid from the master cylinder.
7   Pull back the dust cover on the pushrod and remove the circlip.
8   Remove the retaining washer and the pushrod from the cylinder.
9   Tap the master cylinder on a block of wood to eject the plunger assembly from inside the bore. Carefully lay out all of the components in order and in the correct orientation as they are removed.
10   Lift up on the spring retainer, which will release the spring from the plunger.
11   Compress the spring and release the valve stem from the spring retainer.
12   Remove the spring, valve spacer and spring washer from the valve stem.
13   Remove the valve seal from the end of the valve stem.
14   Carefully remove the seal from the plunger.
15   Inspect the inside bore of the master cylinder for deep scratches, scoring or ridges. The surface must be smooth to the touch. If the bore is not perfectly smooth, the master cylinder must be replaced with a new or factory rebuilt unit.
16   If the cylinder will be rebuilt, use the new parts contained in the rebuild kit and follow any specific instructions that may have accompanied the rebuild kit.
17   Fit the plunger seal to the plunger and the valve seal to the head of the valve stem. Note that the smaller diameter of the seals face the plunger.
18   Assemble the spring washer, valve spacer and spring onto the valve stem.
19   Install the spring retainer to the spring and compress the spring until the valve stem passes through the key hole slot and engages in the center.
20   Fit the spring to the plunger and lock it in place with the retainer.
21   Lubricate the inside bore of the cylinder and the seals with plenty of fresh brake fluid (DOT 3).
22   Carefully guide the plunger assembly into the bore, being careful not to damage the seals. Make sure the valve end is installed first, with the plunger closest to the opening.
23   Position the pushrod and retaining

washer into the bore and install a new circlip.
24   Smear a liberal amount of Girling Rubber Grease or equivalent onto the inside of the dust cover and attach it to the master cylinder.

Installation

25   Install the master cylinder onto the firewall, installing the two mounting bolts finger tight.
26   Connect the hydraulic line to the master cylinder, moving the cylinder slightly as necessary to thread the fitting properly into the bore. Do not cross-thread the fitting as it is installed.
27   Tighten the two mounting bolts to the specified torque.
28   Inside the vehicle, connect the pushrod to the clutch pedal and install the hush panel.
29   Fill the clutch master cylinder reservoir with brake fluid conforming to DOT 3 specifications and bleed the clutch system as outlined in Section 10.

---

8   Neutral start switch (manual transmission) - removal and installation

---

1   The Neutral start switch is mounted on the clutch pedal support and allows the vehicle to be started only with the clutch pedal fully depressed.
2   Disconnect the negative cable at the battery. Place the cable out of the way so it cannot accidentally come in contact with the negative terminal of the battery, as this would once again allow power into the electrical system of the vehicle.
3   Remove the hush pad from under the dashboard to gain access to the top of the clutch pedal.
4   At the top of the clutch pedal is a small rod which passes through the pedal. Remove the clip from the end of this rod.
5   Remove the bolt that secures the Neutral safety switch to the clutch pedal support bracket.
6   Disconnect the electrical lead to the switch and remove the switch.
7   Install the new switch in the reverse order of removal. Check to be sure that the vehicle can be started only when the clutch pedal is fully depressed.

---

9   Clutch pedal - removal and installation

---

1   Disconnect the negative cable at the battery. Place the cable out of the way so it cannot accidentally come in contact with the negative terminal of the battery, as this would once again allow power into the electrical system of the vehicle.
2   Remove the hush pad from under the dashboard.

10.5 When bleeding the hydraulic clutch system, place the slave cylinder as shown in order to get the bleeder valve at the highest position. The clear tube shown here runs into a container of fresh brake fluid. Any air will show up as bubbles in the tube or container

3   If equipped with cruise control, remove the anticipate switch from the clutch pedal support.
4   Disconnect and remove the Neutral start switch (see Section 8).
5   Disconnect the pushrod that goes into the master cylinder. It is held by a spring clip.
6   Remove the nut from the clutch pedal pivot bolt and pull the bolt out to remove the pedal.
7   Wipe clean all parts; however, do not use cleaning solvents on the bushings. Replace all worn parts with new ones.
8   Installation is the reverse of removal; however, note that the bolt must be installed in the same direction. Check that the Neutral safety switch allows the vehicle to be started only with the clutch pedal fully depressed.
9   To set the cruise control switch, depress the clutch pedal and push the switch all the way through the retainer, then lift up on the pedal until it reaches the pedal stop. As this is done, the pedal will force the switch back through the retainer to the proper point. While driving, make sure that the cruise control disengages when the clutch pedal is depressed.

10   Clutch hydraulic system - bleeding

Refer to illustration 10.5
1   The hydraulic system should be bled of all air whenever any part of the system has been removed or if the fluid level has been allowed to fall so low that air has been drawn into the master cylinder. The procedure is very similar to bleeding a brake system.
2   Fill the master cylinder with new brake fluid conforming to DOT 3 specifications. **Caution:** *Do not re-use any of the fluid coming from the system during the bleeding operation or use fluid which has been inside an open container for an extended period of time.*
3   Raise the vehicle and place it securely on jackstands to gain access to the slave

cylinder, which is located on the left side of the clutch housing.
4   On 1984 through 1987 models, with your fingers, pry the rubber dust cover out of the clutch cover and the remove the attaching bolts. 1988 through 1996 models can be bled without removing the slave cylinder from the clutch cover.
5   Remove the dust cap that fits over the bleeder valve and push a length of plastic hose over the valve. Place the other end of the hose into a clear container with about two inches of brake fluid. The hose end must be in the fluid at the bottom of the container. Hold the slave cylinder up at a 45¡ angle with the bleeder valve at its highest point **(see illustration)**.
6   Have an assistant depress the clutch pedal and hold it. Open the bleeder valve on the slave cylinder, allowing fluid to flow through the hose. Close the bleeder valve when your assistant signals that the clutch pedal is at the bottom of its travel. Once closed, have your assistant release the pedal.
7   Continue this process until all air is evacuated from the system, indicated by a full, solid stream of fluid being ejected from

the bleeder valve each time and no air bubbles in the hose or container. Keep a close watch on the fluid level inside the master cylinder; if the level drops too low, air will be sucked back into the system and the process will have to be started all over again.
8   Install the slave cylinder and lower the vehicle. Check carefully for proper operation before placing the vehicle in normal service.

11   Driveshaft and universal joints - description and check

Refer to illustrations 11.3 and 11.5
1   The driveshaft (or prop shaft as it is often called) is a tube running between the transmission and the rear end. It should not be confused with the similar shafts at the rear of the vehicle which are the driveaxles.
2   Universal joints are located at either end of the driveshaft and permit power to be transmitted to the rear wheels at varying angles.
3   The driveshaft features a splined yoke at the front, which slips into the rear extension of the transmission or overdrive unit **(see illustration)**. This arrangement allows the driveshaft to slide back and forth within the transmission/overdrive unit as the vehicle is in operation. An oil seal is used to prevent leakage of fluid at this point and to keep dirt and contaminants from entering the transmission or overdrive unit. If leakage is evident at the front of the driveshaft, replace this oil seal referring to the procedures in Chapter 7 Part A (manual transmission) or Chapter 7 Part B (automatic transmission).
4   The driveshaft assembly requires very little service. The universal joints are lubricated for life and require replacement should a fault develop. The driveshaft must be removed from the vehicle for this procedure.
5   Since the driveshaft is a balanced unit, it is important that no undercoating, mud, etc. be allowed to stay on it. When the vehicle is raised for service it is a good idea to clean the driveshaft and inspect for any obvious damage. Also check that the small weights

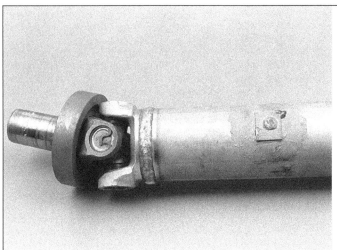

11.3 The transmission end of the driveshaft has a sliding yoke which fits into the transmission (automatic transmission) or overdrive unit (manual transmission)

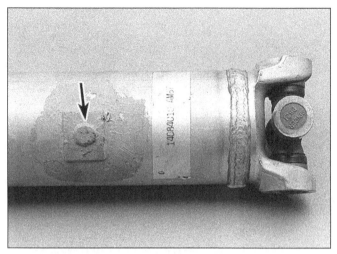

11.5 The rear axle end of the driveshaft has a universal joint which clamps into the rear axle pinion yoke. Note the weight (arrow) which is used to balance the driveshaft

12.7 Before removing the driveshaft, carefully scribe alignment marks across the driveshaft components (arrows)

used to originally balance the driveshaft **(see illustration)** are in place and securely attached. Whenever the driveshaft is removed it is important that it be reinstalled in the same relative position to preserve this balance.

6   Problems with the driveshaft are usually indicated by a noise or vibration while driving the vehicle. A road test should verify if the problem is the driveshaft or another vehicle component:

a) *On an open road, free of traffic, drive the vehicle and note the engine speed (rpm - tachometer) at which the problem is most evident.*

b) *With this noted, drive the vehicle again, this time manually keeping the transmission in 1st, then 2nd, then 3rd gear ranges and running the engine up to the engine speed noted.*

c) *If the noise or vibration occurs at the same engine speed regardless of which gear the transmission is in, the driveshaft is not at fault because the speed of the driveshaft varies in each gear.*

d) *If the noise or vibration decreased or was eliminated, visually inspect the driveshaft for damage, material on the shaft that would effect balance, missing weights or damaged universal joints. Another possibility for this condition would be tires that are out of balance.*

7   To check for worn universal joints:

a) *On an open road, free of traffic, drive the vehicle slowly until the transmission is in high gear. Let off on the accelerator, allowing the vehicle to coast, then accelerate. A clunking or knocking noise will indicate worn universal joints.*

b) *Drive the vehicle at a speed of about 10 to 15 mph and then place the transmission in Neutral, allowing the vehicle to coast. Listen for abnormal driveline noises.*

c) *Raise the vehicle and support it securely on jackstands. With the transmission in*

Neutral, manually turn the driveshaft, observing the universal joints for excessive play.

12   Driveshaft - removal and installation

Refer to illustration 12.7

Removal

1   Disconnect the negative cable at the battery. Place the cable out of the way so it cannot accidentally come in contact with the negative terminal of the battery, as this would once again allow power into the electrical system of the vehicle.

2   Raise the vehicle on a hoist or support it securely on jackstands. Place the transmission in Neutral with the parking brake off.

3   If equipped with a convertible top, remove the underbody braces.

4   Remove the complete exhaust system, referring to Chapter 4.

5   Support the transmission with a transmission jack.

6   Remove the bolts and nuts that attach the driveline support beam (Section 18) and push the beam as far to the right (passenger) side as possible.

7   Using a sharp scribe or a hammer and punch, place marks on the driveshaft and the differential flange in line with each other **(see illustration)**. This is to make sure the driveshaft is reinstalled in the same position to preserve the balance of the system.

8   Remove the rear universal joint strap bolts and the straps. Turn the driveshaft (or tires) as necessary to bring the bolts into the most accessible position.

9   Tape the bearing caps to the spider to prevent the caps from coming off during removal.

10   Lower the rear of the driveshaft and then slide the front from the transmission/overdrive unit.

11   To prevent loss of fluid and protect against contamination while the driveshaft is out, wrap a plastic bag over the transmission/overdrive housing and hold it in place with a rubber band.

Installation

12   Remove the plastic bag on the transmission/overdrive unit and wipe clean the area. Inspect the oil seal carefully. Procedures for replacement of this seal can be found in Chapters 7 Part A or 7 Part B, depending on transmission type.

13   Slide the front of the driveshaft into the transmission/overdrive unit.

14   Raise the rear of the driveshaft into position, checking to be sure that the marks are in perfect alignment. If not, turn the rear wheels to match the pinion flange and the driveshaft.

15   Remove the tape securing the bearing caps and install the straps and bolts. Tighten to the proper torque specification.

16   The remainder of the procedure is the reverse of the removal procedure. **Caution:** *Do not overtighten the driveline support beam bolts, as serious damage can occur to the transmission housing.*

13   Universal joints - replacement

Refer to illustrations 13.2a, 13.2b and 13.4
Note: *A press or large vise will be required for this procedure. It may be advisable to take the driveshaft to a local dealer or machine shop where the universal joints can be replaced for you, normally at a reasonable charge.*

1   Remove the driveshaft as discussed in the previous Section.

2   Using small pliers, remove the snaprings from the spider **(see illustrations)**.

3   Supporting the driveshaft, place it in position on either an arbor press or on a workbench equipped with a vise.

4   Place a piece of pipe (about 1-1/4 inch

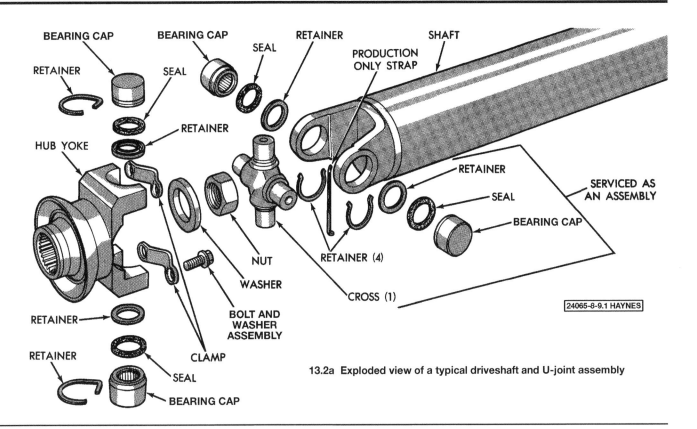

13.2a Exploded view of a typical driveshaft and U-joint assembly

inside diameter) or a socket with the same inside diameter over one of the bearing caps. Position a socket which is of slightly smaller diameter than the cap (about 11/16-inch) on the opposite bearing cap **(see illustration)** and use the vise or press to force the cap out (inside the pipe or large socket), stopping just before it comes completely out of the yoke. Use the vise or large pliers to work the cap the rest of the way out.

5   Transfer the sockets to the other side and press the opposite bearing cap out in the same manner.

6   Pack the new universal joint bearings with grease. Ordinarily, specific instructions for lubrication will be included with the universal joint servicing kit and should be followed carefully.

7   Position the spider in the yoke and partially install one bearing cap into the yoke.

8   Start the spider into the bearing cap and then partially install the other cap. Align the spider and press the bearing caps into position, being careful not to damage the dust seals.

9   Install the snap-rings.

10   Follow the same procedures to replace the rear universal joint, noting that only one-half of the spider needs to be pressed out, since the other two ends are held to the pinion flange with straps and bolts.

11   Install the driveshaft and all components previously removed.

14   Rear axle - description and check

Refer to illustrations 14.3 and 14.4

1   There are two types of rear axles incorporated on Corvettes. The Dana Model 36 is used on early models and vehicles equipped

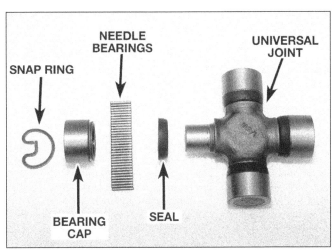

13.2b Exploded view of the universal joint components

13.4 To press the universal joint out of the driveshaft, set it up in a vise as shown. The small socket (at left) will push the joint and bearing cap into the large socket (at right)

14.3 Underside view of the rear axle assembly and components

1 Driveaxle
2 Driveaxle universal joint
3 Cam bolt and mounting bracket
4 Leaf spring
5 Differential carrier
6 Driveaxle universal joint
7 Spindle support rod
8 Driveaxle
9 Knuckle

with an automatic transmission, and the Model 44 is used with the manual transmission.

2   The Model 36 has a 7.875-inch ring gear, and the Model 44 features a larger 8.5-inch ring gear. This is the major difference between the two, and for purposes of this manual, all procedures included in this Chapter will apply to both types.

3   Aluminum is used extensively in the rear axle assemblies, with the axle carriers, differential carrier housing and cover beam all being manufactured from this material (see illustration).

4   Information regarding the rear axle can be found etched onto the bottom of the differential carrier (see illustration).

5   Many times a fault is suspected in the rear axle area when, in fact, the problem lies elsewhere. For this reason, a thorough check should be performed before assuming a rear axle problem.

6   The following noises are those commonly associated with rear axle diagnosis procedures:

a) Road noise is often mistaken for mechanical faults. Driving the vehicle on different surfaces will show whether the road surface is the cause of the noise. Road noise will remain the same if the vehicle is under power or coasting.

b) Tire noise is sometimes mistaken for mechanical problems. Tires that are worn or low on pressure are particularly susceptible to emitting vibrations and noises. Tire noise will remain about the same during varying driving situations,

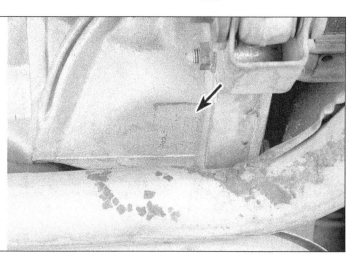

14.4 Rear axle identification information is etched into the differential carrier - later models (shown) have this information on the underside, while some earlier models will have it on the side of the carrier

where rear axle noise will change during coasting, acceleration, etc.

c) Engine and transmission noise can be deceiving because it will travel along the driveline. To isolate engine and transmission noises, make a note of the engine speed at which the noise is most pronounced. Stop the vehicle, place the transmission in Neutral and run the engine to the same speed. If the noise is the same, the rear axle is not at fault.

7   Overhaul and general repair of the rear axle components is beyond the scope of the home mechanic due to the many special tools and critical measurements required for the many parts to operate properly. Thus, the procedures listed here will involve seal replacements to stop oil leakage and removal of the entire unit for repair or replacement.

15 Rear axle oil seals - replacement

Refer to illustration 15.3

1   Three external seals are used on the differential unit. A pinion seal is used at the front of the differential, where the driveshaft yoke attaches to the differential. Seals are also used at the sides of the differential where the driveaxles enter the differential. Replacement of these seals is a somewhat difficult task, and except for experienced home mechanics, these procedures should be performed by a dealer or auto repair shop. Read through the procedures before beginning.

Pinion seal

2   Make sure the vehicle is well supported and remove the driveshaft as described in

15.3 Before removing the large pinion nut, carefully scribe alignment marks (arrow) on the pinion yoke and the differential case

16.6 To retain correct rear end alignment, scribe alignment marks (arrow) on the cam bolt and the mounting bracket. Do this on both sides of the assembly

Section 12.

3   Make alignment marks on the differential and the pinion yoke in order to reinstall the yoke in the same position **(see illustration)**.

4   Remove the pinion nut and washer. Remove the pinion yoke by pulling it from the differential.

5   Remove the dust shield.

6   Using a seal removal tool or a long screwdriver, pry the old seal from the differential.

7   Lubricate the lips of the new seal with gear oil.

8   To install the new seal, use a soft-face hammer and a large piece of pipe or socket to force the new seal into place. Do not hammer directly on the seal.

9   Install the dust shield and the yoke, aligning the marks to return it to its original position. Place the pinion washer into position and install a new pinion nut. Tighten this nut to the proper torque, which is 190 to 210 Ft-lbs (258 to 285 N-m). This torque is important for proper operation, so obtain a torque wrench and a pinion yoke holding tool that will allow you to tighten the nut to this specification.

10   Install the driveshaft.

Driveaxle seals

11   Remove the driveaxle(s) as described in Section 16.

12   Remove the differential carrier and cover beam assembly as described in Section 17.

13   Remove the bolts that attach the cover beam to the differential carrier and allow the lubricant to drain.

14   Separate the cover from the carrier. At the center of the differential assembly, remove the snap-rings that secure the yoke shafts using pliers. Tag each snap-ring so that it can be reinstalled in the same position (the snap-rings are different thicknesses to set the proper end play).

15   Remove the yoke shafts by pulling them out.

16   Remove the dust shield and then pry the oil seal out of the differential carrier.

17   To install the new seal, use a large piece of pipe or a socket to force it into place. Do not hammer directly on the seal.

18   Lubricate the lips of the seal with gear oil.

19   Install the dust shield and the yokes, making sure the snap-rings are installed in the same positions from which they were removed.

20   Install the differential carrier assembly and the remaining parts in the reverse order of removal, tightening all components to the proper specifications.

16 Driveaxles - removal and installation

Refer to illustration 16.6

1   Raise the vehicle and support it firmly on a hoist or on jackstands. Put the transmission in Neutral with the parking brake off.

2   Place a floor jack under the end of the leaf spring to support it, then remove the cotter key and the nut to disconnect the spring from the knuckle (for more information see Chapter 10).

3   Disconnect the tie-rod end from the knuckle.

4   Scribe alignment marks on the universal joints at both ends of the driveaxle, in line with marks made on the yoke and the spindle. This is to ensure the driveaxle is reinstalled in the same relative position.

5   Remove the driveaxle spider strap bolts at the differential end of the driveaxle, followed by the spindle end. Turn the driveaxle (or tires) as necessary to bring the strap bolts into an accessible position.

6   Carefully scribe a line on the cam bolt and the mounting bracket **(see illustration)**

so that the two can be realigned during installation. This will preserve rear end alignment.

7   Remove the cam bolt and separate the spindle support rod from the mounting bracket at the carrier. Be careful that the driveaxle does not fall at this point.

8   Pull out on the tire, then remove the driveaxle.

9   At this time the universal joints can be serviced. The universal joint servicing is the same as that for the driveshaft (see Section 13).

10   Place the driveaxle into position and install the spider straps. Tighten the bolts to the proper torque specification.

11   Install the spindle support rod and tighten the securing bolt.

12   Align the marks on the cam bolt and bracket, then tighten it securely.

13   Install the remaining components, tightening all fasteners to the proper specification. Have the rear suspension alignment checked and adjusted if necessary.

17 Differential carrier and cover - removal and installation

Refer to illustration 17.17

1   Disconnect the negative cable at the battery. Place the cable out of the way so it cannot accidentally come in contact with the negative terminal of the battery, as this would once again allow power into the electrical system of the vehicle.

2   Remove the air cleaner assembly (1984 models).

3   Remove the distributor cap. This is to prevent possible damage to the cap as the engine may move somewhat as the carrier is removed.

4   Raise the vehicle on a hoist or support it securely on jackstands.

5   Remove the spare tire.

17.17 The differential carrier-to-body mounting bolt (arrow)

18.6 Remove the support beam bolts and nuts at transmission end (arrows) and at the differential

6   Remove the spare tire cover by disconnecting the support hooks.

7   If equipped with a convertible top, remove the upper and lower underbody braces.

8   Remove the entire exhaust system as a unit (refer to Chapter 4).

9   Remove the rear leaf spring (see Chapter 10).

10   Scribe a mark on the cam bolts and the mounting bracket so that the two can be replaced in the exact same location upon installation.

11   Remove the cam bolts and the bracket from the carrier.

12   Disconnect both tie-rod ends from the knuckles.

13   Disconnect the driveaxles from the differential. Do this by removing the spider straps and then pulling out on the tire to disengage them (see the previous Section).

14   Support the transmission with a transmission jack or floor jack.

15   Place a floor jack under the differential unit and support it.

16   Remove the driveshaft and the driveline support beam (see Sections 12 and 18).

17   Remove the differential cover beam attaching bolts at the frame brackets **(see illustration)**.

18   Lower and remove the differential carrier assembly.

19   Installation is essentially the reverse of removal. However, note the following:

   a) *Tighten all fasteners to the proper torque specification.*

   b) *Apply thread sealant to the bolts that secure the driveline support beam to the differential carrier.*

   c) *Fill the rear axle with the proper oil (see the information in Chapter 1).*

   d) *If necessary, have the rear suspension alignment checked and adjusted.* **Caution:** *Do not overtighten the bolts securing the driveline support beam to the transmission, as damage to the transmission can occur.*

18   Driveline support beam - removal and installation

Refer to illustration 18.6

1   Disconnect the negative cable at the battery. Place the cable out of the way so it cannot accidentally come in contact with the negative terminal of the battery, as this would once again allow power into the electrical system of the vehicle.

2   Raise the vehicle and support it securely on jackstands.

3   On models equipped with a convertible top, remove the upper and lower braces.

4   Remove the complete exhaust system as a unit (refer to Chapter 4).

5   Support the transmission.

6   Remove the bolts and nuts that attach the support beam at the transmission extension housing and at the differential **(see illustration)**. One wrench is required on the nut above while the bolt is turned with a second wrench.

7   Push the beam as far to the right (passenger) side of the vehicle as possible and remove the driveshaft (see Section 12).

8   The support beam can then be completely removed by gently prying between it and the transmission.

9   Installation is the reverse of removal. However, note the following:

   a) *Apply sealant to the mating surfaces of the driveline support beam at the transmission and the differential.*

   b) *Tighten all fasteners to the proper torque.* **Caution:** *Do not overtighten the bolts which secure the support beam to the transmission extension housing, as damage may result.*

Chapter 9   Brakes

Contents

Specifications

Brake disc
    Parallelism .... 0.0005 in maximum
    Runout .... 0.006 in maximum
    Minimum thickness .... 0.724 in
Parking brake freeplay clearance (1988 and later) .... 0.024 to 0.028 in

Torque specifications
Ft-lbs

Caliper self-locking pin bolts
    1984 through 1987 (front and rear) 24
    1988 and later (rear only)
        Upper 26
        Lower 16
Caliper mounting bracket bolts
    1984
        Front 70
        Rear 44
    1985 and later
        Front 137
        Rear 70
Master cylinder mounting nuts 15

1   General information

General

The braking system used on the Corvette is a four-wheel disc, power-assisted system. Models manufactured in 1986 and later feature an Anti-lock Brake System (ABS).

The system is a split design, meaning there are separate components for the front and rear brakes. If one circuit fails, the other circuit will remain functional and a warning indicator will light up on the dashboard, showing that a failure has occurred.

The ABS is designed to maintain vehicle steerability, directional stability, and optimum deceleration under severe braking conditions and on most road surfaces. It does so by monitoring the speed of each wheel and controlling the brake pressure to each wheel during braking. This prevents the wheel from locking-up and provides maximum vehicle controllability. All parts and procedures listed in this Chapter also apply to models with ABS. More information on the ABS, and the components that are unique to it, can be found in Section 14.

In 1992 Acceleration Slip Regulation (ASR) was added to the Corvette. This system automatically controls the amount of wheel spin and chooses the optimum wheel speed for the traction conditions. ASR uses inputs from the anti-lock brake system and engine control computers to determine the amount of wheel speed correction necessary. Wheel speed reduction is accomplished several ways depending on the severity of unwanted wheel spin, they are spark retard, rear brake application and throttle close down. The later, throttle close down can actually be felt in the throttle pedal as the event occurs. Due to the complexity of the ASR system any concerns should be directed to the dealer service department.

Master cylinder

The master cylinder is located on the engine firewall, under the hood, and is best recognized by the fluid reservoir(s) on top. Since power brakes are standard on all Corvettes, a large power booster is located just behind the master cylinder. Connected to the top of the brake pedal is a pushrod, which runs through the power booster and operates the components inside the master cylinder.

The master cylinder is designed for the "split system" mentioned above and has separate primary and secondary pistons. Earlier models feature two separate reservoirs to hold the brake fluid, while later models have a single reservoir that feeds both circuits.

Proportioning valve

A proportioning valve is built into the master cylinder. Its purpose is to regulate pressure to the front and rear systems to eliminate rear wheel lock-up.

A bypass feature assures full system pressure to the rear brakes in the event of a front brake failure, or to the front brakes should a problem arise in the rear system.

A pressure differential warning switch constantly compares front and rear brake pressures. If a problem is experienced, the dashboard-mounted warning light will come on to warn the driver of the problem.

Disc brake assemblies

Each of the wheels has an aluminum caliper with a cylindrical piston inside it. Hydraulic pressure created by pushing on the brake pedal forces the piston and caliper body to move, in effect clamping against the moving brake disc.

Replaceable pads - two per wheel - are the components that actually press against the brake disc and will, after a time, wear out.

Parking brake

On 1984 through 1987 models, the parking brake is located inside the rear brake disc assembly and somewhat resembles conventional drum brakes. When the parking brake level is applied, the linings are forced against the inside of the brake disc housing preventing movement of the rear wheels. The parking brake on 1988 and later models are basically the same as earlier models instead of brake linings being forced against an inner drum, the parking brake cables actuate the levers on the rear caliper, moving them out and causing the calipers to slide in, forcing the brake pads against the disc, preventing movement of the rear wheels.

Unless the parking brake system has been abused (as in driving with the brake partially engaged), the system components do not need periodic inspection and maintenance. If, however, the vehicle does not hold firmly on a hill with the parking brake alone, the system components should be carefully inspected.

Precautions

There are some general notes and cautions involving the brake system on this vehicle:

a) *Use only DOT 3 or Delco Supreme II brake fluid in this system.*
b) *The brake pads and linings contain asbestos fibers that are hazardous to your health if inhaled. Whenever you work on brake system components, carefully wipe all parts clean with a water-dampened cloth. Do not allow the fine dust to become airborne.*
c) *Safety should be paramount whenever any servicing of the brake components is performed. Do not use parts or fasteners that are not in perfect condition, and be sure that all clearances and torque specifications are adhered to. If you are at all unsure about a certain procedure, seek professional advice. Upon completion of any brake system work, test the brakes carefully in a controlled area before putting the vehicle into normal service. If a problem is suspected in the brake system, do not drive the vehicle until the fault is corrected.*

2   Brake pads - replacement

Note: *The following information applies to both the front and rear brakes.*
Warning: *Disc brake pads must be replaced on both wheels at the same time - never replace the pads on only one wheel. Also, brake system dust contains asbestos, which is harmful to your health. Never blow it out with compressed air and do not inhale any of it. Do not, under any circumstances, use petroleum-based solvents to clean brake parts. Use brake cleaner or denatured alcohol only.*

1984 through 1987 models

Refer to illustrations 2.4a through 2.4g

1   Remove about two-thirds of the fluid from the master cylinder reservoir(s).
2   Raise the vehicle and support it securely on jackstands. Remove the wheels.
3   Inspect the brake disc carefully as outlined in Section 4. If machining is necessary, follow the information in that Section to remove the brake disc, at which time the pads can be removed from the calipers as well.
4   Follow the accompanying photos, beginning with illustration 2.4a, for the actual pad replacement procedure. Be sure to stay in order and read the information in the caption under each illustration.

2.4a Using a large C-clamp, push the piston back into the caliper bore. Note that one end of the clamp is on the flat area near the fluid inlet fitting and the other is pressing against the flat, raised portion of the outboard pad. Moderate force is all that is required

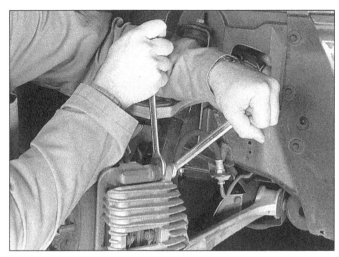

2.4b In order to remove the upper self-locking bolt, two wrenches are necessary. Place one wrench on the flats of the guide pin while the other wrench is used to remove the bolt

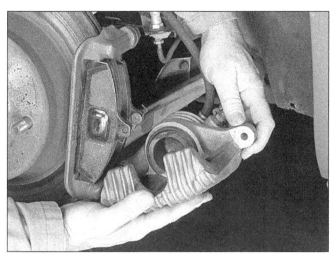

2.4c With the upper self-locking bolt removed, the caliper will swing down, exposing the brake pads. Do not put excessive strain on the rubber hose or damage could result

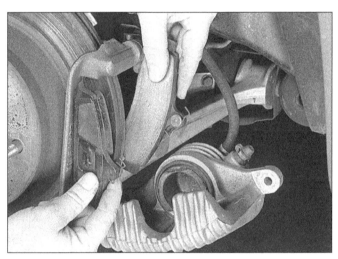

2.4d Separate the two pads from the brake disc. As this is done, note that the pads differ slightly, with the inboard pad having a small, bent piece of metal (wear sensor) and the outboard pad having the raised section (where the C-clamp was placed earlier). Make careful note of this because the new pads will have to go in the same way

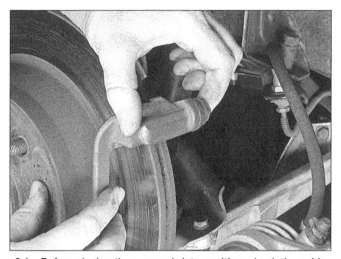

2.4e Before placing the new pads into position, check the guide pins at either end of the caliper mounting bracket. The pins should move freely in and out and the rubber boots should be in good condition

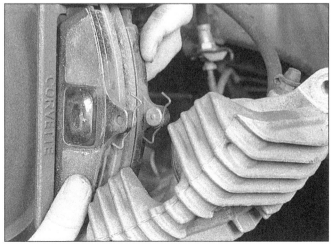

2.4f Place the new pads into position against the brake disc. The pad with the wear sensor goes on the inboard side

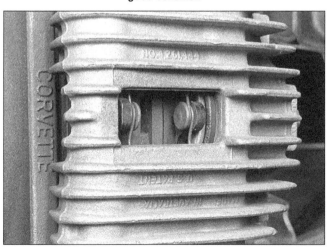

2.4g Swing the caliper back into position and check that the springs are within the cut-out section of the caliper as shown. Use a new self-locking bolt, tightened to the proper torque specification

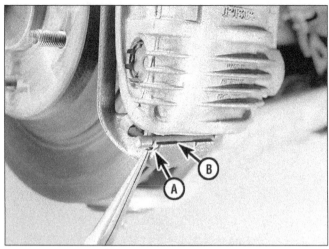

2.13 Use needle-nose pliers to pull the circlip (A) off the retainer pin (B)

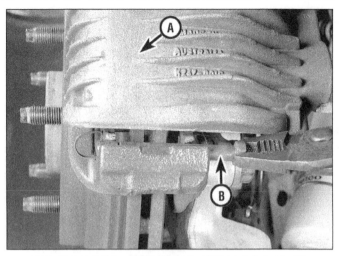

2.14 Support the caliper (A) and pull the retainer pin (B) out with pliers

2.15 Lift the caliper (A) off the disc (B) and bracket (C)

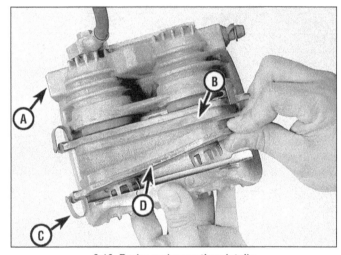

2.16 Brake pad mounting details

A	Caliper	C	Outer pad
B	Inner pad	D	Bias spring

5   Once the new pads are in place and a new self-locking bolt has been installed and properly tightened, install the wheel and lower the vehicle to the ground. **Note:** *If the fluid inlet fitting was disconnected from the caliper for any reason, the brake system must be bled to remove all air as described in Section 13.*
6   Fill the master cylinder reservoir(s) with new brake fluid and slowly pump the brakes a few times to seat the pads against the brake disc.
7   Check the fluid level in the master cylinder reservoir(s) one more time and then road test the vehicle carefully before placing it into normal service.

1988 and later models

Front brakes

Refer to illustrations 2.13, 2.14, 2.15, 2.16, 2.17 and 2.18

8   The two-piston front disc brake calipers

used on 1988 and later models are similar to the single piston unit used on earlier models except the caliper is retained to the mounting bracket by a pin and circlip and integral clips on the back side locate the pads in the caliper.
9   Remove about two-thirds of the fluid from the master cylinder reservoirs.
10   Raise the front of the vehicle and support it securely on jackstands. Apply the parking brake and block the rear wheels to keep the vehicle from rolling. Mark the relationship of the wheels to the hubs and remove the front wheels.
11   Install two of the lug nuts to retain the brake disc.
12   If necessary, depress the caliper pistons into the bores with a C-clamp
13   Use needle-nose pliers to remove the circlip on the inner end of the caliper retainer pin **(see illustration)**.
14   Withdraw the retainer pin with pliers **(see illustration)**.

15   Lift the caliper off the bracket **(see illustration)**.
16   Detach the pads from the caliper **(see illustration)**.
17   Use a C-clamp or large pliers to depress the pistons all the way into the bores **(see illustration)**.
18   Install the new pads, making sure the bias springs are secure **(see illustration)**.
19   Install the caliper. Install a NEW retainer pin and circlip.

Rear brakes

20   The rear disc brakes on 1988 and later models are basically the same as earlier models except the parking brake is a new design. On later models, the parking brake cables actuate the levers on the rear caliper, moving them out and causing the calipers to slide in, forcing the brake pads against the disc.
21   Replacement of the brake pads can be accomplished without disconnecting the

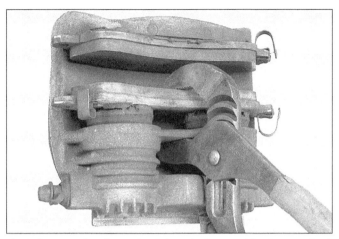

2.17 Squeeze the pistons back into the bores with a large pair of adjustable pliers

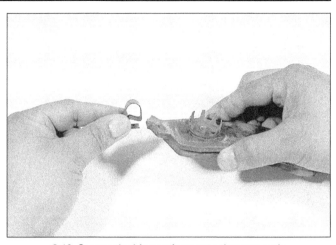

2.18 Secure the bias springs onto the new pads

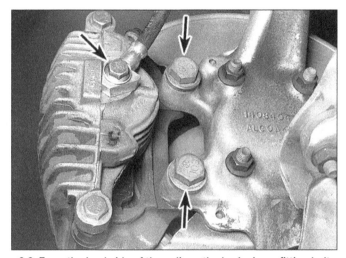

3.2 From the backside of the caliper, the brake hose fitting bolt and the two caliper mounting bracket bolts (arrows) can be seen

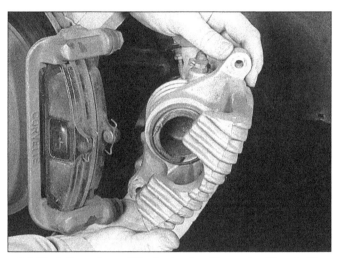

3.4 Work the caliper off the brake disc. The pads will probably stay with the brake disc; however, be careful that they do not fall and become damaged

parking brake cables at the levers on the calipers. Caliper removal, however, does require disconnection of the levers, which means disabling the automatic adjuster (see Section 7).

3   Caliper - removal, overhaul and installation

Note: *The following procedure applies to both the front and rear calipers. If an overhaul is indicated (usually because of fluid leakage) explore all options before beginning the job. New and factory-rebuilt calipers are available on an exchange basis, which makes this job quite easy. If it is decided to rebuild the calipers, make sure that a rebuild kit is available before proceeding.*

Removal
Refer to illustrations 3.2 and 3.4
1   Raise the vehicle and support it securely on jackstands. Remove the wheel.

2   Disconnect the brake hose fitting from the back of the caliper by removing the bolt **(see illustration)**. Have a rag handy to catch fluid spills and wrap a plastic bag around the end of the hose to prevent fluid loss and contamination. Discard the inlet fitting washers; use new ones during reassembly. **Note:** *If the caliper will not be completely removed from the vehicle - as for a pad inspection or brake disc removal- leave the hose connected and suspend the caliper with heavy wire or a coat hanger. This will save the trouble of bleeding the brake system.*
3   On 1988 and later rear calipers, the parking brake automatic adjuster must be disabled prior to removing the caliper, or anytime the parking brake cables or parking brake lever return springs are disconnected. **Warning:** *Failure to disable the parking brake automatic adjuster before disconnecting the parking brake system cables or parking brake lever return springs can cause severe personal injury.* Disable the automatic adjuster (see Section 7), remove the lever return spring and disconnect the cable from the caliper

lever and bracket.
4   Remove the two self-locking bolts that secure the caliper. Discard these also, as new bolts should be used during reassembly. Work the caliper assembly out of the mounting bracket on the brake disc **(see illustration)**. The pads will probably stay with the brake disc, but don't allow them to fall and be damaged.

Overhaul
1984 through 1987 (front and rear calipers) and 1988 and later front caliper
5   Place numerous shop towels in the center of the caliper and then force the piston out of its bore by directing compressed air into the inlet opening. Use care when doing this, as the piston could be ejected with some force.
6   Using a wood or plastic tool, remove the rubber piston seal and boot from the caliper bore. Do not use a metal tool, as the bore can be easily damaged.

3.19 Remove the anchor bracket

3.20 Remove the springs from each end of the parking brake collar

3.21 Place a wood block or shop rag in the caliper as a cushion, then use compressed air to remove the piston from the caliper (use no more air pressure than necessary to ease the piston out of the bore)

3.22 Remove the clamp rod, actuating collar boot retainers, actuating collar boots, actuating collar and piston as a single assembly

7   Carefully inspect the piston and the bore for score marks, nicks, corrosion and damage of any kind. If any of the above are found, the caliper should be replaced with a new or rebuilt unit.

8   Remove the bleeder valve and rubber cap.

9   Inspect the guide pins for corrosion and damage. Replace them with new ones if necessary.

10   Use clean brake fluid or denatured alcohol to clean all the parts. **Warning:** *Do not, under any circumstances, use petroleum-based solvents to clean brake parts. Allow all parts to dry, preferably using compressed air to blow out all passages. Make sure the compressed air is filtered, as a harmful lubricant residue will be present in unfiltered systems.*

11   Check the fit of the piston in the bore by sliding it into the caliper. The piston should move easily.

12   Install the rubber cap over the bleeder

valve and thread the valve into the caliper. Tighten it securely.

13   Lubricate the new piston seal with silicone grease or clean brake fluid. Position the seal in the caliper bore groove, making sure that the seal does not twist.

14   Using silicone grease or clean brake fluid, lubricate the piston and the caliper bore.

15   Install the piston boot over the end of the piston with the fold of the boot facing out. Push the piston into the caliper and seat the boot in the groove in the bore. Push the piston all the way to the bottom of the bore. Check to make sure the boot is properly seated in the groove around the piston and in the caliper bore groove.

16   Lubricate the guide pins with silicone grease and place new boots over them.

17   Install the guide pins in the mounting bracket, seating the boots properly in the grooves in the pins and bracket.

1988 and later rear caliper

Refer to illustrations 3.19, 3.20, 3.21, 3.22, 3.23, 3.24, 3.25, 3.30, 3.31, 3.32, 3.34, 3.35 and 3.36

Note: *Purchase a brake caliper overhaul kit for your particular vehicle before beginning this procedure.*

18   Clean the exterior of the brake caliper with brake system cleaner (never use gasoline, kerosene or any petroleum-based cleaning solvents), then place the caliper on a clean workbench.

19   Remove the anchor bracket **(see illustration)**.

20   Remove the parking brake collar return springs **(see illustration)**.

21   Place a wood block or shop rag in the caliper as a cushion, then use compressed air to remove the piston from the caliper **(see illustration)**. Use only enough air pressure to ease the piston out of the bore. If the piston

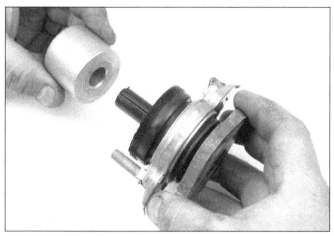

3.23 Slide the piston off the clamp rod bushing

3.24 Separate the clamp rod from the actuating collar boot retainers and actuating collar boots

3.25 Use a wood or plastic tool to remove the piston seal from the bore

3.30 Lubricate the clamp rod bushing with the supplied lubricant, then slide the piston onto the clamp rod bushing

is blown out, even with the cushion in place, it may be damaged. **Warning:** *Never place your fingers in front of the piston in an attempt to catch or protect it when applying compressed air - serious injury could occur.*

22   Remove the clamp rod, actuating collar boot retainers, actuating collar boots, actuating collar and piston as a single assembly **(see illustration)**.

23   Slide the piston off the clamp rod bushing **(see illustration)**.

24   Separate the clamp rod from the actuating collar boot retainers and actuating collar boots **(see illustration)**.

25   Use a wooden or plastic tool to remove the piston seal from the bore **(see illustration)**.

26   Carefully check the caliper bore for score marks, nicks, corrosion and excessive wear. Light corrosion may be removed with crocus cloth; otherwise, replace the caliper housing with a new one.

27   Clean all parts not included in the caliper repair kit with clean brake fluid or brake system cleaner. Do not, under any circumstances, use petroleum-based solvents.

28   Use compressed air to dry the parts and

3.31 Lubricate the piston bore with brake fluid, then place the assembled piston, actuating collar retainers and dust boots and clamp rod into the caliper assembly

blow out all the passages in the caliper housing and bleeder valve.

29   Lubricate the new piston seal with clean brake fluid and install it in the caliper bore groove. Make sure that the seal is not twisted.

30   Lubricate the clamp rod bushing and clamp rod with the liquid lubricant supplied in the overhaul kit, then slide the piston onto the

clamp rod bushing **(see illustration)**.

31   Also lubricate the bead of the actuating collar, actuating collar boot and the boot's grove in the caliper with the same liquid lubricant. Lubricate the piston bore with brake fluid, then place the assembled piston, actuating collar retainers and dust boots and clamp rod into the caliper assembly **(see illustration)**.

3.32 With the piston, actuating collar retainers and dust boots and clamp rod in position, square to the bore, use your thumbs to push the piston into the caliper

3.34 Lubricate the caliper guide pins with high-temperature grease

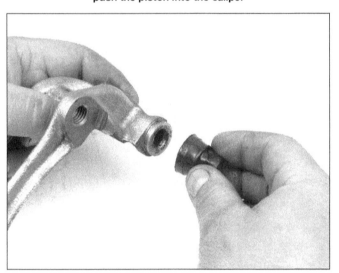

3.35 Inspect the guide pin boots for cracks and replace as necessary

3.36 Install the caliper anchor bracket

32   With the piston, actuating collar retainers and dust boots and clamp rod in position, square to the bore, use your thumbs to push the piston into the caliper **(see illustration)**.
33   Install the parking brake collar return springs.
34   Lubricate the caliper guide pins with high temperature grease **(see illustration)**.
35   Inspect the guide pin boots for cracks. Replace as necessary **(see illustration)**.
36   Install the caliper anchor bracket **(see illustration)**.
37   Install the brake pads (see Section 3).
38   Install the bleeder valve.

Installation

39   Install the caliper over the brake disc and into the mounting bracket. The brake pads should be in place against the brake disc at this time. Refer to Section 2 for more information and illustrations concerning proper pad positioning.
40   Install new self-locking bolts and tighten them a little at a time until the proper torque specification is reached. On 1988 and later model rear brakes, attach the parking brake cable to the caliper lever and bracket, install the lever return spring and restore the automatic adjuster (see Section 7).
41   Install the inlet fitting bolt (with new washers) and tighten it to the proper torque.
42   Pump the brake pedal a few times to bring the pads into contact with the brake disc.
43   Bleed the brakes as described in Section 13. This is not necessary if the inlet fitting was left connected to the caliper. On 1988 and later model rear brakes, adjust the parking brake clearance.
44   Install the wheel and lower the vehicle. Test the brake operation carefully before placing the vehicle into normal service.

4   Brake disc - inspection, removal and installation

Refer to illustrations 4.3 and 4.4
Note: *The following information applies to both the front and rear brakes.*

Inspection

1   Raise the vehicle and support it securely on jackstands. Remove the wheel and hold the brake disc in place with two lug nuts.
2   Visually inspect the brake disc surface for score marks and other damage. Light scratches and shallow grooves are normal after use and are not detrimental to brake operation. Deep scoring - over 0.015-inch (0.38 mm) - requires brake disc removal and refinishing by an automotive machine shop. Be sure to check both sides of the brake disc.

4.3 To check runout, a dial indicator is attached to the caliper with the indicator just touching the brake disc. Note the two lug nuts (arrows) securing the brake disc

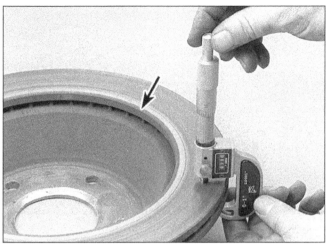

4.4 A micrometer is used to measure the thickness of the brake disc. This can be done on the bench (as shown) or on the vehicle. The minimum thickness is stamped on the brake disc (arrow)

3   To check brake disc runout, position a dial indicator at a point about 1/2-inch from the outer edge **(see illustration)**. Set the indicator to zero and turn the brake disc. The indicator reading should not exceed 0.006-inch (0.15 mm). If it does, the brake disc should be refinished by an automotive machine shop.

4   It is absolutely critical that the brake disc not be machined to a thickness under 0.724-inch (18.4 mm) **(see illustration)**. This minimum thickness should also be stamped on the brake disc itself.

Removal

5   With the vehicle raised and the wheel off, use a C-clamp to push the piston back into the bore. See the Section 2 illustrations if necessary.

6   On the rear brakes, carefully note the parking brake cable bracket, which is on the backside of the caliper mounting bracket. Upon installation, this bracket will have to be relocated in the same position.

7   Have a piece of heavy wire or a length of coat hanger ready, and remove the two large bracket bolts located on the backside of the

caliper mounting bracket. Lift the bracket (with the caliper attached) away from the brake disc and use the wire or coat hanger to suspend the caliper assembly. **Warning:** *Do not allow the caliper to hang by the brake hose and do not disconnect the hose from the caliper.*

8   Remove the two lug nuts that were put on to hold the brake disc in place and detach the brake disc. **Note:** *If the rear brake disc cannot be removed easily, check to make sure the parking brake is completely released.*

Installation

9   Clean off any residual hardened bolt adhesive from the mounting bracket threads, mounting bracket mating surface and knuckle mating surface.

10   Place the brake disc in position over the threaded studs.

11   Install the caliper assembly over the brake disc and secure it with two new bracket bolts, using the special adhesive which will come with the bolts from the Chevrolet dealer. On the rear, properly position the parking brake cable bracket tang in the recess in the mounting plate. Tighten the

bolts to the proper torque.

12   Recheck the torque on each of the bolts.

13   Allow the adhesive to cure for at least two hours before driving the vehicle.

14   Install the wheel, then lower the vehicle to the ground. Depress the brake pedal a few times to bring the brake pads into contact with the brake disc. Bleeding of the system will not be necessary unless the brake hose was disconnected from the caliper. Check the operation of the brakes carefully before placing the vehicle into normal service.

---

5   Master cylinder - removal, overhaul and installation

Refer to illustrations 5.4 and 5.6
Note: *Before deciding to overhaul the master cylinder, check on the availability and cost of a new or factory-rebuilt unit and also the availability of a rebuild kit.*

Removal

1   Place rags under the fluid fittings and prepare caps or plastic bags to cover the ends of the lines once they are disconnected. **Caution:** *Brake fluid will damage paint. Cover all body parts and be careful not to spill fluid during this procedure.*

2   Loosen the tube nuts at the ends of the brake lines where they enter the master cylinder. To prevent rounding off the flats on these nuts, the use of a special flare-nut wrench, which wraps around the nut, is preferred.

3   Pull the brake lines slightly away from the master cylinder and plug the ends to prevent contamination.

4   Disconnect the electrical connector at the master cylinder, then remove the two nuts attaching the master cylinder to the power booster **(see illustration)**. Pull the master cylinder off the studs and out of the engine compartment. Again, be careful not to spill the fluid as this is done.

5.4 The master cylinder is attached to the power booster with two nuts, one of which is shown here with an arrow. Two fluid lines are located on the other side and one line at the bottom of the master cylinder

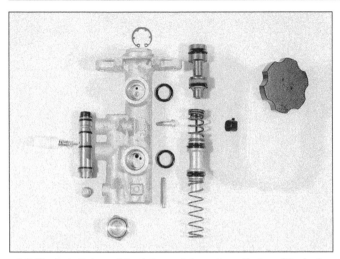

5.6 Exploded view of the late model master cylinder with single fluid reservoir

6.6 To detach the power brake booster from the firewall, remove the four nuts from the mounting studs (two shown of the left side of the booster pushrod)

Overhaul

5   Before attempting the overhaul of the master cylinder, obtain the proper rebuild kit which will contain the necessary replacement parts and also any instructions which may be specific to your model.

6   Inspect the reservoir grommet(s) for indications of leakage near the base of the reservoir. If necessary, remove the reservoir cap(s) and diaphragm(s), inspecting for further damage **(see illustration)**.

7   Using a wooden dowel or a brass rod, fully depress the pistons until they bottom against the other end of the master cylinder. Hold the pistons in this position and remove the stop bolt on the side of the master cylinder.

8   Carefully release the pistons and remove the snap-ring at the end of the master cylinder. Special snap-ring pliers should be used for this.

9   The various internal components can now be removed from the cylinder bore. To dislodge the secondary piston assembly, you may have to gently bump the open end of the master cylinder against a piece of wood. Make a note of the installed order of the components so they can be returned to their original locations.

10   Carefully inspect the bore of the master cylinder. Any deep scoring or other damage will mean a new master cylinder is required.

11   Carefully note the positions of all components, making sketches if necessary. Pay particular attention to the seals and the direction of the seal lips so the new ones can be installed in the same position and direction.

12   Replace all parts included in the rebuild kit, following any instructions in the kit. Clean all re-used parts with clean brake fluid or denatured alcohol. **Warning:** *Do not, under any circumstances, use petroleum-based solvents to clean brake parts. During assembly, lubricate all parts liberally with clean brake fluid.*

13   Use the wood dowel or brass rod to push the assembled components into the bore, bottoming them against the end of the master cylinder, then install the stop bolt.

14   Install the new snap-ring, making sure it is seated properly in the groove.

15   Whenever the master cylinder is removed, the complete hydraulic system must be bled. The time required to bleed the system can be reduced if the master cylinder is filled with fluid and bench bled as follows before it's installed on the vehicle.

16   Insert threaded plugs of the correct size into the brake line outlet holes and fill the reservoirs with brake fluid. The master cylinder should be supported so brake fluid won't spill during the bench bleeding procedure.

17   Loosen one plug at a time and push the piston assembly into the bore to force air from the master cylinder. To prevent air from being drawn back in, the appropriate plug must be replaced before allowing the piston to return to its original position.

18   Stroke the piston three or four times for each outlet to ensure that all air has been expelled.

19   Since high pressure isn't involved in the bench bleeding procedure, there is an alternative to the removal and replacement of the plugs with each stroke of the piston assembly. Before pushing in on the piston assembly, remove one of the plugs completely. Before releasing the piston, however, instead of replacing the plug, simply put your finger tightly over the hole to keep air from being drawn back into the master cylinder. Wait several seconds for the brake fluid to be drawn from the reservoir into the piston bore, then repeat the procedure. When you push down on the piston it'll force your finger off the hole, allowing the air inside to be expelled. When only brake fluid is being ejected from the hole, replace the plug and go on to the other port.

20   Refill the master cylinder reservoirs and install the diaphragm and cover assembly.

Installation

21   Install the master cylinder over the studs on the power brake booster and tighten the nuts only finger tight at this time.

22   Using your fingers, thread the brake line fittings into the master cylinder. Since the master cylinder is still a bit loose, it can be moved slightly in order for the fittings to thread in easily. Do not strip the threads as the fittings are tightened.

23   Tighten the brake line fittings and the two mounting nuts.

24   Fill the master cylinder reservoir(s) with fluid and bleed the brake system as described in Section 13. Test the operation of the brake system carefully before placing the vehicle in normal service.

---

6   Power brake booster - inspection, removal and installation

Refer to illustration 6.6

1   The power brake booster unit requires no special maintenance apart from periodic inspection of the vacuum hose and the case. Early models will have an in-line filter, which should be inspected periodically and replaced if clogged or damaged.

2   Dismantling of the power unit requires special tools and is not ordinarily done by the home mechanic. If a problem develops, it is recommended that a new or factory rebuilt unit be used.

3   Remove the nuts attaching the master cylinder to the booster and carefully pull the master cylinder forward until it clears the mounting studs. Use caution so as not to bend or kink the brake lines. **Note:** *On 1990 and later models, the ECM and cruise control servo/bracket may have to be unbolted and moved aside to allow removal of the brake booster. Disconnect the battery first on these models, then refer to Chapter 6 for ECM*

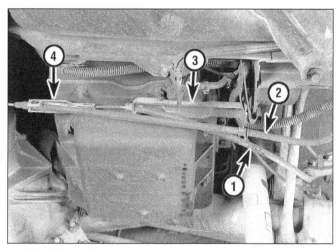

7.2a The parking brake cable system is composed of three separate cables all working together. Each cable can be replaced individually

| 1 | Front cable | 3 | Bracket |
| 2 | Left rear cable | 4 | Equalizer |

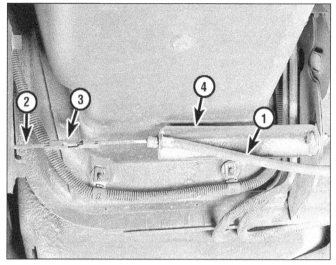

7.2b 1988 and later model parking brake cable details

| 1 | Right rear cable | 3 | Retainer clips |
| 2 | Left rear cable | 4 | Bracket |

removal, and Chapter 12 for the cruise control.

4   Disconnect the vacuum hose where it attaches to the power brake booster.

5   From the passenger compartment, disconnect the power brake pushrod from the top of the brake pedal.

6   Also from this location, remove the nuts attaching the booster to the firewall **(see illustration)**.

7   Carefully lift the booster unit away from the firewall and out of the engine compartment.

8   Upon installation, place the booster into position and tighten the retaining nuts. Connect the brake pedal.

9   Install the master cylinder and vacuum hose.

10   Carefully test the operation of the brakes before placing the vehicle in normal service.

7   Parking brake cable - replacement

Warning: *On 1988 and later rear calipers, the parking brake automatic adjuster must be disabled anytime the parking brake cables or parking brake lever return springs are disconnected. Failure to disable the parking brake automatic adjuster before disconnecting the parking brake system cables or parking brake lever return springs can cause severe personal injury.*

Front cable

Refer to illustrations 7.2a and 7.2b

1   On 1988 and later models, disable the automatic adjuster (see below).

2   Raise the vehicle and support it securely on jackstands. Disconnect the left rear cable

at the equalizer, followed by the right rear cable at the equalizer **(see illustrations)**.

3   To gain access to the cable where it attaches to the handle, remove the lower door sill molding.

4   Remove the cable nut, cable guide and then the cable itself.

5   Install the new cable and adjust it as outlined in the next Section.

Left rear cable

Refer to illustration 7.8

6   On 1988 and later models, disable the automatic adjuster (see below).

7   Raise the vehicle and support it securely on jackstands. Disconnect the left rear cable where it attaches to the equalizer.

8   Disconnect the cable mounting at the frame, caliper mounting bracket and the parking brake lever at the wheel **(see illustration)**.

9   Install the new cable and adjust as outlined in the next Section.

Right rear cable

10   On 1988 and later models, disable the automatic adjuster (see below).

11   Raise the vehicle and support it securely on jackstands. Remove enough tension at the equalizer to enable the cable to be removed from the retainer.

12   With the cable disconnected from the retainer, remove the cable at the frame, caliper mounting bracket and the parking brake lever at the wheel.

13   Install the new cable and adjust it as outlined in the next Section.

Disabling the parking brake automatic adjuster

Refer to illustration 7.16

14   Remove the driver's seat cushion.

15   Remove the screws and detach the parking brake lever cover.

16   Fabricate a hook out of 0.080-inch diameter wire, use it to keep the drive pawl

7.8 At the wheel, the parking brake cable attaches to a lever (arrow). Once slack is achieved in the cable, it can be unhooked at this point

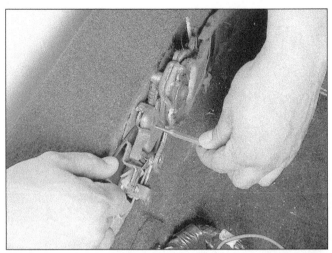

7.16 Use a hook tool and a small punch to pull up and hold the drive pawl in place

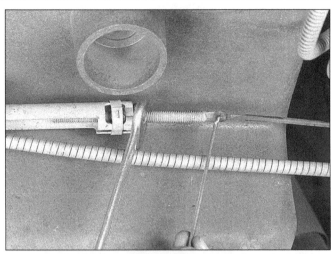

8.3 The equalizer is the point where the parking brake system is adjusted. Use a small wrench to keep the threaded rod from turning while another wrench is used to turn the adjusting nut

8.5 With the disc turned to this position, the access hole will be directly over the star wheel inside. A standard screwdriver is then used to move the star wheel and adjust the brake

8.13 If the gap is too small (insufficient cable free travel), turn the adjustment screw counterclockwise; if the gap is too big (too much cable free travel), turn the adjustment screws clockwise (caliper removed for clarity)

disengaged from the drive sector, then insert a nail or small punch through the hole in the anchor plate to hold the pawl in place **(see illustration)**.

17   Move the parking brake lever to align it with the lock pawl, depress the button, then move the lever to the down position.

18   Check to make sure the anchor plate is against the stud; if it isn't, repeat the procedure.

19   The front parking brake cable can now be pulled to the rear to provide enough slack so the parking brake system can be serviced. Be very careful when handling the parking brake cables because the surface can be easily damaged, which could cause binding when the parking brake is applied.

Restoring the parking brake automatic adjuster

20   Remove the nail or small punch from the

anchor plate, then apply and release the parking brake three times. Pull up on the parking brake handle and make sure it moves between 7 and 9 clicks when applied.

21   Release the parking brake and check to make sure the rear brakes don't drag and there is no gap between the caliper housing and the parking brake levers.

---

8   Parking brake - adjustment

1984 through 1987 models

Refer to illustrations 8.3 and 8.5

1   Raise the vehicle and support it securely on jackstands. Remove the rear wheels and then reinstall two lug nuts at each wheel to ensure correct disc/drum alignment.

2   Using a C-clamp, press the caliper piston into the bore (see Section 2 for illustrations).

3   At the equalizer, loosen the parking brake cable so there is no tension on the shoes **(see illustration)**.

4   Turn the disc so the hole in the drum aligns with the star wheel adjuster inside.

5   Use a standard screwdriver to turn the star wheel **(see illustration)**. On the driver's side, move the screwdriver handle up to adjust the shoes out and down to adjust them in. The passenger's side will be the opposite.

6   Keep adjusting until the disc/drum unit will not rotate, then back off the adjuster 5 to 7 notches. Do the same for the other side.

7   Install the rear wheels and apply the parking brake lever two notches.

8   At the equalizer, adjust the cables so the wheels have a slight drag.

9   Release the parking brake lever and check that the wheels rotate freely. Adjust the equalizer as necessary for free movement of the wheels.

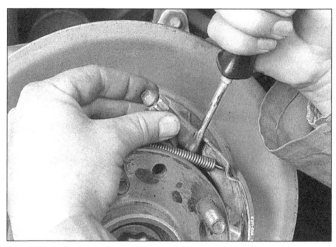

9.5 A screwdriver is used to pry the shoe back so the adjuster assembly can be lifted out of place

9.6 Needle-nose pliers are used to disconnect the adjuster return spring from each of the shoes

1988 and later models

Refer to illustration 8.13

10   The parking brake freeplay must be adjusted after the rear calipers are over-hauled. The rear brake pads must be new or parallel to 0.008-inch for the adjustment to be valid. If the pads are worn or tapered, replace them with new ones before adjusting the freeplay.

11   Raise the rear of the vehicle, support is securely on jackstands and block the front wheels. Remove the rear wheels.

12   Have an assistant apply the brakes lightly (just enough so the brake disc can't be turned by hand); this will take up the clearance and align the components.

13   Working under the vehicle, push on the parking brake lever and measure the freeplay between the lever and the caliper housing **(see illustration)**. To adjust it, remove the adjustment screws, clean the threads, apply new thread locking compound and install the screw far enough to achieve the specified clearance.

14   Have the assistant release the brake pressure and pump the brakes three times, then recheck the freeplay, adjusting as necessary.

---

9   Parking brake assembly (1984 through 1987 models only) - check, removal and installation

Refer to illustrations 9.5, 9.6 and 9.8

Warning: *Some of the parking brake components are made of asbestos fibers that may cause serious harm if inhaled. When servicing these components, do not create dust by grinding or sanding the linings or by using compressed air to blow away dust; use a water-dampened cloth to wipe away the residue.*

Caution: *During this procedure, be careful not to damage the wear shims. If damaged, the entire mounting plate assembly will require replacement.*

9.8 The wear bracket at the bottom of the assembly is removed to provide additional room to remove the shoe assemblies and the return spring. A screwdriver used to pry the shoe away from the wear bracket will help. At this stage, look carefully at how the lever and strut assembly are positioned

Check

1   The parking brake system can and should be checked as a normal part of driving. With the vehicle parked on a hill, apply the brake, place the transmission in Neutral and check that the parking brake alone will hold the vehicle. This is a simple check that should be performed regularly. However, every 24 months or so (or whenever a fault is suspected), the assembly itself should be visually inspected as discussed below.

2   With the vehicle raised and supported on jackstands, remove the rear wheels.

3   Remove the rear brake discs as outlined in Section 4. Support the caliper assemblies with a coat hanger or heavy wire and do not disconnect the brake line from the caliper.

4   With the brake disc removed, the parking brake components are visible and can be inspected for wear and damage. Most critical are the linings, which can wear after a time, especially if the parking brake system has been improperly adjusted. There should be at least 1/8-inch of lining on top of the metal backing. Also check the springs and adjuster mechanism for damage. Finally, inspect the inside of the brake disc for deep scratches and other damage.

Removal

5   Spread the shoes apart at the top and remove the adjuster assembly **(see illustration)**.

6   Note how the adjuster return spring is attached to each shoe and then remove the spring using needle-nose pliers **(see illustration)**.

7   Using a screwdriver, push down on the hold-down spring and then turn the end of the hold-down pin until it can be removed through the hole at the end of the spring. Remove the hold-down springs and pins.

8   Loosen each of the two small bolts at the wear bracket (at the bottom of the assembly) and then remove the bracket **(see illustration)**. Carefully note how the lever and strut assembly positions against the forward shoe and the wear shim. Upon installation this assembly must be positioned the same way.

9   With the shoe return spring still attached to the shoes, work the shoes and spring off the backing plate. The spring should come through the opening where the wear bracket was. It may help to rotate the hub as the spring comes through this opening (some models will have a small notch in the back-

side of the hub for extra clearance).

10   Detach the shoe assemblies and disengage the shoe return spring from each shoe.

11   Disassemble the adjuster and inspect it for damage. Apply multipurpose grease to the threads and confirm that the adjuster screw turns easily in the socket.

12   Inspect all parts and replace them with new ones if damage is found.

Installation

13   Noting the Warning at the beginning of this Section, clean all parts (except the new linings) with denatured alcohol.

14   Place a thin layer of brake lube on the wear bracket and wear shims.

15   Engage the shoe return spring in the shoes and work the new shoes into position on the backing plate, making sure the operating lever and strut assembly is positioned correctly.

16   Install the hold-down pins and springs by depressing the springs and fitting the pin through the slot in the backing plate. There are a couple different methods for doing this. First, rotate the hub until the large hole is directly over the spring. Working through this hole, depress the spring with either a screwdriver or needle-nose pliers and attempt to move the other end of the pin into the slot. You may find a second screwdriver is necessary to move the pin into the slot.

17   Install the adjuster return spring.

18   Spread the shoes and install the adjuster assembly, making sure the star wheel is positioned towards the front.

19   Slide the brake disc into position, turning the adjuster as necessary for the brake disc to clear the shoes.

20   Install the caliper bracket, referring to Section 4. Install the wheels. Adjust the parking brake assemblies and cable as outlined in Section 8.

---

10   Brake pedal - removal, installation and adjustment

Removal

1   Disconnect the battery ground cable. Place the cable out of the way so that it cannot come into contact with the battery terminal.

2   Remove the screws that secure the cover under the dashboard, lower the cover and disconnect the wiring to the courtesy lamp.

3   Disconnect the brake pedal from the power brake booster pushrod by removing the retainer and washer.

4   Remove the nut and pull out the bolt that travels through the top of the pedal.

5   The brake pedal can now be removed from the bracket.

Installation

6   Upon installation, use new bushings and lubricate the bushings, pivot bolt and all fric-

tion parts with brake lube.

7   Place the pedal in position and slide the bolt into place. Note that it should be installed in the direction shown in the accompanying illustration.

8   Tighten the nut and attach the pedal to the booster pushrod, using the washer and a new retainer.

9   Operate the brake pedal several times to ensure proper operation.

10   Connect the wiring to the courtesy lamp and install the under dash cover. Connect the battery cables.

Adjustment

11   Although the brake pedal does not ordinarily require adjustment, there is a brake pedal free play check which should be performed as a part of routine maintenance (see Brake check in Chapter 1).

12   Excessive brake pedal travel is an indication of worn brake components (usually disc brake pads) and indicates a thorough brake system inspection is required.

13   The power brake booster does have a fitting for adjustment on the pushrod that attaches to the brake pedal. This adjustment, however, should not be performed until all remaining brake components have been found to be in proper working order and a dealer has been consulted.

---

11   Brake light switch - removal, installation and adjustment

Removal

1   The brake light switch, or stop light switch as it is sometimes called, is located on a bracket at the top of the brake pedal. The switch activates the brake lights at the rear of the vehicle whenever the pedal is depressed.

2   Remove the under dash cover and disconnect the wiring to the courtesy lamp in this panel.

3   Locate the switch at the top of the brake pedal. If equipped with cruise control, there will be another switch very similar in appearance. The brake light switch is the one towards the end of the bracket.

4   Disconnect the negative battery cable and secure it out of the way so that it cannot come into contact with the battery terminal.

5   Disconnect the wiring connectors at the brake light switch.

6   Depress the brake pedal and pull the switch out of the clip. The switch appears to be threaded, but it is designed to be pushed into and out of the clip and not turned.

Installation and adjustment

7   With the brake pedal depressed, push the new switch into the clip. Note that audible clicks will be heard as this is done.

8   Pull the brake pedal all the way to the rear against the pedal stop until the clicking sounds can no longer be heard. This action will automatically move the switch the proper

amount and no further adjustment will be required. **Note:** *Do not apply excessive force during this adjustment procedure, as power booster damage may result.*

9   Connect the wiring at the switch and the battery. With an assistant, check that the rear brake lights are functioning properly. If so, install the under dash cover.

---

12   Brake hoses and lines - inspection and replacement

Refer to illustrations 12.4a and 12.4b

1   About every six months, with the vehicle raised and placed securely on jackstands, the rubber hoses which connect the steel brake lines with the front and rear brake assemblies should be inspected for cracks, chafing of the outer cover, leaks, blisters and other damage. These are important and vulnerable parts of the brake system and inspection should be complete. A light and mirror will prove helpful for a thorough check. If a hose exhibits any of the above conditions, replace it with a new one immediately.

Flexible (rubber) hose replacement

2   Clean all dirt away from the ends of the hose.

3   Disconnect the brake line from the hose fitting using a back-up wrench on the fitting. Be careful not to bend the frame bracket or line. If necessary, soak the connections with penetrating oil.

4   Remove the U-clip from the female fitting at the bracket **(see illustrations)** and remove the hose from the bracket.

5   Disconnect the hose from the caliper, discarding the copper washers on either side of the fitting block.

6   Using new copper washers, attach the new brake hose to the caliper.

7   Pass the female fitting through the frame or frame bracket. With the least amount of twist in the hose, install the fitting in this position (use the paint stripe on the hose to help determine twist). **Note:** *The weight of the vehicle must be on the suspension, so the vehicle should not be raised while positioning the hose.*

8   Attach the U-clip to the female fitting at the frame bracket.

9   Attach the brake line to the hose fitting using a back-up wrench on the fitting.

10   Carefully check to make sure the suspension or steering components do not make contact with the hose. Have an assistant push on the vehicle and also turn the steering wheel from lock-to-lock during inspection.

11   Bleed the brake system as described in Section 13.

Metal brake line replacement

12   When replacing brake lines it is important that the proper replacements be purchased. Do not use copper tubing for any

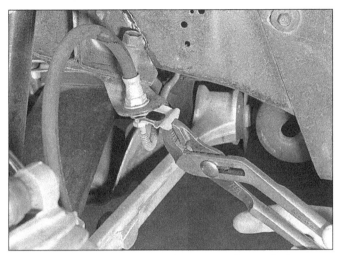

12.4a On the front brakes, the flexible hose is attached to the steel line at this bracket. Pliers are used to remove the U-clip that secured the two components

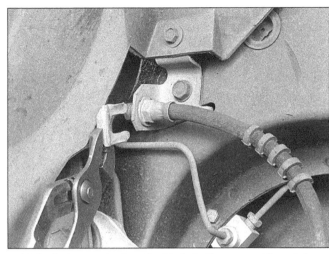

12.4b At the rear brakes, the U-clip is the same, although the bracket differs slightly

brake system connections. Purchase steel brake line from a dealer or brake specialist.

13   Prefabricated brake line, with the tube ends already flared and connectors installed, is available at auto parts stores and dealers. These lines are also bent to the proper shapes if necessary.

14   If prefabricated lengths are not available, obtain the recommended steel tubing and fittings to match the line to be replaced. Determine the correct length by measuring the old brake line (a piece of string can usually be used for this) and cut the new tubing to length, allowing about 1/2-inch extra for flaring the ends.

15   Install the fitting onto the cut tubing and flare the ends of the line with an ISO flaring tool.

16   If necessary, carefully bend the line to the proper shape. A tube bender is recommended for this. **Warning:** *Do not crimp or damage the line.*

17   When installing the new line make sure it is well supported in the brackets and has plenty of clearance between moving or hot components.

18   After installation, check the master cylinder fluid level and add fluid as necessary. Bleed the brake system as outlined in the next Section and test the brakes carefully before placing the vehicle into normal service.

13   Brake system bleeding

Refer to illustration 13.9

Note: *Bleeding the hydraulic system is necessary to remove any air that manages to find its way into the system when it has been opened during removal and installation of a hose, line, caliper or master cylinder.*

1   It will probably be necessary to bleed the system at all four brakes if air has entered the system due to low fluid level, or if the brake lines have been disconnected at the master cylinder.

2   If a brake line was disconnected at only one wheel, then only that caliper must be bled.

3   If a brake line is disconnected at a fitting located between the master cylinder and any of the brakes, that part of the system served by the disconnected line must be bled.

4   If the rear brakes are being bled, raise the front of the vehicle, which will position the bleeder valve at the 12 o'clock position and prevent air from being trapped in the caliper.

5   Remove any residual vacuum from the brake power booster by applying the brake several times with the engine off.

6   Remove the master cylinder reservoir cover(s) and fill the reservoir(s) with brake fluid. Reinstall the cover. **Note:** *Check the fluid level often during the bleeding operation and add fluid as necessary to prevent the fluid level from falling low enough to allow air bubbles into the master cylinder.*

7   Have an assistant on hand, as well as a supply of new brake fluid, an empty clear plastic container, a length of 3/16-inch plastic, rubber or vinyl tubing to fit over the

bleeder valve and a wrench to open and close the bleeder valve.

8   Beginning at the left rear wheel, loosen the bleeder valve slightly, then tighten it to a point where it is snug but can still be loosened quickly and easily.

9   Place one end of the tubing over the bleeder valve and submerge the other end in brake fluid in the container **(see illustration)**.

10   Have the assistant pump the brakes a few times to get pressure in the system, then hold the pedal firmly depressed.

11   While the pedal is held depressed, open the bleeder valve just enough to allow a flow of fluid to leave the valve. Watch for air bubbles to exit the submerged end of the tube. When the fluid flow slows after a couple of seconds, close the valve and have your assistant release the pedal.

12   Repeat Steps 10 and 11 until no more air is seen leaving the tube, then tighten the bleeder valve and proceed to the right rear wheel, the left front wheel and the right front wheel, in that order, and perform the same procedure. Be sure to check the fluid in the

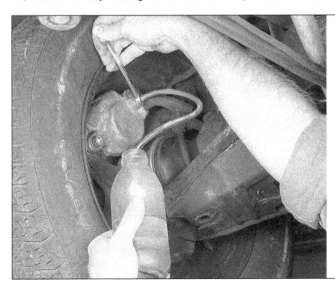

13.9 When bleeding the brakes, a hose is connected to the bleeder valve at the caliper and then submerged in brake fluid. Air will be seen as bubbles in the container or as bubbles inside the tube. All air must be removed before continuing to the next wheel

14.9a The front wheel speed sensors (arrow) are bolted to the spindles

14.9b Rear wheel speed sensor location (arrow)

master cylinder reservoir frequently.

13   Never use old brake fluid. It contains moisture that will deteriorate the brake system components.

14   Refill the master cylinder with fluid at the end of the operation.

15   Check the operation of the brakes. The pedal should feel solid when depressed, with no sponginess. If necessary, repeat the entire process. **Warning:** *Do not operate the vehicle if you are in doubt about the effectiveness of the brake system.*

16   If any difficulty is experienced in bleeding the hydraulic system, or if an assistant is not available, a pressure bleeding kit is a worthwhile investment. If connected in accordance with the instructions, each bleeder valve can be opened in turn to allow the fluid to be pressure ejected until it is clear of air bubbles without the need to replenish the master cylinder reservoir during the process.

---

14   Anti-lock brake system (ABS) - general information

Refer to illustrations 14.9a and 14.9b

The anti-lock braking system (ABS) was introduced in 1986 and is designed to maintain vehicle control under severe braking conditions. The system performs this function by monitoring the rotational speed of each wheel and then controlling the brake line pressure to prevent wheel lock-up.

Operation

Each time the vehicle is started, the amber "anti-lock" warning light on the dashboard illuminates for a brief time and then goes out. This indicates that the system is operating properly. Additionally, since the ABS functions may not be used every day, there is a test that actually runs the modulator valve to ensure that the system is totally functional. This test occurs when the vehicle

is first started and when it reaches about four mph. This test can sometimes be heard and felt (if the driver's foot is on the brake pedal). During vehicle operation the system is constantly monitored by the control module. If a fault occurs in any part of the ABS system, the dashboard warning light will illuminate to indicate that repair is necessary. **Note:** *If a fault in the ABS system should occur, the conventional brake system will remain fully operational provided it is not faulty.*

Components

Modulator valve

The modulator valve is located in the storage compartment directly behind the driver's seat. The function of the modulator valve is to maintain or reduce the brake fluid pressure to the wheel calipers. It cannot increase the pressure above that transmitted to the master cylinder, and can never apply the brakes by itself. The modulator valve receives all its instructions from the control unit.

If a problem occurs in the modulator valve, it must be replaced as a unit because it is not serviceable in any way.

Control module

The control module is also located in the rear storage compartment behind the driver's seat and is essentially the "brain" for the ABS system. The function of the control module - made up of resistors, diodes, transistors and large integrated circuits - is to accept and process information received from the wheel speed sensors. Wheel acceleration, deceleration and slip values are calculated to produce control commands for the modulator valve.

If a problem is suspected in the control module, it must be replaced as an assembly.

Lateral acceleration switch

The lateral acceleration switch is located on the floorboard, just under the air condi-

tioning control head. This switch is used to detect if the vehicle is cornering faster than a given curve speed. If so, a signal is sent to the control module indicating this hard cornering situation.

Wheel speed sensors

These sensors are located at each wheel and send an electric signal to the control module indicating rotational speed.

The wheel speed sensors are installed in the knuckles and have toothed rings pressed onto the front hub and bearing assemblies and the rear driveaxle spindles **(see illustrations)**. The sensors do not require any adjustment, nor are repairs possible.

Diagnosis and repair

If the dashboard warning light comes on and stays on while the vehicle is in operation, the ABS system requires attention. Although a special electronic ABS diagnostic tester is available, the home mechanic can perform a few preliminary checks before taking the vehicle to a dealer who is equipped with this tester.

a) *Open the rear compartment lid and check that the control module electrical connector is securely connected.*

b) *Follow the wiring harness to each wheel and check that all connections are secure and that the wiring is not damaged.*

c) *Check the condition of the Brake fuse at the fuse box.*

d) *Check the condition of the Gauge fuse at the fuse box.*

If the above preliminary checks do not rectify the problem, the vehicle should be diagnosed by a dealer service department. Due to the rather complicated nature of this system, all actual repair work must be done by a dealer.

Chapter 10
Steering and suspension systems

Contents

Specifications

Torque specifications

Ft-lbs

Steering linkage

Tie-rod-to-knuckle nut	33
Tie-rod-to-steering gear nut	70
Right hand steering gear clamp bolt	18
Steering gear-to-crossmember bolt	
1984 through 1987	25
1988 on	30

Intermediate shaft

Steering gear pinch bolt	
1984 only	46
1885 on	44
Steering column pinch bolt	
1984 only	35
1985 on	44

Power steering pump

Pump bracket-to-engine bolt	
1984 through 1987	25
1988 on	18
Pump-to-bracket bolt	
1984 through 1987	35
1988 on	18

Steering wheel and column

Steering wheel-to-shaft nut	30
Steering column bracket-to-frame bolt	20

Torque specifications

Ft-lbs

Front suspension

Upper control arm-to-balljoint nut	32
Lower control arm-to-balljoint nut	48
Upper balljoint-to-knuckle nut	33 to 63
Lower balljoint-to-knuckle nut	50 to 88
Upper control arm shaft-to-crossmember bolt	
1984 through 1987	63
1988 on	37
Lower control arm shaft-to-crossmember bolt	
1984 through 1987	96
1988 on	82
Spring protector-to-crossmember bolt	18
Spring retainer bolt	46*
Hub to bearing-to-knuckle bolt and nut	46
Shock absorber-to-lower control arm bolt	22
Shock absorber-to-frame nut	19
Shock absorber bracket-to-lower control arm	22
Stabilizer clamp-to-frame bolt	40*
Stabilizer link nut	35*
Stabilizer bracket-to lower control arm	22 to 35

Rear suspension

Driveshaft trunnion joint strap bolt	22 to 29
Spindle nut	164
Hub-to-knuckle bolt	59 to 73
Shock absorber stud-to-knuckle	
1984 through 1987	51 to 66
1988 on	89
Shock absorber-to-body bolt	
1984 through 1987	55 to 73
1988 on	61
Control arm-to-chassis bolt	55 to 70
Control arm-to-knuckle bolt	125 to 154
Transverse spring-to-differential carrier bolt	
1984 through 1987	29 to 54
1988 on	37
Cam bolt-to-support rod	158 to 213
Support rod-to-knuckle	95 to 118
Support rod bracket-to-differential carrier bolt	47 to 62
Tie-rod end-to-knuckle bolt	29 to 36
Tie-rod housing-to-differential carrier bolt	47 to 61
Stabilizer bar-to-chassis bolt	14 to 22
Stabilizer link-to-bracket and bar	33 to 44
Stabilizer link bracket-to-knuckle	14 to 22

* Component must be at normal ride height before tightening

Miscellaneous

Wheel lug nuts	100

---

1   General information

Refer to illustrations 1.1 and 1.2

The front suspension is independent, allowing each wheel to compensate for road surface changes without appreciably affecting the opposite wheel **(see illustration)**. Each wheel is independently connected to the chassis by unequal length forged aluminum control arms and a steering knuckle incorporating balljoints. The front wheels are held in proper relationship to each other by tie-rods, which are connected to the steering knuckle and the power-assisted rack and pinion steering gear. A transverse fiberglass single leaf spring is used, along with a stabilizer bar connected between the lower control arms. Shock absorbing is provided by double-acting shock absorbers mounted between the chassis and the lower control arms. Gas-charged shock absorbers are available as an option at all four wheels.

Rear suspension is independent as well and also features a transverse fiberglass spring and double-acting shock absorbers **(see illustration)**. Each wheel is connected to the chassis by forged aluminum trailing upper and lower control arms, a rear knuckle and a spindle and wheel bearing assembly. The rear wheels are located by a spindle support rod at the bottom of the knuckles and a tie-rod and a stabilizer bar at the top. The spring is pivoted at the knuckles and is clamped to the differential carrier assembly.

The rack-and-pinion steering gear is located in front of the engine and actuates steering arms, which are integral with the steering knuckles. Power assist is standard and the steering column is designed to collapse in the event of an accident.

1.1 Front suspension components

1	Steering knuckle	5	Steering gear	8	Lower control arm
2	Outer tie-rod end	6	Sway bar	9	Lower shock absorber mount
3	Tie-rod	7	Spring retainer	10	Lower balljoint
4	Transverse spring				

1.2 Rear suspension components

1	Rear transverse spring	3	Rear knuckle	5	Control rods
2	Spindle support rod	4	Shock absorber	6	Driveaxle

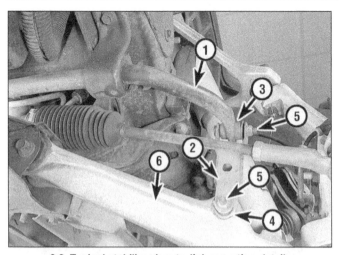

2.2 Typical stabilizer bar-to-link mounting details

1   Stabilizer bar
2   Link assembly
3   Stabilizer bar-to-link
    bushing
4   Link-to-control arm
    bushing
5   Link assembly mounting
    bolts/nuts
6   Lower control arm

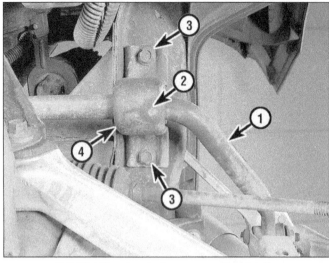

2.3 Stabilizer bar-to-frame mounting details

1   Stabilizer bar
2   Frame clamp
3   Mounting bolts
4   Insulator bushings

2   Front stabilizer bar - removal and installation

Refer to illustrations 2.2, 2.3 and 2.5

1   Raise the front of the vehicle, support it securely on jackstands and remove the front wheels.
2   Remove the stabilizer bar nuts, bolts and bushing retainers at the control arm **(see illustration)**. Make careful note of the directions they are installed, they must be installed in their original directions.
3   Unbolt the frame clamps and remove the stabilizer bar from the vehicle **(see illustration)**.
4   To install it, place the bar in position on the frame rails and then install the clamps and bolts.
5   Press the upper bushing into the stabi-lizer bar **(see illustration)**. Install the bushing retainers, nuts and bolts at the lower control arm.
6   Raise the lower control arms to normal ride height and tighten the nuts to the specified torque.
7   Install the wheels, lower the vehicle weight onto the suspension and tighten frame clamp bolts to the specified torque.

3   Front transverse spring - removal and installation

Refer to illustrations 3.2 and 3.7

1   Raise the vehicle, support it securely on jackstands and remove the front wheels.
2   Remove the spring protectors and spring mounting nuts **(see illustration)**.

3   Place a jack under the lower control arm and raise it slightly to compress the spring. **Warning:** *The jack must be situated firmly under the control arm, such that the jack can't slip out.*
4   Disconnect the lower balljoints with a balljoint-separating tool. The balljoints can be separated by removing the cotter pins and nuts, then tapping each balljoint with a hammer as the jack under the control arm is lowered, allowing the spring to push the balljoint out of the knuckle. Using this procedure, the spring can be decompressed and the balljoints disconnected in one operation.
5   Remove the bolts and disconnect the lower shock absorber mount.
6   Remove the spring bolts and shims by pushing them up and over the spring.
7   Disengage the spring insulators from the lower control arms by carefully prying them

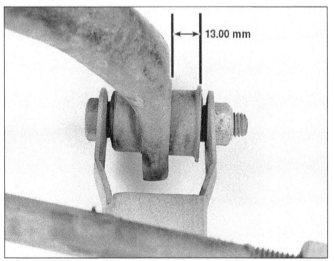

2.5 The stabilizer bar upper bushing must be pressed into the eye until it protrudes the amount indicated

13.00 mm

3.2 Spring protector bolt (A) and spring retaining nut (B) locations

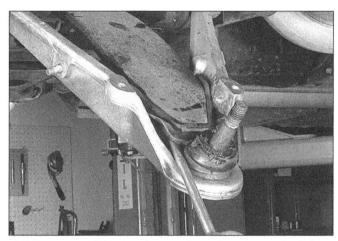

3.7 Be very careful not to damage the spring when prying it free from the lower control arm

4.6 The balljoints can be separated with a two-jaw puller

free **(see illustration)**.

8   Pull the left side lower control arm down to the limit of its travel and remove the right side lower control arm.

9   Remove the spring by carefully withdrawing it from the right side of the vehicle. Be very careful not to damage the spring surface.

10   To install, place the spring in position from the right side of the vehicle. **Note:** *The spring number must be on the right side.*

11   Insert the end of the spring into the left lower control arm and then install the right side lower control arm. Raise each control arm with the jack and connect the balljoints and the lower shock absorber mount. Tighten the shock absorber mount and lower control arm bolts to the specified torque.

12   Install the spring mount and protector nuts and bolts. Tighten the nuts and bolts to the specified torque.

13   Install the wheels and lower the vehicle.

14   The wheel alignment should be checked by a dealer service department or alignment shop.

---

4   Balljoints - check and replacement

Refer to illustration 4.6

Check

Upper balljoint

1   Raise the front of the vehicle and support it securely under the lower control arms on jackstands placed as close as possible to the balljoints.

2   Mount a dial indicator against the lower wheel rim, grasp the top and bottom of the wheel and move the top of the wheel in and the bottom out while measuring the deflection of the balljoint. Repeat the procedure, moving the bottom of the wheel out and the top in to get an accurate reading of the horizontal deflection caused by balljoint wear. If the horizontal deflection of the balljoint

exceeds 0.125-inch (3.18 mm) the balljoint must be replaced with a new one.

Lower balljoint

3   With the vehicle weight resting on the suspension, inspect the balljoint to make sure the wear indicator extends 0.050-inch (1.27 mm) above the balljoint surface (Refer to Chapter 1, Section14). Replace the balljoint with a new one if it is worn.

Replacement

4   Raise the vehicle, support it securely on jackstands and remove the front wheels.

5   Place a floor jack under the lower control arm. **Note:** *The jack must remain under the lower control arm during removal and installation of the balljoint to hold the spring and control arm in position.*

6   Remove the cotter pins and upper and lower balljoint nuts. Separate the balljoints, using a balljoint separator to press the balljoints out of the steering knuckle **(see illustration)**.

7   To remove the upper balljoint from the control arm, drill out the securing rivets, remove the balljoints and clean the control arm. Install the new upper balljoint and secure it with the supplied nuts and bolts.

8   On lower balljoints take the control arms to an automotive machine shop to have the balljoints pressed out and new ones pressed in.

9   Inspect the tapered holes in the steering knuckle, removing any accumulated dirt first. If out-of-roundness, deformation or other damage is noted, the knuckle must be replaced with a new one (Section 19).

10   Reconnect the balljoints to the steering knuckle and tighten the nuts to the specified torque.

11   If the cotter pin does not line up with the opening in the castellated nut, tighten (never loosen) the nut just enough to allow installation of the cotter pin.

12   Install the lubrication fittings and lubricate the new balljoints.

13   Install the wheels and lower the vehicle.

14   The wheel alignment should be checked

by a dealer service department or alignment shop.

---

5   Front lower control arm - removal and installation

Caution: *On ABS-equipped vehicles, use care when working in the vicinity of the brake sensing components.*

1   Raise the vehicle, support it securely on jackstands and remove the front wheel.

2   Remove the spring protector and then compress the spring with a jack.

3   Unbolt and remove the lower shock absorber bracket.

4   Remove the cotter pin and nut and tap the lower control arm balljoint lightly while lowering the jack, using the spring pressure to disconnect the balljoint by pushing it out of the steering knuckle.

5   Remove the bolts and lower control arm from the crossmember.

6   To install, place the lower control arm in position and install the bolts. Raise the control arm to normal ride height and tighten the bolts to the specified torque.

7   Install the shock absorber bracket and tighten the nuts and bolts to the specified torque.

8   Raise the spring with the jack and connect the balljoint to the steering knuckle.

9   Release the spring tension and install the spring protector.

10   Install the wheel and lower the vehicle.

11   Have the wheel alignment checked by a dealer service department or an alignment shop.

---

6   Front upper control arm - removal and installation

Refer to illustration 6.4

Caution: *On ABS-equipped vehicles, use care when working in the vicinity of the brake sensing components.*

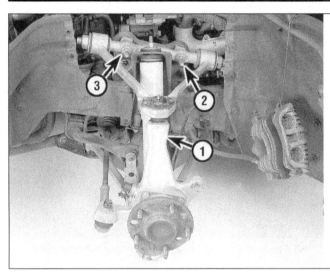

6.4 Upper control arm mounting details - tag the location of all shims and washers

1 Upper balljoint stud/nut
2 Front control arm shaft mounting stud
3 Rear control arm shaft mounting stud

1   Raise the vehicle, support it securely on jackstands and remove the wheel.
2   Use a jack or jackstand to support the lower control arm.
3   Disconnect the upper balljoint (Section 4).
4   Remove the nuts, bolts, shims (noting their number and location) and detach the control arm from the vehicle **(see illustration)**.
5   To install, place the arm, bolts and shims in position and install the nuts. Tighten the nuts to the specified torque.
6   Connect the balljoint.
7   Remove the jack or jackstand from under the lower control arm, install the wheel and lower the vehicle.
8   Have the wheel alignment checked by a dealer service department or an alignment shop.

7   Front hub and bearing - removal and installation

Refer to illustrations 7.4 and 7.5
Caution: *On ABS equipped vehicles, use care when working in the vicinity of the brake*

sensing components.
1   Raise the front of the vehicle, support it securely on jackstands and remove the front wheel.
2   Remove the brake caliper assembly and hang it out of the way on a piece of wire so there is no strain on the brake hose.
3   Remove the brake disc.
4   Remove the bolts and nuts and detach the hub and bearing assembly **(see illustration)**. On ABS-equipped models, take care not to damage the toothed sensor ring when withdrawing the assembly from the knuckle.
5   To install, place the assembly in position, insert the bolts through the access holes in the hub and install the nuts **(see illustration)**. Tighten the nuts and bolts to the specified torque.
6   Install the brake disc and caliper and lower the vehicle.

8   Shock absorbers - inspection, removal and installation

Refer to illustrations 8.4 and 8.8
Caution: *On ABS-equipped vehicles, use care when working in the vicinity of the brake*

sensing components.
Note: *Some models are equipped with Selective Ride Control. This system incorporates four adjustable shock absorbers, four electrical actuator motors, a center console mounted ride select switch and a ride control module. Due to the complexity of the system, extent of diagnosis and high expense of the components, any concerns of the Selective Ride Control system should be directed to the dealer.*

Inspection

1   On conventional air/oil shock absorbers, oil on the outside of the shock absorber bodies is an indication that the seals have started to leak and the units must be replaced. If the shock absorber has failed internally, the vehicle ride will be affected, particularly over uneven surfaces. If a shock absorber is suspected to have failed, remove it from the vehicle and holding it in a vertical position operate it for the full length of travel several times. Any lack of resistance in either direction will indicate the need for replacement.
2   On the optional gas-charged shock absorber units, a certain amount of oil film is acceptable, but the appearance of actual oil drops means the unit has failed. If the gas charge has leaked out of the shock absorber, the appearance may be normal but the ride will deteriorate considerably. **Caution:** *Because gas-charged shock absorbers contain gas and oil under high pressure, they should be taken to a dealer to be depressurized before they are disposed of.*

Removal and installation

3   Raise the vehicle, support it securely on jackstands and remove the wheels.

Front shock absorber

4   Use an open end wrench to prevent the upper (squared) end from turning, then remove the upper stem retaining nut. On some models it may be necessary to remove the wheel well panel for better access. Remove the two bolts retaining the lower

7.4 The front hub and bearing assembly can be detached after removing the mounting nuts/bolts

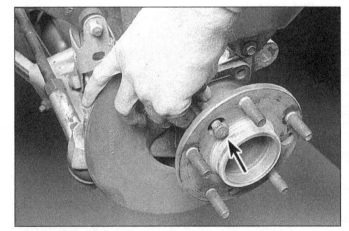

7.5 The front hub and bearing bolts are installed through the access hole in the hub flange

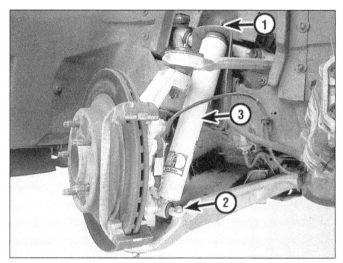

8.4 Front shock absorber mounting details

1   *Shock stem mounting nut*
2   *Shock pivot-to-control arm mounting bolts*
3   *Shock absorber*

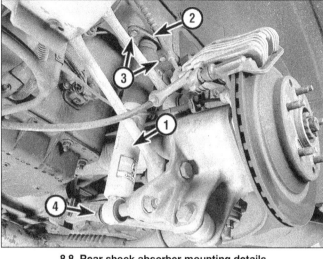

8.8 Rear shock absorber mounting details

1   *Shock absorber*   3   *Mounting bolts*
2   *Shock retaining bracket* 4   *Lower mounting nut*

9.3 Rear spring-to-knuckle mount components

1   *Spring*   3   *Insulators*
2   *Bolt*

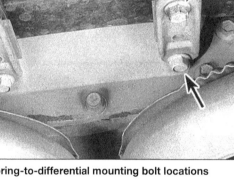

9.4 Rear spring-to-differential mounting bolt locations

shock absorber pivot to the control arm **(see illustration)**.

5   Prior to installation, place the grommet and washer in position on the upper stem. Insert the stem through the upper mount, push the grommet and washer into place and install the retaining nut. Tighten the nut to the specified torque.

6   Place the lower end of the shock absorber in position and install the bolts. Tighten the bolts to the specified torque.

Rear shock absorber

7   Use a backup wrench on the stud on the knuckle and remove the shock absorber nut, then remove the upper through bolt and nut and detach the shock absorber.

8   To install, place the shock absorber in position and install the retaining nuts and bolts. Using a backup wrench on the stud, tighten the shock absorber-to-knuckle nut to the specified torque **(see illustration)**.

9   Raise the knuckle to normal ride height with a jack and tighten the upper shock absorber through bolt and nut to the specified torque.

---

9   Rear transverse spring - removal and installation

Refer to illustrations 9.3 and 9.4

Caution: *On ABS-equipped vehicles, use care when working in the vicinity of the brake sensing components.*

1   Raise the vehicle and support it securely on jackstands.

2   Remove the rear wheels.

3   Support the end of the spring with a jack, remove the cotter pins, nuts, bushings and link bolts which attach the spring to the knuckle **(see illustration)**. Repeat the operation at the other end of the spring.

4   Remove the spring-to-differential carrier bolts, mark the positioning of the spacers, remove the spacers and insulators and lower the spring from the vehicle **(see illustration)**.

5   To install, place the spring, insulators and spacers in position on the carrier and install the bolts. Tighten the bolts to the specified torque.

6   Raise each end of the spring with a jack, connect the spring to the knuckles, tighten each of the nuts until the slots align with the link bolt holes and then install the cotter pins.

7   Install the wheels and lower the vehicle.

---

10   Rear knuckle - removal and installation

Caution: *On ABS-equipped vehicles, use care when working in the vicinity of the brake sensing components.*

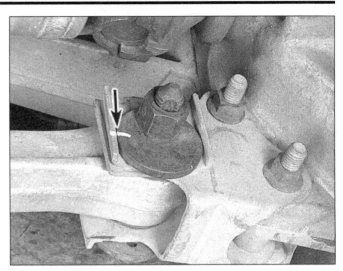

11.3 Rear control arm mounting components

1	*Control rods*	*2*	*Mounting bolts*

12.2 Use white paint (arrow) to mark the relationship of the cam bolt and bracket before removing the bolt

1   Raise the vehicle, support it securely on jackstands and remove the wheel. Remove the center cap, reinstall the wheel and lower the vehicle.

2   Remove the cotter pin, spindle nut and washer.

3   Raise the vehicle, support it securely on jackstands and remove the wheel.

4   Remove the brake caliper and brake disc (Chapter 9).

5   Disconnect the parking brake cable at the backing plate.

6   Disconnect the tie-rod end, transverse spring and stabilizer bar from the knuckle.

7   Disconnect the shock absorber (Section 8) and the spindle control rods (Section 11) at the knuckle.

8   Lower the knuckle, slide the driveshaft and spindle out and remove the knuckle assembly from the vehicle. Remove the hub and bearing and then remove the splash shield from the knuckle.

9   Installation is the reverse of removal, remembering to tighten all fasteners to the specified torque.

10   Have the wheel alignment checked by a dealer or properly equipped alignment shop.

11   Rear control rods - removal and installation

Refer to illustration 11.3

1   Raise the vehicle, support it securely on jackstands and remove the rear wheels.

2   Disconnect the transverse spring from the knuckle (Section 9).

3   Remove the attaching bolts and nuts at the knuckle and the body bracket and remove the control rods **(see illustration)**.

4   Installation is the reverse of removal.

12   Rear spindle support rod - removal and installation

Refer to illustrations 12.2 and 12.3
Caution: *On ABS-equipped vehicles, use care when working in the vicinity of the brake*

sensing components.

1   Raise the vehicle and support it securely on jackstands.

2   Scribe or paint across the cam bolt and mounting bracket so they can be in reinstalled in the same position **(see illustration)**.

3   Remove the cam bolt and detach the spindle support rod from the mounting bracket **(see illustration)**. Remove the spindle support rod bolt from the knuckle and detach the rod from the vehicle.

4   To install, place the rod in position and install the bolts finger tight. Line up the cam bolt with the scribe or paint mark and tighten the cam bolt and bracket bolt to the specified torque. Tighten the support rod-to-knuckle bolt and nut to the specified torque.

13   Rear tie-rod assembly - removal and installation

Refer to illustrations 13.3 and 13.4
Caution: *On ABS-equipped vehicles, use care when working in the vicinity of the brake sensing components.*

1   Raise the vehicle and support it securely on jackstands.

2   Remove the cotter pins and nuts from the tie-rod ends at the knuckles.

3   Disconnect the tie-rod ends from the knuckle with a two-jaw puller **(see illustration)**.

4   Remove the tie-rod assembly-to-differential carrier bolts and detach the assembly from the vehicle **(see illustration)**.

5   To replace the tie-rod ends, loosen the jam nut sufficiently to mark the position of the tie-rod end on the tie-rod then remove the tie-rod end.

6   Thread the tie-rod end onto the rod to the marked position and tighten the jam nut securely.

7   To install the tie-rod assembly, place it in position and install the retaining bolts.

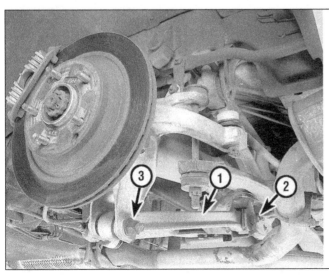

12.3 Spindle support rod details

1   *Knuckle support rod*
2   *Cam bolt*
3   *Knuckle-to-support rod mounting bolt*

13.3 To avoid damage to the threads, disconnect the tie-rod end with a puller only

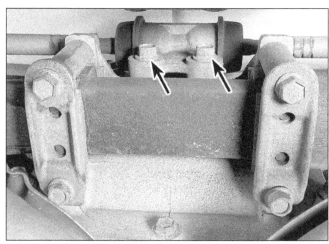

13.4 The rear tie-rod center mounting bolts are located above the leaf spring mount (arrows indicate two of the three)

15.7 Use a block of wood to support the driveaxle while the hub is removed

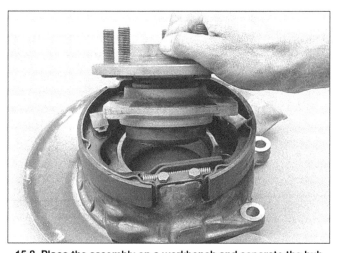

15.8 Place the assembly on a workbench and separate the hub and bearing assembly from the caliper mounting bracket

Tighten the bolts to the specified torque.

8   Connect the tie-rod end to the knuckle, install the nut and tighten it to the specified torque. Install a new cotter pin.

9   Have the wheel alignment checked by a dealer service department or an alignment shop.

14 Rear wheel spindle - removal and installation

Caution: *On ABS-equipped vehicles, use care when working in the vicinity of the brake sensing components.*

1   Raise the vehicle, support it securely on jackstands and remove the wheel. Remove the center cap, reinstall the wheel, lower the vehicle and remove the cotter pin, nut lock, spindle nut and washer.

2   Raise the vehicle, support it securely on jackstands and remove the wheel.

3   Disconnect the tie-rod end and the transverse spring from the knuckle.

4   Scribe or paint across the spindle support rod cam bolt and mounting bracket and remove the cam bolt and bracket assembly (Section 12).

5   Remove the driveaxle trunnion straps at the spindle, push out on the knuckle to separate the driveaxle from the spindle and then remove the spindle from the hub and bearing.

6   Installation is the reverse of removal, replacing the spindle seal with a new one if necessary and tightening all fasteners to the specified torque. Apply the parking brake when tightening the spindle nut.

7   Have the wheel alignment checked by a dealer service department or an alignment shop.

15 Rear hub and bearing - removal and installation

Refer to illustrations 15.7, 15.8 and 15.11

Caution: *On ABS-equipped vehicles, use care when working in the vicinity of the brake sensing components.*

1   Raise the vehicle, support it securely on jackstands and remove the wheel. Remove

the center cap and reinstall the wheel.

2   Lower the vehicle and remove the cotter pin, nut lock and spindle nut, then again raise the vehicle and support it on jackstands.

3   Remove the brake caliper, adapter and disc and hang the caliper out of the way on a piece of wire. Do not let it hang by the hose.

4   Disconnect the tie-rod at the knuckle.

5   Remove the three Torx head bolts retaining the hub and bearing assembly. Turn the driveaxle universal joint as necessary for access to the upper bolt.

6   Lift the hub and bearing and caliper mounting bracket assembly from the knuckle as a unit (don't lose the spindle nut washer). It may be necessary to use a puller to press the spindle in to disengage the hub from the spindle splines.

7   Support the driveaxle with a block of wood placed between the driveaxle and the support rod **(see illustration)**.

8   Place the assembly on a workbench. Noting the direction in which it is installed, separate the hub and bearing assembly from the caliper mounting bracket **(see illustration)**.

9   To install, place the hub and bearing assembly in the caliper mounting bracket and backing plate assembly.
10   Place the assembly in position on the axle spindle and remove the wood block.
11   Line up the bolt holes in the hub and knuckle by inserting a punch through from the hub side **(see illustration)**.
12   Install the hub retaining bolts. Tighten the bolts to the specified torque.
13   Install the spindle washer and nut. Tighten the nut to the specified torque and install the nut lock and cotter pin.
14   Connect the tie-rod.
15   Install the brake caliper and disc.
16   Install the wheel and lower the vehicle.

16   Rear stabilizer bar - removal and installation

Refer to illustration 16.3
1   Raise the vehicle and support it securely on jackstands.
2   Remove the spare tire and carrier assembly.
3   Disconnect the stabilizer bar at the knuckles, remove the bracket nuts and lower the bar from the vehicle **(see illustration)**.
4   To install, place the bar in position and install the brackets and nuts. Tighten the nuts to the specified torque. **Note:** *The bolts are installed from the inside, with the nuts on the wheel side of the assembly.*
5   Connect the bar ends to the links and attach them to the knuckles. Raise the each knuckle to normal ride height with a jack before tightening the link bolts and nuts to the specified torque.
6   Install the spare tire and carrier assembly.
7   Install the wheels and lower the vehicle.

17   Steering system - general information

Warning: *Later models are equipped with airbags. Always disable the airbag system before working in the vicinity of the steering column, instrument panel or impact sensors to avoid the possibility of accidental deployment of the airbag, which could cause personal injury* (see Chapter 12).

All models are equipped with power assisted rack-and-pinion steering. The steering gear is bolted to the chassis directly in front of the engine and operates the steering arms via tie-rods. The inner ends of the tie-rods are protected by rubber boots, which should be inspected periodically for secure attachment, tears and leaking lubricant.

The power assist system consists of a belt-driven pump and associated lines and hoses. The power steering pump reservoir fluid level should be checked periodically (Chapter 1).

The steering wheel operates the steer-

15.11 Use a punch or similar tool to line up the hub and knuckle bolt holes

ing shaft, which actuates the steering gear through universal joints, and the intermediate shaft. Looseness in the steering can be caused by wear in the steering shaft universal joint, the steering gear, the tie-rod ends and loose retaining bolts.

18   Steering tie-rod ends - removal and installation

Caution: *On ABS-equipped vehicles, use care when working in the vicinity of the brake sensing components.*
1   Raise the front of the vehicle, support it securely on jackstands, block the rear wheels and set the parking brake. Remove the front wheels.
2   Disconnect the tie-rod from the steering knuckle arm with a puller.
3   Loosen the jam nut sufficiently to mark the position of the tie-rod end on the tie-rod threads then remove the tie-rod end.
4   Thread the tie-rod end onto the rod to the marked position and tighten the jam nut securely.
5   Connect the tie-rod end to the steering knuckle arm, install the nut and tighten it to the specified torque. Install a new cotter pin.

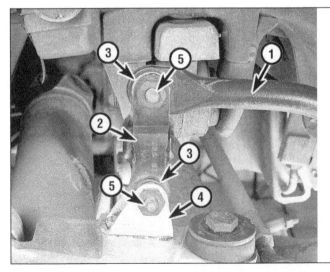

6   Have the wheel alignment checked by a dealer service department or an alignment shop.

19   Steering knuckle - removal and installation

Refer to illustration 19.8
Caution: *On ABS-equipped vehicles, use care when working in the vicinity of the brake sensing components.*
1   Raise the vehicle, support it securely on jackstands and remove the front wheels.
2   Remove the brake caliper retaining bolts and hang the caliper out of the way on a piece of wire. Do not allow the caliper to hang by the brake hose.
3   Remove the hub and bearing assembly (Section 7).
4   Unbolt and remove the splash shield.
5   Disconnect the tie-rod from the steering knuckle.
6   Support the lower control arm with a jack.
7   On ABS-equipped models, disconnect the brake sensor.
8   Disconnect the balljoints from the steering knuckle (refer to Section 4) and remove

16.3 The rear stabilizer bar linkage and bolts must be installed exactly as shown

1   Rear stabilizer bar
2   Link flange
3   Link bushings
4   Bracket
5   Link flange mounting bolts

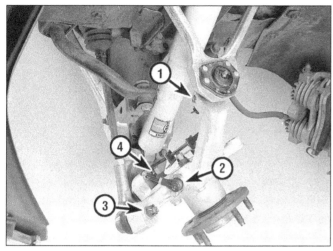

19.8 Front steering knuckle assembly

1	Upper balljoint stud/nut	3	Tie-rod end nut
2	ABS wheel sensor	4	Lower balljoint stud/nut

20.3 Steering column adjuster retaining screws (A), retainer (B) and lever (C) locations

20.5 Use a puller to remove the steering wheel - do not hammer on the shaft!

20.6 Thread a bolt into the steering shaft, pull the shaft out and then tighten the bolt to lock the shaft in place

the knuckle from the vehicle **(see illustration)**.

9   Installation is the reverse of removal. Tighten all bolts to the specified torque.

10   Have the wheel alignment checked by a dealer service department or an alignment shop.

---

20   Steering wheel - removal and installation

Warning: *Later models are equipped with airbags. Always disable the airbag system before working in the vicinity of the steering column, instrument panel or impact sensors to avoid the possibility of accidental deployment of the airbag, which could cause personal injury (see Chapter 12).*

Models without airbags

Refer to illustrations 20.3, 20.5 and 20.6

1   Disconnect the negative cable at the battery. Place the cable out of the way so it cannot accidentally come in contact with the negative terminal of the battery, as this would once again allow power into the electrical system of the vehicle.

2   Pry the horn cap off with a small screwdriver, disconnect the horn contact and remove the cap assembly.

3   Remove the two column adjusting lever-to-lock screws, the clip (if equipped), unscrew the retainer and lift the adjusting lever out through the slot in the steering wheel **(see illustration)**.

4   Mark the relationship of the steering shaft and hub to simplify installation and remove the steering wheel retaining nut.

5   Use a puller to detach the steering wheel **(see illustration)**. Remove the spring and spacers.

6   Because the steering shaft will fall into the column after the adjuster lever and steering wheel are removed, it will be necessary to pull it out and lock it in place before the wheel is reinstalled. This is accomplished by screwing a 5/16 - 18 bolt into the end of the shaft, pulling the shaft out to a position where the steering wheel can be installed and then further tightening the bolt to lock the shaft in place **(see illustration)**. It may be necessary to grind the head of the bolt down sufficiently to allow the steering wheel nut to pass over it.

7   To install the wheel, align the mark on the steering wheel hub with the mark made on the shaft during removal and slip the wheel onto the shaft. Install the hub nut and tighten it to the specified torque. Remove the bolt from the steering column shaft. Insert the telescoping column lever through the slot in the steering wheel, lubricate the threads of the adjuster with a few drops of oil and assemble the adjuster mechanism.

8   Connect the horn lead wire.

9   Line up the horn cap with the openings in the contact plate and press the cap evenly into place until it locks.

10   Connect the negative battery cable.

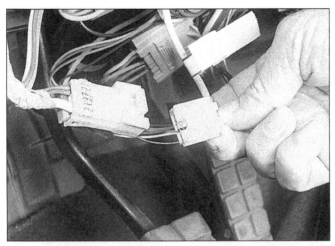

20.12 Disconnect the driver's side airbag inflator module wiring connector

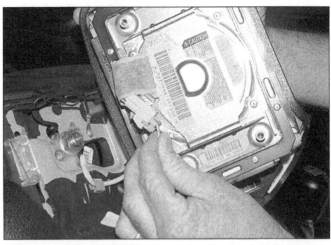

20.14 Disconnect the inflator module wiring lead. Always point the inflator module in a safe direction and store with the airbag unobstructed, facing up

Models equipped with airbags

Refer to illustrations 20.12 and 20 14

11 Disable the inflatable restraint system by first turning the ignition off, then remove the AIRBAG fuse from the fuse box.

12 Remove the driver side lower trim panel, locate the bright yellow airbag lead and connector at the base of the steering column.

13 Remove the connector lock and disconnect the airbag lead from the vehicle harness. The steering wheel is now safe to service, however caution should be used in handling the airbag module, static electricity can cause unwanted bag deployment.

14 Remove the screws attaching the inflator module from the back of the steering wheel, the yellow connector from the inflator module and remove the airbag. Carry the airbag with the trim side facing AWAY from your body and place it in a safe place with the trim side facing UP.

15 Remove the steering wheel retaining nut.

16 Disconnect the horn connector.

17 Attach a steering wheel puller and remove the steering wheel.

18 Assembly is the reverse of removal. Turn the ignition switch to the run position and check that the INFL REST warning light flashes 7 to 9 times then goes off. This will verify the airbag system is operational.

---

21 Power steering system - bleeding

1 The power steering system must be bled whenever a line is disconnected. Bubbles can be seen in power steering fluid that has air in it and the fluid will often have a tan or milky appearance. On later models, low fluid level can cause air to mix in with the fluid resulting in a noisy pump as well as foaming of the fluid.

2 Open the hood and check the fluid level in the reservoir, adding the specified fluid necessary to bring it up to the proper level (see Chapter 1).

22.4a Steering gear fastener locations (arrows)

1984 models

3 With the power steering pump and fluid at normal operating temperature, start the engine and slowly turn the steering wheel several times from left-to-right. Do not turn the wheel fully from lock-to-lock. Check the fluid level, topping it up as necessary until it remains steady and no more bubbles appear in the reservoir.

1985 on

4 Start the engine and run it for approximately ten seconds, then shut if off and refill the reservoir.

5 Run the engine for a few seconds while turning the steering wheel from lock-to-lock several times.

6 Shut the engine off and refill the reservoir.

---

22 Steering gear - removal and installation

Refer to illustrations 22.4a and 22.4b
Warning: *On models equipped with airbags, do not allow the steering shaft to rotate with*

the steering gear removed or damage to the airbag coil under the steering wheel could occur. Lock the steering wheel with the front wheels pointing straight ahead and remove the ignition key before beginning work.
Caution: *On ABS-equipped vehicles, use care when working in the vicinity of the brake sensing components.*

1 Raise the vehicle, support it securely on jackstands and remove the front wheels.

2 Disconnect the power steering fluid hoses from the steering gear and plug them.

3 Disconnect the tie-rod ends from the steering knuckles (Section 18).

4 Remove the upper and lower mounting bolts on the passenger side and the mounting bolt from the driver's side **(see illustration)**. Mark the relationship of the intermediate shaft universal joint sections and detach them from the steering gear **(see illustration)**.

5 On some models it may be necessary to remove the drivebelt, cooling fan and the stabilizer bar.

6 Detach the steering gear and lift it from the vehicle.

7 To install the steering gear, position it in the crossmember, install the bolts, tighten

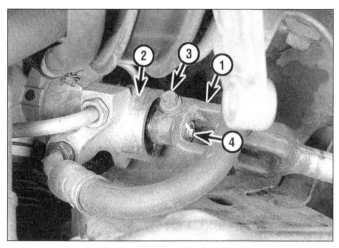

22.4b Steering installation details

1 Intermediate shaft U-joint	4 Area which must be
2 Steering gear	marked prior to
3 Pinch bolt	disassembly

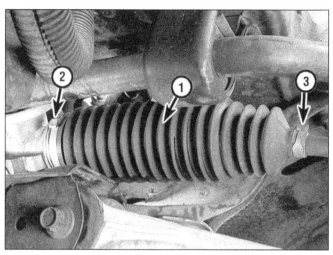

23.3 Steering gear boot components

1 Boot	3 Outer clamp
2 Inner clamp	

them securely and reconnect the intermediate shaft universal joint.

8   Attach the tie-rod ends to the steering knuckles.

9   Reconnect the hoses.

10   Start the engine and bleed the steering system (Section 21). While the engine is running, check for leaks at the hose connections.

11   Have the wheel alignment checked by a dealer service department or an alignment shop.

23 Steering gear boots - replacement

Refer to illustration 23.3

1   Raise the vehicle and support it securely on jackstands.

2   Mark the jam nut position on the tie-rod threads with paint. Loosen the jam nut and disconnect the tie-rod end from the steering knuckle. Unscrew the tie-rod end and jam nut from the steering rod.

3   Cut the inner clamp and slide the boot off the steering arm **(see illustration)**.

4   Before installing the new boot, it is a good idea to wrap the threads and serrations on the end of the steering rod with a layer of tape so the small end of the new boot is not damaged.

5   Slide the new boot into position on the steering gear until it seats in the groove in the steering rod and install a new inner clamp.

6   Remove the tape, install the tie-rod end and tighten the locknut, making sure it is at the marked position.

7   Lower the vehicle.

8   Have the wheel alignment checked by a dealer service department or alignment shop.

24 Power steering pump - removal and installation

Refer to illustration 24.4

1   Remove the serpentine drivebelt (Chapter 1).

2   Disconnect the power steering pump

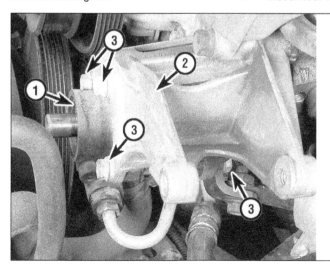

24.4 Power steering pump and brackets

1   Power steering pump

2   Bracket

3   Mounting bolts

hoses and plug the hoses and pump ports.

3   Using a power steering pump pulley removal tool (available at most auto parts stores), remove the pump pulley. On 1991 and later models, remove the drivebelt idler pulley.

4   Remove the retaining bolts and lift the pump and bracket from the engine compartment **(see illustration)**.

5   To install, place the pump and bracket in position and install the bolts.

6   Install the pump pulley using the installation tool.

7   Attach the hoses to the pump.

8   Install the serpentine drivebelt (Chapter 1).

9   Fill the pump reservoir with the specified fluid, bleed the system and check the fluid level.

25 Steering column - removal and installation

Refer to illustration 25.2

Warning 1: *Some of the models covered by this manual are equipped with Supplemental Inflatable Restraints (SIR), more commonly known as airbags. Always disable the airbag system before working in the vicinity of any airbag system components to avoid the possibility of accidental deployment of the airbag, which could cause personal injury (see Chapter 12).*

Warning 2: *On models equipped with airbags, do not allow the steering shaft to rotate with the steering column removed or damage to the airbag coil under the steering wheel could occur. Lock the steering wheel with the front wheels pointing straight ahead and remove the ignition key before beginning work.*

1   Disconnect the negative battery cable from the battery. If equipped with airbags, disable the airbag system (see Chapter 12).

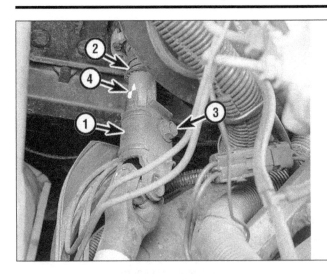

25.2 Steering shaft pinch bolt details

1   U-joint
2   Shaft
3   Pinch bolt
4   Mark components here prior to disassembly

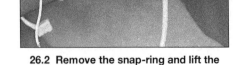

26.2 Remove the snap-ring and lift the airbag coil off

2   Mark the intermediate shaft-to-steering column relative positions with paint and remove the pinch bolt universal joint coupling in the engine compartment **(see illustration)**.
3   Remove the steering wheel (Section 20).
4   Remove the steering column-to-instrument panel and lower support retaining nuts and bolts.
5   Unplug the steering column wiring connectors.
6   Grasp the steering column assembly firmly and pull it toward the rear so the lower stub shaft is disconnected from the intermediate shaft universal joint coupling.
7   Remove the assembly carefully from the vehicle.
8   To install, place the steering column in position with the stub shaft paint mark aligned with the mark on the intermediate shaft universal joint and insert the shaft into the joint coupling. Install the pinch bolt and tighten it to the specified torque.
9   Install the retaining nuts and bolts finger tight. After all the nuts and bolts have been installed, tighten them to the specified torque.
10   Install the steering wheel.

11   Plug in the wiring connectors.
12   Connect the negative battery cable and check the operation of the horn and lights.

---

26   Turn signal switch - removal and installation

Refer to illustrations 26.2, 26.3, 26.5a, 26.5b, 26.7, 26.12 and 26.13
Warning: *Later models are equipped with airbags. Always disable the airbag system before working in the vicinity of the steering column, instrument panel or impact sensors to avoid the possibility of accidental deployment of the airbag, which could cause personal injury* (see Chapter 12).
1   Disconnect the negative cable at the battery. If equipped with airbags, disable the airbag system (see Chapter 12).
2   Remove the steering wheel (Section 20). If equipped with airbags, Remove the airbag coil retaining snap-ring and remove the coil assembly and wave washer; let the coil hang by the wiring harness **(see illustration)**.
3   Pry the plastic horn spring insulator off

the lock plate **(see illustration)**.
4   Lock the steering shaft in position.
5   Install the lockplate depressor tool, press the lockplate down and pry the retaining clip out of the groove with a screwdriver **(see illustrations)**.
6   Remove the depressor and withdraw the lockplate. Remove the canceling cam and spring.
7   Remove the hazard warning switch knob. Remove the turn signal switch arm screw and remove the arm, then remove the three switch retaining screws **(see illustration)**.
8   Remove the under-dash panels and disconnect the turn signal switch wiring harness at the base of the steering column. Remove the wiring harness protector. Remove the switch, pulling the harness wires and connector up through the steering column.
9   Feed the new connector and wires down through the column and plug the connector into the harness. Use a section of mechanics wire to pull it through, if necessary. Plug in the connector and replace the wiring harness protector.

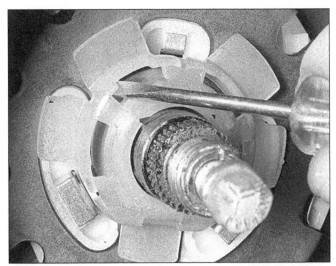

26.3 Use a small screwdriver to pry the spring insulator off

26.5a Compress the lockplate with the tool until . . .

26.5b ... the retaining clip can be pried out of the steering shaft groove

26.7 Column switch retaining screw locations (arrows)

26.12 To center the SIR coil, hold it with its underside facing up, depress the spring lock and rotate the hub in the direction of the arrow on the coil assembly until it stops

26.13 When properly installed the airbag coil will be centered with the marks aligned (circle) and the tab fitted between the projections on the top of the steering column

10   Place the turn signal switch in position and install the screws. Install the hazard warning switch knob.

11   Install the spring, cancel cam and lock plate. Depress the lock plate, install the retaining ring and insulator.

12   If necessary, center the airbag coil as follows (it will only become uncentered if the spring lock is depressed and the hub rotated with the coil removed) **(see illustration)**:

   a) *Turn the coil over and depress the spring lock.*
   b) *Rotate the hub in the direction of the arrow until it stops.*
   c) *Rotate the hub in the opposite direction 2-1/2 turns and release the spring lock.*

13   Install the wave washer and the airbag coil **(see illustration)**. Pull the slack out of the airbag coil lower wiring harness to keep it tight through the steering column, or it may be cut when the steering wheel is turned. Install the airbag coil retaining snap-ring.

14   The remainder of installation is the reverse of removal.

27   Intermediate steering shaft - removal and installation

Warning 1: *Some of the models covered by this manual are equipped with Supplemental Inflatable Restraints (SIR), more commonly known as airbags. Always disable the airbag system before working in the vicinity of any airbag system components to avoid the possibility of accidental deployment of the airbag, which could cause personal injury (see Chapter 12).*

Warning 2: *On models equipped with airbags, do not allow the steering shaft to rotate with the intermediate shaft removed or damage to the airbag coil under the steering wheel could occur. Lock the steering wheel with the front wheels pointing straight ahead and remove the ignition key before beginning work.*

1   Remove the bolts, disengage the retaining hooks and remove the upper and lower intermediate shaft shields.

2   Mark the relative position of the intermediate shaft and steering gear. Remove the pinch bolts and detach the intermediate shaft from the steering gear and column bearing **(see illustration 25.2)**.

3   Installation is the reverse of removal.

28   Wheel alignment - general information

Refer to illustration 28.1

A wheel alignment refers to the adjustments made to the wheels so they are in proper angular relationship to the suspension and the ground. Wheels that are out of proper alignment not only affect steering control, but also increase tire wear. Toe-in can be adjusted on the front and rear wheels. The front and rear camber and caster angles should be checked to determine if any of the suspension components are worn out or bent **(see illustration)**.

Getting the proper wheel alignment is a very exacting process, one in which complicated and expensive machines are necessary to perform the job properly. Because of this, you should have a technician with the proper equipment perform these tasks. We will, however, use this space to give you a basic idea of what is involved with wheel alignment so you can better understand the process and deal intelligently with the shop that does the work.

Toe-in is the turning in of the wheels. The purpose of a toe specification is to ensure parallel rolling of the wheels. In a vehicle with zero toe-in, the distance between the front edges of the wheels will be the same as the distance between the rear edges of the wheels. The actual amount of toe-in is normally only a fraction of an inch. At both the front and rear wheels, toe-in is controlled by the tie-rod end position on the tie-rod. Incorrect toe-in will cause the tires to wear improperly by making them scrub against the road surface.

Camber is the tilting of the wheels from the vertical when viewed from the front or rear of the vehicle. When the wheels tilt out at the top, the camber is said to be positive (+). When the wheels tilt in at the top the camber is negative (-). The amount of tilt is measured in degrees from the vertical and this measurement is called the camber angle. This angle affects the amount of tire tread which contacts the road and compensates for changes in the suspension geometry when the vehicle is cornering or traveling over an undulating surface. At the front, camber is adjusted by adding or subtracting an equal amount of shims between the control arm and the frame bracket at the two upper control arm-to-frame mounting bolts. At the rear, camber is adjusted by turning an adjuster cam bolt on the inner end of the spindle support rod.

Caster is the tilting of the top of the steering axis from the vertical. A tilt toward the rear is positive caster and a tilt toward the front is negative caster. At the front, caster is adjusted by adding or subtracting shims between the control arm and the frame bracket at one of the upper control arm mounting bolts, but not the other. Caster isn't adjustable on the rear of these vehicles.

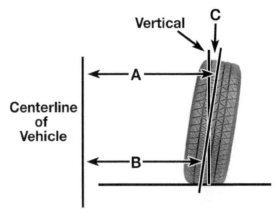

CAMBER ANGLE (FRONT VIEW)

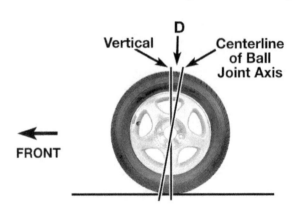

CASTER ANGLE (SIDE VIEW)

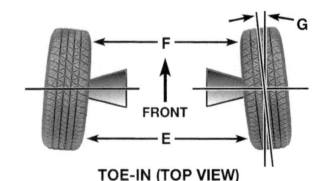

TOE-IN (TOP VIEW)

28.1 Wheel alignment details

A minus B = C (degrees camber)
D - Degrees caster
E minus F = toe-in (measured in inches)
G - toe-in (expressed in degrees)

---

29   Wheel stud - replacement

1   Raise the vehicle, support it securely on jackstands and remove the wheel.
2   Remove the brake caliper and disc and the splash shield.
3   Rotate the hub until the stud is positioned and press it out of the hub with a tool or a large C-clamp and socket. Thread a lug nut partway over the stud, leaving a "pocket" that will keep the tool centered over the stud being pressed out.

4   Insert the new stud into the hub, install four washers and the wheel bolt and then tighten the bolt until the stud is seated.
5   Install the brake caliper, disc, splash shield and wheel and lower the vehicle.

Chapter 11   Body

Contents

Specifications

Torque specifications

	Ft-lbs
Front bumper impact bar-to-skid plate	15 to 25
Front bumper impact bar-to-front crossmember	15 to 25
Roof locator bracket adjusting bolts	15

1   General information

The Corvette utilizes a unique blending of unibody construction with a multi-part perimeter frame. The main frame assembly is made of galvanized steel, with the major fiberglass body panels bonded directly to the frame to form a rigid structure. An aluminum rear frame assembly supporting the rear axle assembly and rear suspension is bolted to the unibody.

2   Body - maintenance

1   The condition of your vehicle's body is very important, because it is an important component of the resale value of the vehicle. It is much more difficult to repair a neglected or damaged body than it is to repair mechanical components. And the condition of the hidden areas of the body (the fender wells, the frame, and the engine compartment, to name a few) - even though they don't require as much attention as the rest of the body - is just as important as the condition of more obvious exterior panels.

2   Once a year, or every 12,000 miles, it is a good idea to have the underside of the body and the frame steam cleaned. All traces of dirt and oil will be removed and the underside can then be inspected carefully for damaged brake lines, frayed electrical wiring, damaged cables, and other problems. The suspension components should be greased after completion of this job.

3   At the same time, clean the engine and the engine compartment using either a steam cleaner or a water soluble degreaser.

4   The body should be washed once a week (or when dirty). Wet the vehicle thoroughly to soften the dirt, then wash it down with a soft sponge and plenty of clean soapy water. If the surplus dirt is not washed off very carefully, it will eventually wear down the paint.

5   Spots of tar or asphalt coating thrown up from the road should be removed with a cloth soaked in solvent.

6   Once every six months, give the body a thorough waxing.

3   Vinyl trim - maintenance

1   Under no circumstances try to clean any vinyl trim with detergents, caustic soap or petroleum-based cleaners. Plain soap and water is all that is required, with a soft brush to clean dirt that may be ingrained. Wash the vinyl as frequently as the rest of the vehicle.

2   After cleaning, application of a high quality rubber and vinyl protectant will help prevent oxidation and cracking. This protectant can also be applied to weatherstripping, vacuum lines and rubber hoses, which often fail as a result of chemical degradation, and to the sidewalls of the tires.

4   Upholstery and carpets - maintenance

1   Every three months remove the carpets or mats and clean the interior of the vehicle (more frequently if necessary). Vacuum the upholstery and carpets to remove loose dirt and dust.

2   Leather upholstery requires special care. Stains should be removed with warm water and a very mild soap solution. Use a clean, damp cloth to remove the soap, then wipe again with a dry cloth. Never use alcohol, gasoline, nail polish remover or thinner to clean leather upholstery.

3   After regular cleaning, treat leather upholstery with a leather wax. Never use car wax on leather upholstery.

4   In areas where the interior of the vehicle is subject to bright sunlight, it is advisable to cover leather seats with a sheet or other protective layer if the vehicle is to be left out for any length of time.

5   Convertible top - general information and operation

General information

Some Corvette models are equipped with a full convertible top made of cambria cloth or vinyl. The top is manually operated and stows in an enclosed compartment behind the seat. The linkage members are made of aluminum and pivot on shoulder bolts passing through Teflon bushings for ease of operation.

Operation

1   To lower the convertible top, lower the left and right door windows then loosen the headliner tape fasteners on the rear rails at the back of each door window opening.

2   Reach under the stowage compartment lid and pull sideways on the release lever for the bottom (number 5) bow.

3   Grasp the bottom bow and raise the rear of the top to the vertical position.

4   Unlock the stowage lid. This is accomplished either by pushing the button located at the rear edge of either door, the button inside the center console, or by pulling the right and left manual release rings attached directly to the latches.

5   Raise the stowage lid and lower the rear of the top to its original position.

6   Turn the left and right sun visors down and unlock the right and left top levers.

7   Grasp the front edge of the top and lift it up and to the rear, folding it into the stowage compartment.

8   Close the stowage compartment lid by pressing down lightly on the front edge at each side until the latches catch.

6   Body damage - minor repair

Refer to illustration 6.7

1   In most cases, repair of a scratched panel will involve only refinishing the paint in the area of the scratch.

2   If the scratch is superficial, lightly rub the scratched area with a fine rubbing compound to remove loose paint and built-up wax. Rinse the area with clean water.

3   Apply touch-up paint to the scratch, using a small brush. Continue to apply thin layers of paint until the surface of the paint is level with the surrounding paint. Allow the new paint at least two weeks to harden, then blend it into the surrounding paint by rubbing with a very fine rubbing compound. Finally, apply a coat of wax to the scratch area.

4   For deeper scratches, such as those which extend all the way through the paint, but which have not penetrated into the fiberglass, clean and sand the area as above, then use a rubber or nylon applicator to coat the scratched area with a glaze-type filler. If required, the filler can be mixed with thinner to provide a very thin paste, which is ideal for filling very narrow scratches. Before the glaze filler in the scratch hardens, wrap a small piece of cotton around the tip of a finger. Dip the cloth in thinner and quickly wipe it along the surface of the scratch. This will ensure that the surface of the filler is slightly concave. The scratch can now be painted as described earlier.

5   Where slightly more repair is required, such as where the scratch extends into the surface of the fiberglass, remove the paint completely from around the scratch, sanding into the fiberglass to provide a clean surface for the repair material.

6   After sanding clean the area thoroughly with solvent, then use a tack rag to ensure that no fiberglass particles remain on the surface.

7   Use epoxy solder/adhesive, mixed with hardener in the ratio recommended by the manufacturer, and apply the mixture to the repair area with a putty knife or rubber squeegee **(see illustration)**.

8   Allow the filler to cure at room temperature for 8 to 10 hours before sanding. If heat lamps are available, they can be placed a minimum of 12-inches from the surface to speed curing. In most cases the repair will cure with heat lamps in between one and two hours.

9   After curing the repair area can be sanded to the proper contour with 150 grit sandpaper, then finish sanded with 220 grit wet-or-dry paper, making sure the original paint is carefully feather-edged into the repair area.

10   Temporary protection of the repaired

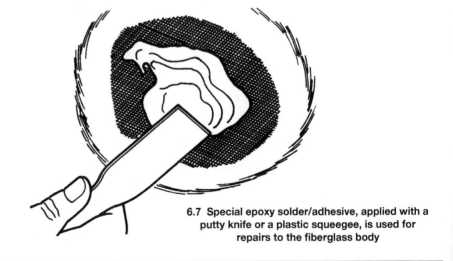

6.7 Special epoxy solder/adhesive, applied with a putty knife or a plastic squeegee, is used for repairs to the fiberglass body

9.3 Electrical connectors on each side provide power to the underhood service lights

9.5 Before removing the hood scribe the position of the hinge bolt washers to aid in alignment when reinstalling the hood

area can be accomplished with a spray-can primer. Apply several light coats, allowing each coat to dry thoroughly before applying the next. If desired, the final primer coat can be wet sanded with 220 or 320 grit wet-or-dry sandpaper.

11 Applying paint to the repaired area that will correctly match the original finish should be left to a professional paint shop. After painting, follow carefully the instructions of the paint shop regarding washing and waxing of the new finish.

7 Body damage - major repair

1 The multi-part frame used on the Corvette demands that the components be properly aligned to assure correct suspension location and body panel fit. In the event of a collision, it is essential that the frame and underbody be thoroughly checked and, if necessary, realigned by a body repair shop with the proper equipment.

2 Because each component contributes directly to the overall strength of the vehicle,

proper materials and techniques must be used during body repair service. Again, because of the specialized materials, machinery, tools and techniques required for this kind of work, it should be performed by a professional paint and body shop.

3 Because many of the exterior panels are separate and replaceable units, they should be replaced rather than repaired if major damage occurs. Some of these components can be found in a wrecking yard that specializes in used vehicle components, often at considerable savings over the cost of new parts.

8 Hinges and locks - maintenance

Once every 3000 miles, or every three months, the hinges, locks and latch assemblies on the doors and the hood and rear hatch should be given a few drops of light oil or lock lubricant. The door latch strikers should also be lubricated with a thin coat of grease to reduce wear and ensure free movement.

9 Hood - removal, installation and adjustment

Refer to illustrations 9.3, 9.5 and 9.7

Removal

1 Open the hood and disconnect the negative cable from the battery.

2 Place a fender cover in the opening between the hood and nose to protect the paint.

3 Disconnect the connectors for the underhood service lights **(see illustration)**.

4 Disconnect the headlight electrical connectors and remove the headlights from the hood (refer to Section 10).

5 Scribe the location of the hinge bolt washers for alignment on reassembly **(see illustration)**.

6 Remove the hood support rod by unbolting it from the hood.

7 Remove the hinge-to-frame bolts **(see illustration)**, including the lower bolt on the side of the frame rail.

8 With the aid of an assistant, lift off the hood. Note the number of shims used on each hinge.

Installation

9 With the aid of an assistant, put the hood in place, install any shims that were removed earlier and loosely install the hinge-to-frame bolts.

10 Align the hinges with the scribe marks made earlier and tighten the hinge bolts.

11 Install the hood support rod.

12 Referring to Section 10 if necessary, replace the headlights and reconnect the wires.

13 Reconnect the service lights.

14 Connect the ground cable to the battery.

Adjustment

15 The front of the hood is adjusted later-

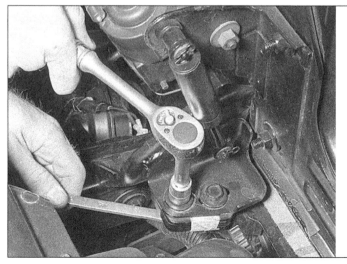

9.7 With an assistant supporting the hood, remove the hinge-to-frame bolts

10.3 Six nuts (arrows) attach the headlight assembly to the hood

10.4 Remove the carriage bolt and nut holding the headlight assembly to the hood hinge

ally by loosening the hood hinge nuts, moving the hood, then retightening the nuts.

16   Fore-and-aft adjustment of the hood is accomplished by loosening bolts that attach the hood hinges to the frame, repositioning the hood, then tightening the bolts.

17   Lateral adjustment of the rear of the hood (at the windshield) is accomplished by loosening either the lock and bolt assemblies or the latch assemblies, shifting the position of the hood as necessary, then retightening the lock and bolt or latch.

10   Headlight assembly - removal and installation

Refer to illustrations 10.3, 10.4 and 10.6

1   Scribe the nut position on the headlight assembly mounting brackets to assure proper realignment.

2   Disconnect the negative cable from the battery.

3   Remove the six nuts holding the headlight assembly to the hood **(see illustration)**.

4   Remove the carriage bolt and nut holding the headlight assembly to the hood hinge **(see illustration)**.

5   Disconnect the headlight and electric motor connectors.

6   Detach the headlight assembly from the hood **(see illustration)**.

7   Installation is the reverse of the removal procedure.

11   Front bumper and energy pad - removal and installation

Warning: *Later models are equipped with airbags. Always disable the airbag system before working in the vicinity of the steering column, instrument panel or impact sensors to avoid the possibility of accidental deployment of the airbag, which could cause personal injury* (see Chapter 12).

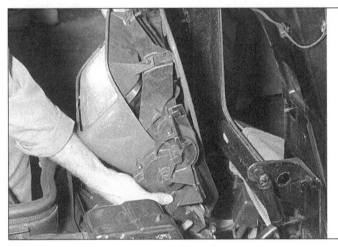

10.6 Disconnect the electrical connector and detach the headlight assembly from the hood

1   Open the hood to gain working clearance.

2   Disconnect the negative cable from the battery.

3   Remove the bulb holders from the side marker and parking lights (refer to Chapter 12 if necessary).

4   Remove the wheel well seal from the upper edge of the fascia and remove the four fascia screws on each side.

5   Remove the 12 screws attaching the top of the fascia to the impact bar reinforcement.

6   Remove the screws attaching the rear vertical edge of the fascia to the splash shield and the two inner screws attaching the fascia to the support.

7   Remove the four screws on each side attaching the fascia to the outer air dams.

8   Remove the screws attaching the fascia to the lower edge of the impact bar.

9   Remove the fascia-to-skid plate bolt.

10   Remove the fascia by lifting the upper edge away from the reinforcement and at the same time rotating the fascia down.

11   Disconnect the evaporator control system canister purge lines.

12   Remove the nuts attaching the impact bar assembly to the crossmember. Note the number and location of shims used for

reassembly.

13   Drill out the rivets retaining the energy absorber pad to the impact bar.

14   Installation is the reverse of the removal procedure, noting that new rivets must be used to attach the energy absorber pad to the impact bar.

12   Rear bumper and energy pad - removal and installation

1   Disconnect the negative cable from the battery.

2   Remove the gas filler door-to-body screws.

3   Remove the gas cap and the filler neck seal. Replace the gas cap.

4   Remove the three screws on each side attaching the fascia to the gas door panel. There is a fourth screw on each side, but it will be removed later from underneath.

5   Raise the vehicle and support it securely on jackstands.

6   Remove the screws attaching the bottom edge of the fascia to the energy pad.

7   Remove the screws attaching the rear wheel well panel to the lower side fascia.

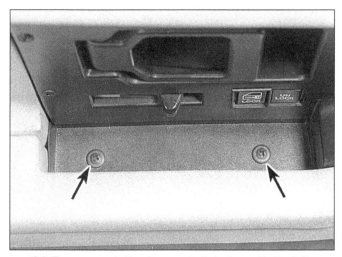

13.2 Two screws in the door control cluster well attach the armrest to the trim panel

13.5 The door panel contains several electrical devices that must be disconnected before the panel can be remove

8   Disconnect the side marker and back-up/turn signal lights.
9   Remove the two fascia-to-body attaching nuts on each side.
10   Remove the license plate light and disconnect the back-up lights.
11   Remove the two bolts attaching the license plate opening to the frame extension.
12   Pull the bottom edge of the fascia backwards to gain access to the upper fascia mounting nuts and remove the nuts.
13   Disconnect the taillights.
14   Pull the fascia back and remove the remaining top fascia-to-body attaching screws, including the gas door panel screws.
15   Remove the fascia and energy absorbing pad from the frame extension.
16   Installation is the reverse of the removal procedure.

13   Door trim panel - removal and installation

1984 through 1989 models

Refer to illustrations 13.2 and 13.5
1   Disconnect the cable from the negative terminal of the battery.
2   Remove the two screws attaching the armrest to the trim panel **(see illustration)**.
3   The armrest is also attached to the door trim panel with front and rear hooks that fit into slots in the trim panel. To release the hooks from the slots, grasp the armrest with both hands, push it towards the door and pull up on the side of the armrest facing the door.
4   Remove the four speaker grille mounting screws and the grille.
5   Remove the door trim panel screws from the bottom edge of the panel, the three remote control plate screws from the plate and the single screw in the rear edge of the door **(see illustration)**.
6   Lift up on the door trim panel to release the molding from the top edge of the door,

then slide the panel forward to clear the latch release handle and down to clear the lock rod knob.
7   When the door trim panel is free of the door, disconnect the electrical control connectors, such as the two wiper switch connectors, the interior light, the power door lock switch and the remote hatch release switch, depending upon which door trim panel (right or left) is being removed. Remove the panel.
8   Installation is the reverse of removal.

1990 and later models

Refer to illustration 13.12
9   Lower the window.
10   Remove the Handle bezel by gently prying down on the top corners approximately 1/2-inch from the edge. Rotate the lamp downward and out.
11   Pry the remote control lock knob from the rod.
12   Remove the bezel screws and bezel. It may be necessary to pull the inside door handle inward as you remove the bezel **(see illustration)**.
13   Remove the courtesy lamp and power door lock switch connectors.

14   Remove the screws from the armrest and remove the armrest.
15   If you are removing the right side trim panel, remove the window switch by prying it away from the trim panel and disconnecting the electrical connector. Remove the hidden trim panel screw that was covered by the switch.
16   Remove the trim panel screws and bolts.
17   If you are removing the left door trim panel, it will be necessary to remove the rear compartment release switch wiring by pulling the trim panel away from the bottom of the door and reaching inside the door and disconnecting the switch.
18   The trim panel can be removed by pulling the panel out at the bottom of the door and pushing up towards the top.
19   Installation is the reverse of removal.

14   Door striker - adjustment

Refer to illustration 14.1
1   The door striker is mounted in the body pillar, threaded into a floating cage. To adjust

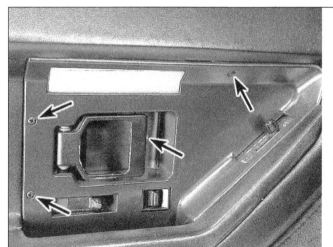

13.12 Typical door handle and bezel details. After all of the mounting screws (arrows) are removed, the assembly can be pried from the door and the wiring disconnected. Some models have screws concealed beneath the light lens and the door lock lever

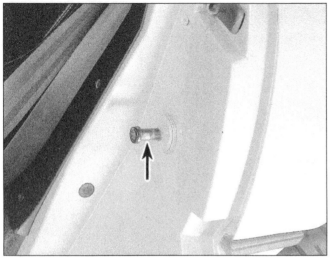

14.1 The door latch striker (arrow) threads into a caged nut in the door pillar

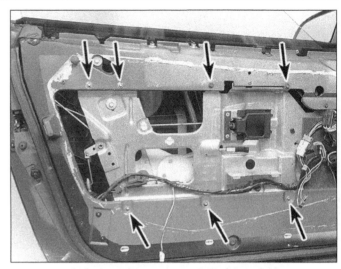

15.3 Perimeter screws attach the inner plate to the door assembly

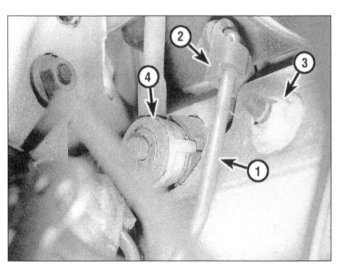

15.4 Inner door lock components

| 1 | Actuator rod | 3 | Door handle nut (one of two) |
| 2 | Clip | 4 | Lock cylinder |

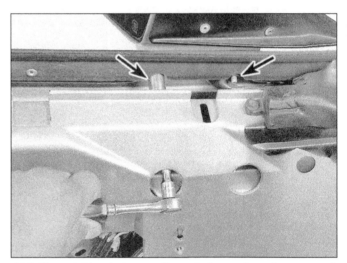

16.3 The outside mirror mounting nuts (arrows) can be accessed through the inner door panel

the striker in an in-out or up-down direction, loosen the striker bolt and shift the striker as required **(see illustration)**.
2   Fore-and-aft adjustment of the striker requires the addition or subtraction of spacers from the striker assembly and the use of modeling clay to determine the amount of adjustment necessary.
3   Apply modeling clay to the lock bolt opening.
4   Close the door only as far as necessary to make an impression in the modeling clay with the striker bolt.
5   Open the door and note the impression made by the striker head. It should be centered in the clay. Spacers to go between the striker and plate are available in graduated sizes from 5/64-inch to 5/16-inch. Use whatever combination of spacers is required to center the striker in the modeling clay, then remove the clay.

15   Door lock and remote controls - removal and installation

Refer to illustrations 15.3 and 15.4
1   Remove the door trim panel (Section 13).
2   Disconnect the wiring harness clip.
3   Remove the screws and bolts attaching the inner plate to the door assembly **(see illustration)**.
4   Disconnect the handle assembly actuator rod **(see illustration)**.
5   Remove the lock remote control rod by pushing the plate down and pulling out slightly to remove the clip.
6   Remove the screws attaching the lock mechanism to the door frame.
7   Disconnect the lock cylinder-to-lock mechanism rod.
8   Lower the lock assembly, pushing the

door lock handle rod back into the door. When the rod and lever are visible, disconnect the clip and rod. Remove the lock assembly from the door.
9   Remove the nuts and backing plate from the outside handle, disconnect the electrical connector and remove the handle.
10   Installation is the reverse of the removal procedure.

16   Outside mirror - removal and installation

Refer to illustration 16.3
1   Roll the window down and remove the door trim panel (refer to Section 13 if necessary).
2   Disconnect the negative cable from the battery.
3   Remove the clips holding the wiring to

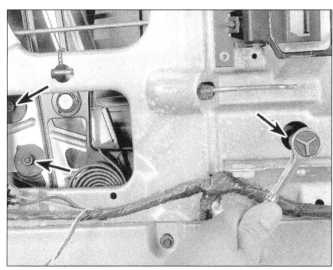

17.2 The fasteners that secure the side window glass to the regulator assembly can be removed through the access holes in the door

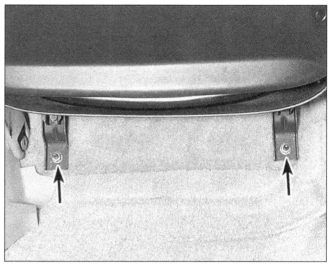

19.2 With the seat in the full forward position and the seat back tilted forward, the two rear mounting nuts are revealed

19.5 Removing the trim cover gives access to the front seat mounting nut

the door panel **(see illustration)**.
4   Remove the attaching nuts and detach the mirror assembly from the door.
5   Installation is the reverse of the removal procedure.

17   Window regulator - removal and installation

Refer to illustration 17.2
Note: *General Motors has replaced the earlier plastic track type regulator with a new style all metal regulator. The new regulators will come with a motor attached. The wire terminals in the passenger side window regulator motor connector may have to be switched in the connector for proper switch operation.*
1   Remove the door trim panel and plastic water deflector sheet (Section 13).
2   Disconnect the window glass from the regulator and rest the window glass on the door side guard beam **(see illustration)**.

3   Lower the regulator 3/4 of the way down.
4   Mark the location of the upper regulator mounting bolts and remove the upper and lower regulator mounting bolts.
5   Disconnect the window motor electrical connector.
6   Rotate the regulator counterclockwise and lift the regulator from the door.
7   If you are replacing the regulator, connect the window motor electrical connector and raise the regulator 1/3 of its travel.
8   The remainder of installation is the reverse of removal.

18   Side window - removal, installation and adjustment

1   Disconnect the negative cable from the battery.
2   Remove the armrest and speaker grille.
3   Remove the door trim panel as

described in Section 13.
4   Mark the location of the anti-rattle pads, loosen them and move them as necessary to provide clearance for window removal.
5   Move the regulator into a position where you can reach through the glass nut holes in the inner mounting plate.
6   Mark the position of the stabilizing guide, then remove the guide.
7   Remove the glass-to-regulator nuts and remove the glass through the top of the door.
8   Installation is the reverse of the removal procedure.

19   Seat - removal and installation

Refer to illustrations 19.2 and 19.5
1   Move the seat to the fully forward position and fold the seat back forward.
2   Remove the nuts from the support bracket rear studs **(see illustration)**.
3   Put the seat back upright and move the seat all the way to the rear.
4   Remove the seat adjuster trim covers by pulling out the retaining pins.
5   Remove the front support nuts **(see illustration)** and remove the seat assembly.
6   Installation is the reverse of the removal procedure.

20   Seat belt and shoulder harness - replacement

Refer to illustrations 20.4 and 20.5
1   Remove the seat as described in Section 19.
2   Remove the bolt retaining the buckle end of the lap belt.
3   Disconnect the belt wiring electrical connector.
4   Remove the seat belt retaining bolt at

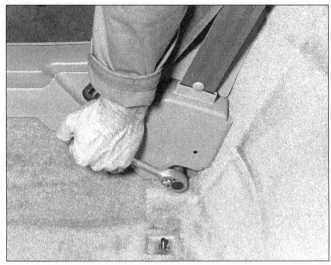

20.4 A large Torx bolt secures the seat belt and retainer assembly to the frame

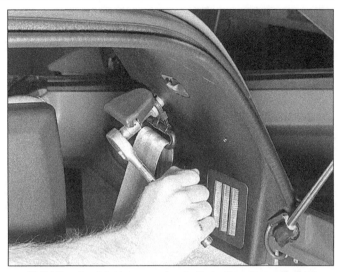

20.5 The shoulder harness ring bolts to the rear roof panel

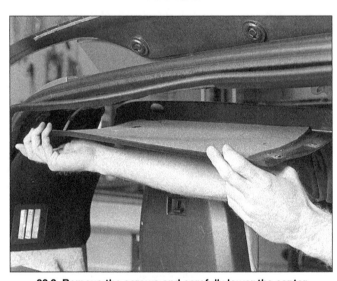

22.3 Remove the screws and carefully lower the center roof trim panel

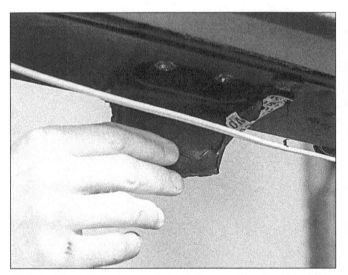

22.4 Peel back the noise control patches covering the hinge bolts

the floor side rail **(see illustration)**.

5   Remove the shoulder harness retaining bolt **(see illustration)** and remove the belt assembly.

6   Installation is the reverse of the removal procedure.

---

21   Roof panel - adjustment

1   Park the vehicle on a level surface and remove the roof panel.

2   Remove the center roof trim panel to gain access to the locator brackets and loosen the adjusting bolts.

3   Remove the visors, mounting covers and header molding to gain access to the front mounting brackets.

4   Loosen the roof panel locating pins.

5   Install the roof panel and snug the rear

bolts lightly. The top should fit level with the rear body panel with approximately a 3/16-inch gap between the roof panel and body. Adjust the locator brackets as necessary to obtain this clearance.

6   Install the front roof panel retaining bolts but do not tighten them at this time.

7   Tighten the rear roof panel retaining bolts securely.

8   Tighten the roof panel locating pins.

9   Tighten the front roof panel retaining bolts securely.

10   Check the front roof panel fit to the windshield molding. The panel should compress the weatherstrip evenly. Adjust the mounting bracket as necessary to obtain the proper fit.

11   Remove the roof panel from the vehicle and hold the T-nut to make sure it does not turn, then tighten the locator pins securely.

12   Install the visors, mounting covers and

header molding.

13   Install the center roof trim panel.

---

22   Rear glass and hydraulic struts - removal and installation

Refer to illustrations 22.3, 22.4, 22.5 and 22.8

1   Disconnect the negative cable from the battery.

2   Open the rear hatch and remove the roof panel.

3   Remove the interior center roof trim panel **(see illustration)**.

4   Peel back the noise control patch on both sides at the glass attachment access holes **(see illustration)**.

5   Unscrew the strut bodies from the glass attachments **(see illustration)**.

6   Disconnect the third (high) brake light electrical connector, if equipped.

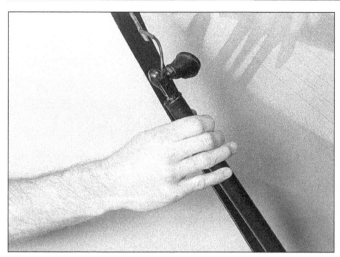

22.5 The hydraulic strut bodies can be unscrewed from the glass attachments

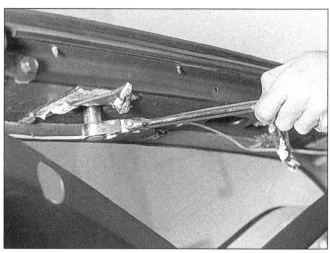

22.8 Remove the glass hinge nuts from underneath and lift off the hatch assembly

7   Close the glass gently. Do not engage the latch.
8   Remove the glass hinge nuts from underneath **(see illustration)** and lift off the glass and hinge.
9   Installation is the reverse of the removal procedure.

23 Battery panel - removal and installation

The battery panel removal is covered in Chapter 5 - refer to the battery removal and installation procedure (Section 2).

24 Gas filler door - removal and installation

The gas filler door is removed and installed as part of Fuel pump - removal and installation. Refer to Chapter 4.

Notes

Chapter 12
Chassis electrical system

Contents

1   General information

The electrical system is a 12-volt, negative ground type. Power for the lights and all electrical accessories is supplied by a lead/acid-type battery that is charged by the alternator.

This Chapter covers repair and service procedures for the various electrical components not associated with the engine. Information on the battery, alternator, distributor and starter motor can be found in Chapter 5.

It should be noted that whenever portions of the electrical system are serviced, the negative battery cable should be disconnected at the battery to prevent electrical shorts and/or fires.

Note: *Information concerning digital instrumentation and dash related accessories*
is not included in this manual. Problems involving these components should be referred to a dealer service department.

2   Electrical troubleshooting - general information

A typical electrical circuit consists of an electrical component, any switches, relays, motors, etc. related to that component and the wiring and connectors that connect the component to both the battery and the chassis. To aid in locating a problem in any electrical circuit, wiring diagrams are included at the end of this book.

Before tackling any troublesome electrical circuit, first study the appropriate diagrams to get a complete understanding of
what makes up that individual circuit. Trouble spots, for instance, can often be narrowed down by noting if other components related to that circuit are operating properly or not. If several components or circuits fail at one time, chances are the problem lies in the fuse or ground connection, as several circuits are often routed through the same fuse and ground connections.

Electrical problems often stem from simple causes, such as loose or corroded connections, a blown fuse or a melted fusible link. Always visually inspect the condition of the fuse, wires and connections in a problem circuit before troubleshooting it.

If testing instruments are going to be utilized, use the diagrams to plan ahead of time where you will make the necessary connections in order to accurately pinpoint the trouble spot.

3.1a The fuse panel is located in the right end of the dashboard - to open the access cover (not shown), rotate the plastic lock knob at the forward end of the cover to the vertical position and pry the cover out from between the weatherstripping of the door frame and the insulating ring around the heating/air conditioning duct

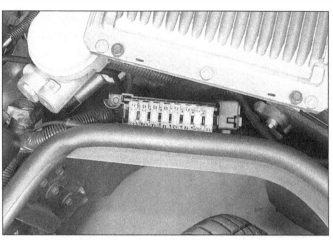

3.1b The auxiliary fuse boxes on 1994 and later models are located in the engine compartment

The basic tools needed for electrical troubleshooting include a circuit tester or voltmeter (a 12-volt bulb with a set of test leads can also be used), a continuity tester, which includes a bulb, battery and set of test leads, and a jumper wire, preferably with a circuit breaker incorporated, which can be used to bypass electrical components.

Voltage checks should be performed if a circuit is not functioning properly.

Connect one lead of a circuit tester to either the negative battery terminal or a known good ground. Connect the other lead to a connector in the circuit being tested, preferably nearest to the battery or fuse. If the bulb of the tester lights up, voltage is present, which means that the part of the circuit between the connector and the battery is problem free. Continue checking the rest of the circuit in the same fashion. When you reach a point at which no voltage is present, the problem lies between that point and the last test point with voltage. Most of the time the problem can be traced to a loose connection. **Note:** *Keep in mind that some circuits receive voltage only when the ignition key is in the Accessory or Run position.*

One method of finding shorts in a circuit is to remove the fuse and connect a test light or voltmeter in its place to the fuse terminals. There should be no voltage present in the circuit. Move the wiring harness from side-to-side while watching the test light. If the bulb goes on, there is a short to ground somewhere in that area, probably where the insulation has rubbed through. The same test can be performed on each component in the circuit, even a switch.

Perform a ground test to check whether a component is properly grounded. Disconnect the battery and connect one lead of a self-powered test light, known as a "continuity tester," to a known good ground. Connect the other lead to the wire or ground connection being tested. If the bulb goes on, the ground is good. If the bulb does not go on, the ground is not good.

A continuity check determines if there are any breaks in a circuit - if it is conducting electricity properly. With the circuit off (no power in the circuit), a self-powered continuity tester can be used to check the circuit. Connect the test leads to both ends of the circuit (or to the "power" end and a good ground), and if the test light comes on the circuit is passing current properly. If the light doesn't come on, there is a break somewhere in the circuit. The same procedure can be used to test a switch, by connecting the continuity tester to the power in and power out sides of the switch. With the switch turned on, the test light should come on.

When diagnosing for possible open circuits, it is often difficult to locate them by sight because oxidation or terminal misalignment are hidden by the connectors. Merely wiggling a connector on a sensor or in the wiring harness may correct the open circuit condition. Remember this if an open circuit is indicated when troubleshooting a circuit. Intermittent problems may also be caused by oxidized or loose connections.

Electrical troubleshooting is simple if you keep in mind that all electrical circuits are basically electricity running from the battery, through the wires, switches, relays, fuses and fusible links to each electrical component (light bulb, motor, etc.) and back to ground, from which it is passed back to the battery. Any electrical problem is an interruption in the flow of electricity to and from the battery.

3   Fuses - general information

Refer to illustrations 3.1a and 3.1b

The electrical circuits of the vehicle are protected by a combination of fuses, circuit breakers and fusible links. The fuse block is located behind the small plastic cover in the right end of the dashboard **(see illustration)**. 1991 and later models use an auxiliary fuse block, located near the multi-use relay center or in an under-hood fuse panel located near

the ECM **(see illustration)**.

Each fuse protects one or even several circuits. The protected circuit is identified on the fuse panel face right above each fuse. Miniaturized fuses are employed in the fuse block. These compact fuses, with blade terminal design, allow fingertip removal and replacement.

If an electrical component fails, always check the fuse first. A blown fuse, which is nothing more than a broken element, is easily identified through the clear plastic body. Visually inspect the element for evidence of damage. If a continuity check is called for, the blade terminal tips are exposed in the fuse body.

Be sure to replace blown fuses with the correct type. Fuses of different ratings are physically interchangeable, but only fuses of the proper rating should be used. Replacing a fuse with one of a higher or lower value than specified is not recommended. Each electrical circuit needs a specific amount of protection. The amperage value of each fuse is molded into the fuse body. **Caution:** *At no time should a fuse be bypassed with pieces of metal or foil. Serious damage to the electrical system could result.*

If the replacement fuse immediately fails, do not replace it again until the cause of the problem is isolated and corrected. In most cases, this will be a short circuit in the wiring caused by a broken or deteriorated wire.

4   Fusible links - general information

Some circuits are protected by fusible links. These links are used in circuits that are not ordinarily fused, such as the ignition circuit.

Although the fusible links appear to be a heavier gauge than the wire they are protecting, the appearance is due to the thick insulation. All fusible links are four wire gauges

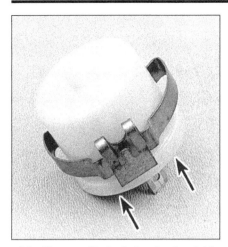

6.6a Because it's buried under the right side of the dash next to the ECM, an installed turn signal flasher on a 1989 and earlier model is impossible to photograph, but this is what it looks like out of the vehicle - to remove it, simply push up on the canister (arrows) to release the spring clip from the bracket

6.6b Turn signal flasher location (arrow) - 1990 and later models

smaller than the wire they are designed to protect. The location of the fusible links on your particular vehicle may be determined by referring to the wiring diagrams at the end of this Chapter.

Fusible links cannot be repaired, but a new link of the same size wire of the old link can be installed in its place. Never use a fusible link longer than 9-inches. The procedure is as follows:

a) *Disconnect the negative cable at the battery.*
b) *Disconnect the fusible link from the wiring harness.*
c) *Cut the damaged fusible link out of the wiring just behind the connector.*
d) *Strip the insulation back approximately 1/2-inch.*
e) *Position the connector on the new fusible link and crimp it into place.*
f) *Use rosin core solder at each end of the new link to obtain a good solder joint.*
g) *Use plenty of electrical tape around the soldered joint. No wires should be exposed.*
h) *Connect the negative battery cable. Test the circuit for proper operation.*

5 Circuit breakers - general information

Circuit breakers protect accessories such as power windows, power door locks, the rear window defogger and the headlights. Some circuit breakers are located in the fuse box. Refer to the wiring diagrams at the end of this book for the location of the circuit breakers located elsewhere on this vehicle.

Because a circuit breaker resets itself automatically, an electrical overload in a cir-

cuit breaker protected system will cause the circuit to fail momentarily, then come back on. If the circuit does not come back on, check it immediately. Once the condition is corrected, the circuit breaker will resume its normal function.

6 Turn signal and hazard flashers - check and replacement

Refer to illustrations 6.6a and 6.6b and 6.9

Turn signal flasher

1   The turn signal flasher is a small canister shaped unit that flashes the turn signals. It's located just to the left of the ECM inside the right side of the dashboard on 1989 and earlier models, or under the instrument panel to the left of the steering column on 1990 and later models.

Check

2   When the flasher unit is functioning properly, an audible click can be heard during its operation. If the turn signals fail on one side or the other and the flasher unit does not make its characteristic clicking sound, a faulty turn signal bulb is indicated.

3   If both turn signals fail to blink, the problem many be due to a blown fuse, a faulty flasher unit, a broken switch or a loose or open connection. If a quick check of the fuse box indicates that the turn signal fuse has blown, check the wiring for a short before installing a new fuse.

Replacement

4   On 1984 through 1989 models, remove the three hex head screws from the dash panel under the right side of the dashboard, then pull down the panel far enough to disconnect the under-dash courtesy light electrical connector from the backside of the light. Remove the under-dash panel. On later models, remove the screws retaining the ALCL connector from the left under-dash panel. Remove the screws holding the trim panel to the knee bolster and the bolt from the trim panel to the steering column. Disconnect the electrical connector from the courtesy lamp and remove the trim panel.
5   Locate the blue and purple wires leading to the flasher unit.
6   Push up on the spring clip that encircles the flasher unit **(see illustrations)** to disconnect the flasher from the dashboard support bracket to which it is attached. Pull the flasher down far enough to disconnect the electrical connector and remove the flasher.
7   Make sure that the replacement unit is identical to the unit it replaces. Compare the old one to the new one before installing it.
8   Installation is the reverse of removal.

Hazard flasher

9   The hazard flasher, a small canister shaped unit located in the multi-use relay center **(see illustration)** behind the driver information function select switch panel, flashes all four turn signals simultaneously when activated.

Check

10   The hazard flasher is checked in a fashion similar to the turn signal flasher (see Steps 2 and 3).

Replacement

11   On 1989 and earlier models:

6.9 The hazard flasher (arrow) is inside the multi-use relay center which resides behind the driver information function select switches and the telltale display screen (removed for this photo); it's mounted - and removed - the same way as the turn signal flasher unit

7.5 The key warning buzzer switch is easily extracted from the slot underneath the key lock cylinder with a pair of needle-nose pliers - be sure to note the orientation of the switch before removing it to facilitate reinstallation

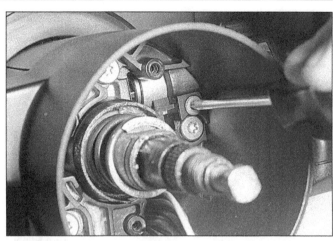

7.6 With a Torx screwdriver of the correct size, remove this key lock cylinder retaining screw before pulling the cylinder from the steering column

a) *Remove the four screws from the left end and the two screws from the right end of the instrument cluster bezel and remove the bezel.*

b) *Remove the two screws from the left side and the two screws from the right side of the accessory console trim plate and remove the plate.*

c) *Remove the three screws from the function select switch panel and the display screen, then remove the function select switch panel/display screen assembly.*

d) *Pry the spring clip surrounding the hazard flasher loose from its bracket and pull the flasher unit far enough out from the multi-use relay center to disconnect the electrical connector. Remove the hazard flasher.*

e) *Make sure that the replacement unit is identical to the one it replaces. Compare the old one to the new one before installing it.*

12   On 1990 and later models:

a) *Remove the screws and bolts from the right side lower trim panel.*

7.10 This tab (arrow) must be retracted prior to installation of the key warning buzzer switch

b) *Remove the fuse box cover and instrument panel side cover.*

c) *Remove the nuts attaching the lower trim panel to the instrument panel.*

d) *Remove the instrument panel air outlet.*

e) *Pull down the trim panel far enough to disconnect the courtesy light.*

f) *Remove the trim panel by guiding it over the floor air outlet.*

g) *Unclip the flasher from its holder located on the multi-use bracket, then disconnect the flasher electrical connector.*

13   Installation is the reverse of removal.

---

7   Ignition key lock cylinder - replacement

Refer to illustrations 7.5, 7.6 and 7.10
Warning: *Later models are equipped with airbags. Always disable the airbag system before working in the vicinity of the steering column, instrument panel or impact sensors to avoid the possibility of accidental deployment of the airbag, which could cause personal injury* (see Section 30).

1   Disconnect the cable from the negative terminal of the battery. If equipped with airbags, disable the airbag system (see Section 30).

2   Remove the steering wheel (Chapter 10).

3   Remove the turn signal switch from the steering column (Chapter 10). Do not disconnect the switch - simply push it aside so that it's out of the way while you replace the key lock cylinder.

4   Turn the key lock cylinder to the On position.

5   Remove the key warning buzzer switch from the steering column with a pair of needle-nose pliers **(see illustration)**.

6   Remove the key lock cylinder retaining screw with a Torx screwdriver **(see illustration)**.

7   Pull the key lock cylinder from the steering column. If equipped with the PASS key security system, disconnect the harness wiring connector and feed the harness through the column.

8   Before installing the new lock cylinder, make sure that it is in the On position or it will not fit into the cavity in the steering column.

9   Tighten the key lock cylinder retaining screw securely.

10   Install the key warning buzzer switch. The tab on the bottom of the key lock cylinder must be retracted **(see illustration)** to install the switch.

11   Installation is otherwise the reverse of removal.

---

8   Ignition switch - replacement

Warning: *Later models are equipped with airbags. Always disable the airbag system before working in the vicinity of the steering column, instrument panel or impact sensors to avoid the possibility of accidental deployment of the airbag, which could cause personal injury* (see Section 30).

1   The ignition switch is located on the top of the steering column, just above the dimmer switch.

2   Disconnect the cable from the negative terminal of the battery. If equipped with airbags, disable the airbag system (see Section 30).

3   Remove all three screws from the under-dash panel and pull the panel down far enough to disconnect the under-dash courtesy light electrical connector. Remove the panel.

4   Disconnect the two electrical connectors from the right side of the switch.

5   Remove the dimmer switch (see Section 10).

6   Remove the stud from the front of the ignition switch.

7   Note how the ignition switch actuator

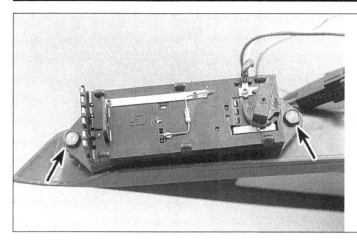

9.4 The windshield wiper switch is mounted to the left door remote control plate with these two screws (arrows)

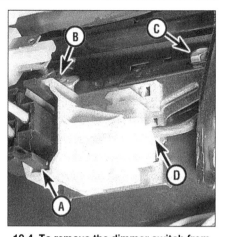

10.4 To remove the dimmer switch from the steering column, disconnect the electrical connector (A) and remove the nut (B), the screw (C) and the rod (D) - to disconnect the switch from the rod, simply slide the switch forward until the two are separated (be sure to note how the rod inserts into the sliding part of the switch before disconnecting them)

rod connects to the sliding part of the switch. Remove the switch.

8   Install the new ignition switch and reconnect the electrical connectors but don't tighten the stud.

9   Turn the ignition key to On and slide the switch (it has slotted mounting holes) until the dashboard lights come on. Then tighten the stud and turn the ignition key to Start. The engine should start. If it doesn't, loosen the stud and slide the switch forward or backward until the engine does start. Tighten the stud securely.

10   Install and adjust the dimmer switch (Section 10).

11   Installation is otherwise the reverse of the removal procedure.

9   Windshield wiper switch (1989 and earlier models) - replacement

Refer to illustration 9.4

Note: *On 1990 and later models the windshield wiper switch is located in the steering column. Replacement of the switch requires disassembly of the steering column and is best left to a dealer service department or other qualified repair facility.*

1   The wiper switch is located in the left door trim panel.

2   Disconnect the cable from the negative terminal of the battery.

3   Remove the armrest and door trim panel (see Chapter 11).

4   Remove the two hex head screws attaching the wiper switch to the remote control plate **(see illustration)**.

5   Installation is the reverse of removal.

10   Dimmer switch - replacement

Refer to illustration 10.4

Warning: *Later models are equipped with airbags. Always disable the airbag system before working in the vicinity of the steering column, instrument panel or impact sensors to avoid the possibility of accidental deploy-*

ment of the airbag, which could cause personal injury (see Section 30).

1   The dimmer switch is located on the steering column, just below the ignition switch.

2   Disconnect the cable from the negative terminal of the battery. If equipped with airbags, disable the airbag system (see Section 30).

3   Remove all three screws from the under-dash panel and pull the panel down far enough to disconnect the under-dash courtesy light electrical connector. Remove the panel.

4   Unplug the electrical connector from the dimmer switch **(see illustration)**.

5   Remove the screw at the rear and the nut at the front of the dimmer switch.

6   Note how the rod connects to the dimmer switch, then slide the assembly forward until it is detached from the rod.

7   Remove the dimmer switch.

8   Install the new dimmer switch but don't tighten the screw and nut. Reconnect the electrical connector to the switch.

9   Reconnect the cable to the negative terminal of the battery.

10   Turn the lights on and activate the dimmer stalk. The lights should alternate between dim and bright. If they don't, slide the switch up the column (toward the steering wheel) until the dimmer stalk activates the switch properly. Tighten the screw and nut and recheck the actuation of the switch to make sure that the dimmer switch mounting location is correct.

11   Installation is otherwise the reverse of the removal procedure.

11   Headlight switch - replacement

Warning: *Some of the models covered by this manual are equipped with Supplemental Inflatable Restraints (SIR), more commonly known as airbags. Always disable the airbag system before working in the vicinity of any airbag system components to avoid the possibility of accidental deployment of the airbag, which could cause personal injury (see Section 30).*

1984 through 1989 models

Refer to illustrations 11.3, 11.7 and 11.9

1   The headlight switch is located in the upper left hand corner of the dashboard, just to the left of the instrument cluster.

2   Disconnect the cable from the negative terminal of the battery.

3   Remove the headlight switch knob **(see illustration)**.

4   Remove the four screws from the left end and the two screws from the right end of the instrument cluster bezel and remove the bezel (see Section 17 for more information).

5   Remove the two screws from the left side and the three screws from the right side of the accessory console trim plate and remove the plate.

6   Remove the two screws from the top center and the five screws from the under-

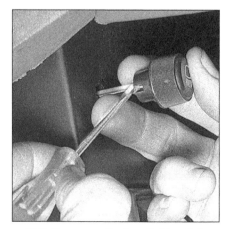

11.3 To remove the headlight switch knob, use a small screwdriver to press the friction clip to the rear until it no longer grips the rod and slide the knob off

11.7 This threaded locking barrel attaches the headlight switch to the backside of the instrument panel carrier assembly - it can be removed by turning it with a couple of screwdrivers

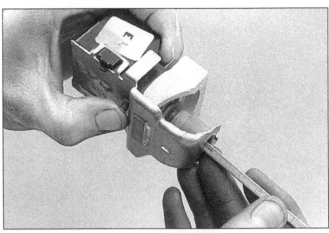

11.9 Before discarding the old headlight switch assembly, depress the button (arrow) on the switch and pull the rod out - install the rod in the new switch

side of the trailing edge of the dashboard upper trim pad and remove the pad.

7   Unscrew the threaded locking barrel **(see illustration)**.

8   Pull the switch from the instrument panel far enough to disconnect the wire harness. Remove the headlight switch assembly.

9   Push the release button on the headlight switch assembly and remove the switch rod **(see illustration)**.

10   Install the rod in the new switch. Installation is the reverse of removal.

1990 and later models

11   Disconnect the cable from the negative terminal of the battery. If equipped with airbags, disable the airbag system (see Section 30).

12   Remove the instrument cluster trim screws and reposition the instrument cluster trim plate to one side allowing access to the right side headlight switch screws.

13   Remove the right headlight switch screws.

14   With the driver side door open remove the switch trim plate.

15   Disconnect the headlight switch and remove the switch.

16   Installation is the reverse of removal.

12   Headlights - removal and installation

1   Turn the headlight switch on to rotate the headlights up into position.

2   Disconnect the cable from the negative terminal of the battery.

3   Remove all four headlight bezel retaining screws (there are two in the front and one on each side). Remove the bezel carefully to avoid scratching the painted surface.

4   Disconnect the electrical connector from the rear of the headlight.

5   On 1984 through 1990 models, using a hooked tool, similar to a cotter pin remover, pull the retaining spring to one side and dis-

connect it. On later models, remove the four retaining ring screws.

6   Rotate the right headlight clockwise to release it from the aiming pins; rotate the left headlight counterclockwise to release it. Remove the headlight assembly.

7   Remove the four retaining ring mounting screws and the retaining ring.

8   Installation is the reverse of removal.

13   Headlights - adjustment

Refer to illustration 13.3

Note: *It is important that the headlights be aimed correctly. If adjusted incorrectly they could blind the driver of an oncoming car and cause a serious accident or seriously reduce your ability to see the road. The headlights*

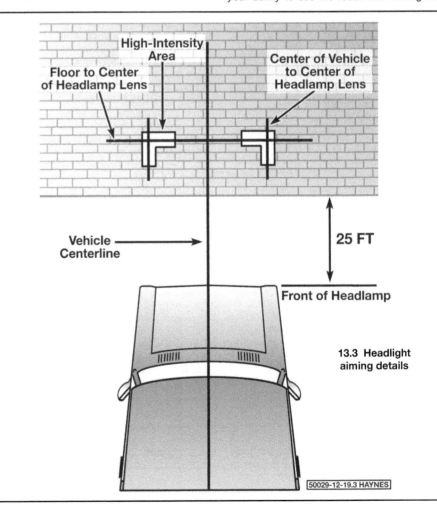

13.3 Headlight aiming details

50029-12-19.3 HAYNES

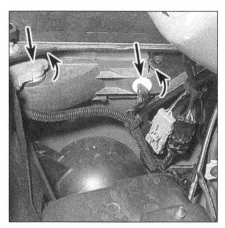

14.5 To replace a front side marker running light bulb, rotate the bulb holder (A) counterclockwise, lift it up and remove the bulb from the holder - to replace a turning light bulb (B), rotate the bulb holder counterclockwise and pull the holder out, then remove the bulb from the holder

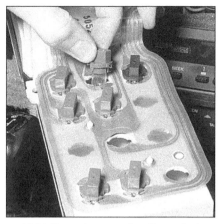

14.11 To replace any function select switch panel or telltale screen bulb, remove the appropriate bulb holder from the backside of the panel by rotating it counterclockwise, pulling it out of the panel and removing the bulb from the holder

should be checked for proper aim every 12 months and any time a new sealed beam headlight is installed or front end body work is performed. It should be emphasized that the following procedure is only an interim step, which will provide temporary adjustment until the headlights can be adjusted by a properly equipped shop.

1   Headlights have two spring loaded adjusting screws, one on the top controlling up-and-down movement and one on the side controlling left-and-right movement.
2   There are several methods of adjusting the headlights. The simplest method requires a blank wall 25-feet in front of the vehicle and a level floor.
3   Park the vehicle on a known level floor 25-feet from the wall **(see illustration)**.
4   Position masking tape vertically on the wall in reference to the vehicle centerline and the centerlines of both headlights.
5   Position a horizontal tape line in reference to the centerline of all the headlights. **Note:** *It may be easier to position the tape on the wall with the vehicle parked only a few inches away.*
6   Adjustment should be made with the vehicle sitting level, the gas tank half-full and no unusually heavy load in the vehicle.
7   Starting with the low beam adjustment, position the high intensity zone so it is two-inches below the horizontal line and two-inches to the right of the headlight vertical line. Adjustment is made by turning the top adjusting screw clockwise to raise the beam and counterclockwise to lower the beam. The adjusting screw on the side should be used in the same manner to move the beam left or right.
8   With the high beams on, the high intensity zone should be vertically centered with the exact center just below the horizontal line. **Note:** *It may not be possible to position the headlight aim exactly for both high and low*

beams. *If a compromise must be used, keep in mind that the low beams are the most used and have the greatest effect on driver safety.*
9   Have the headlights adjusted by a dealer service department or service station at the earliest opportunity.

---

14   Bulb replacement

Refer to illustrations 14.5, 14.11, 14.15 and 14.19

Caution: *Always disconnect the cable from the negative terminal of the battery prior to replacement of any of the following bulbs.*

Fog lights and front park/turn signal lights

1   The fog lights and park/turn signal lights are located in the front bumper. The fog lights are the inboard lights, the park/turn signal lights, the outboard ones.
2   To remove a fog light or park/turn signal light bulb, lift the hood, reach behind the appropriate light assembly and remove the twist-lock bulb holder by turning it counterclockwise and extracting it from the housing.
3   Remove the bulb. **Caution:** *The (inboard) fog lights are quartz halogen bulbs.*

Avoid touching them - even natural skin oils can cause a halogen bulb to explode when suddenly heated to operating temperature.
4   *Installation is the reverse of removal.*

Front side marker lights

5   The front side marker lights are located in the front fender just forward of the front wheels **(see illustration)**. They consist of a front turning light (lights up only when the turn signal switch is activated) and an (amber) rear running light.
6   To remove a turning light, rotate the bulb holder about 1/8-turn counterclockwise and remove the holder. Pull the bulb straight out. Installation is the reverse of removal.
7   To remove a running light, depress the tab on the bulb holder, rotate the holder counterclockwise and lift up. Remove the bulb. Installation is the reverse of removal.

Engine compartment lights

8   There are two engine compartment lights located on the underside of the hood. They come on automatically when the hood is opened.
9   To remove either light, unscrew the lens cover and remove the bulb. Installation is the reverse of removal.

Function select switch panel lights (1984 through 1989 models only)

10   Removal of any of the bulbs that illuminate the function select switch panel (the part of the driver information center located just above the radio) requires removal of the panel itself first (see Section 17).
11   Once the panel is removed, turn the holder of the bulb you are replacing 1/4-turn and remove it from the backside of the panel **(see illustration)**. Remove the bulb from the holder. Installation is the reverse of removal.

Interior courtesy lights

12   The interior courtesy lights are located in the remote control plates of either door.
13   To remove either bulb on a 1984 through 1989 model, remove the armrest and the remote control plate from the door trim panel (Chapter 11).
14   Pop out the interior courtesy light housing from the remote control plate.
15   Pry loose the front end of the lens from the housing with a small screwdriver **(see**

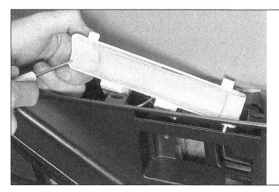

14.15 To replace an interior courtesy light bulb, remove the remote control plate from the door trim panel, pop the courtesy light housing out of the remote control plate, pry the front end of the plastic lens loose from the housing and remove the bulb

14.19 To replace a cargo light bulb, pry the rear edge of the cover loose with a small screwdriver and remove the bulb from the housing

illustration). Remove the bulb.

16   To remove either bulb, on a 1990 and later model, using a small flat bladed screwdriver at the top of the lens, pry down and out until the tabs on the lens are clear of the housing. Rotate the lens away from the door trim panel and remove the bulb from its socket.

17   Installation is the reverse of removal.

Interior cargo light

18   The interior cargo light bulbs are located in the pillar behind either door.

19   To remove a cargo light bulb, pry the rear edge of the cover loose with a small screwdriver **(see illustration)**.

20   Remove the bulb.

21   Installation is the reverse of removal.

Rear side marker lights

22   The rear side marker lights are located in the rear fender behind the rear wheels. They consist of a front (clear) back-up light and a rear (red) running light. Either bulb is replaced by reaching up from under the vehicle between the fuel tank and the inside of the bumper fascia.

23   To remove a back-up light bulb, depress the locking tab on the bulb holder, turn the holder counterclockwise and lift up on it. Remove the bulb from the holder. Installation is the reverse of removal.

24   To remove a running light bulb, simply rotate the bulb holder counterclockwise and remove it from the housing. Remove the bulb. Installation is the reverse of removal.

Taillights

25   The four taillight assemblies are located in the rear bumper fascia. These lights serve as rear running lights, brake lights and turn signal lights.

26   To replace an inboard taillight bulb, remove the license plate and reach through the hole in the bumper fascia. **Note:** *If your arm is too big, also remove the spare tire*

cover support bracket bolts and the bracket. Turn the bulb holder counterclockwise and remove it from the housing. Remove the bulb from the holder. Installation is the reverse of removal.

27   To replace an outboard bulb, reach up between the muffler and the forward part of the bumper fascia. Removal of the bulb holder is similar to that for the inboard bulbs. Installation is the reverse of removal.

Back-up lights/license plate light

28   The back-up lights are located on either side of the license plate. The license plate light is located above the license plate.

29   To remove either back-up light bulb or the license plate light, remove the license plate.

30   To remove a back-up light bulb, depress the locking tab on the bulb holder, rotate it counterclockwise and extract the holder. Remove the bulb. Installation is the reverse of removal.

31   To remove the license plate light, reach up through the hole in the bumper fascia, rotate the bulb holder counterclockwise and remove it from the housing. Remove the bulb. Installation is the reverse of removal.

High level brake light

Coupe

32   Open the rear compartment and remove the two screws securing the lamp assembly.

33   Partially lower the headliner to expose the wiring retainer and connector.

34   Disconnect the wiring connector and remove the wiring retainer nuts.

35   Replace the bulb. Installation is the reverse of removal.

Convertible

36   Remove the fuel tank filler door and filler cap.

37   Remove the fuel tank filler housing and drain hose.

38   Remove the license plate.

39   Remove the two stamped nuts through the upper body panel.

40   Lower the lamp assembly through the license plate opening.

41   Remove the screws from the back of the lamp assembly and replace the bulb. Installation is the reverse of removal.

---

15   Radio and speakers - removal and installation

Warning: *Some of the models covered by this manual are equipped with Supplemental Inflatable Restraints (SIR), more commonly known as airbags. Always disable the airbag system before working in the vicinity of any airbag system components to avoid the possibility of accidental deployment of the airbag, which could cause personal injury (see Section 30).*

Caution: *All radios in this vehicle are the bridge audio type, using two wires to each speaker. It is very important when changing speakers or performing any radio work to avoid pinching any wires, as this will cause damage to the output circuit in the radio.*

1984 through 1989 models

Radio

1   Disconnect the cable from the negative terminal of the battery.

2   Remove the four screws from the left end and the two screws from the right end of the instrument cluster bezel and remove the bezel (see Section 17 for more information).

3   Remove the two screws from the left side and the two screws from the right side of the accessory console trim plate and remove the plate.

4   Remove the four radio mounting screws and pull the radio far enough from the dash to disconnect the five electrical connectors and the antenna lead on the backside.

5   Remove the radio.

6   Installation is the reverse of removal.

Speakers

Note: *There are four speakers. One is mounted in each door and one is mounted in each side of the rear compartment. But they are all installed exactly the same, so the following procedure applies to all four speakers.*

7   Disconnect the cable from the negative terminal of the battery.

8   Remove the four grille retaining screws and the grille. Remove the three speaker retaining screws and pull the speaker far enough from the housing to disconnect the wiring harness electrical connector. Remove the speaker.

9   Installation is the reverse of removal.

1990 and later models

Refer to illustrations 15.11 and 15.17

Radio control

10   Disconnect the cable from the negative terminal of the battery. If equipped with airbags, disable the airbag system (see Section 30).

11   Remove the console trim plate **(see illustration)**.

12   Remove the two screws retaining the center air outlet and remove the outlet.

13   Remove the screws from the accessory trim plate to instrument panel and remove the trim plate.

14   Remove the right side console screws to the instrument panel.

15   Remove the screw on the right side of the driver information center to the upper trim panel.

16   Relocate the trim panel and pad to access the right hand radio control screw and remove the screw.

17   Remove the two screws holding the radio control to the instrument panel then slide the radio control out far enough to disconnect the electrical connector then remove

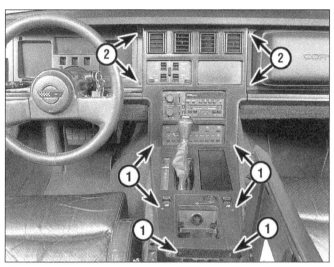

15.11 Center console trim plate and instrument panel trim plate screw locations - typical 1988 vehicle

1   *Console screws*
2   *Instrument panel screws*

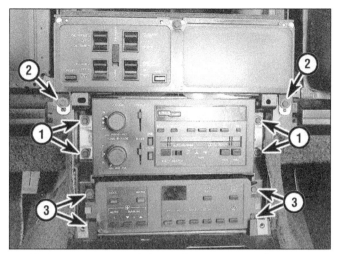

15.17 Instrument panel electronic module mounting screw locations - typical 1988 vehicle

1   *Sound system mounting screws*
2   *Information panel mounting screws*
3   *Climate control panel mounting screws*

the radio control **(see illustration)**.
18   Installation is the reverse of removal.

Radio receiver

19   The radio receiver is located in the right rear cargo compartment and is held in place with Velcro fasteners.
20   Open the cargo hatch and remove the tray.
21   Firmly grasp the receiver and with a pulling-rocking motion separate the receiver from the Velcro fastener.
22   Disconnect the electrical connections and the antenna lead.
23   installation is the reverse of removal.

Speakers

24   The speaker removal and installation is the same as prior model years.

16   Radio antenna - removal and installation

Refer to illustrations 16.10 and 16.11
Note: *If the vehicle is equipped with a convertible top, release the convertible top storage compartment cover before proceeding.*
1   Disconnect the cable from the negative terminal of the battery.
2   Open the rear hatch and remove the trim panel along the rear edge of the cargo area.
3   Disconnect the power antenna motor electrical lead from the relay.
4   Remove the trim panel along the left edge of the cargo area.
5   Remove the cargo shade track.
6   Remove the speaker grille, if equipped. If your vehicle is not equipped with the optional Bose rear speakers, remove the trim panel that covers the area normally occupied by the speaker.

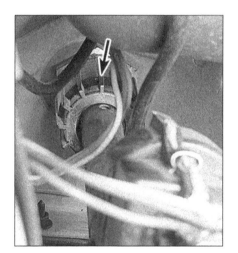

16.10 To remove the antenna, first loosen this threaded locking knob (arrow) . . .

7   Peel back the carpet along the left side of the cargo area until the antenna lead connector is exposed. Disconnect it.
8   Raise the vehicle and secure it on jackstands.
9   Remove the left rear wheel well inner fender panel screws (see Chapter 11) and prop the lower left portion of the bumper fascia open to gain access to the antenna locking knob, which is located at the top of the tube, just under the bodywork.
10   Reach up from below and unscrew the locking knob **(see illustration)** that secures the antenna tube to the body.
11   Remove the two antenna support bracket mounting nuts **(see illustration)**.
12   Pull the antenna lead and power antenna lead through the grommet and remove the antenna.
13   Installation is the reverse of removal.

16.11 . . . then remove these two antenna support bracket mounting nuts and thread the antenna and motor leads through the grommet in the side of the cargo area

17   Instrument panel - removal and installation

Warning: *Some of the models covered by this manual are equipped with Supplemental Inflatable Restraints (SIR), more commonly known as airbags. Always disable the airbag system before working in the vicinity of any airbag system components to avoid the possibility of accidental deployment of the airbag, which could cause personal injury (see Section 30).*
Note: *Because of its solid state design, the instrument cluster assembly is not repairable by nor can the home mechanic disassemble it. It is also extremely expensive to replace. If a malfunction occurs in the instrument cluster assembly, it is recommended that it be taken to a dealer for repairs.*

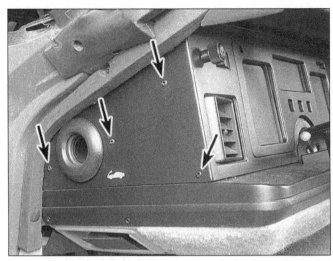

17.2a To remove the instrument cluster bezel, take out these four screws (arrows) from the left end . . .

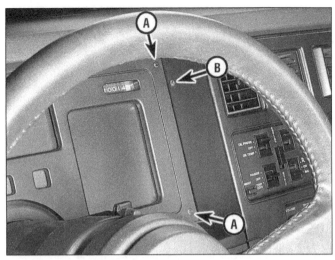

17.2b . . . and these two (A) from the right end - if you are also removing the instrument panel accessory trim plate, remove screw (B) as well

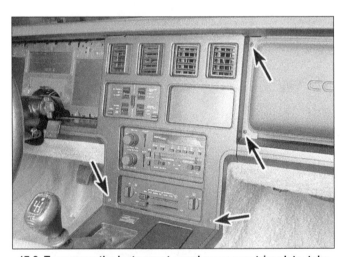

17.9 To remove the instrument panel accessory trim plate, take out five screws - the four screws shown (arrows) and a fifth shown in 17.2b

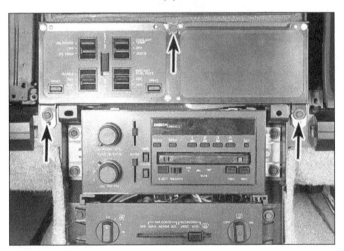

17.10 To remove the driver information function select switch panel and the telltale screen, take out these three screws (arrows)

1984 through 1989 models

Refer to illustrations 17.2a, 17.2b, 17.9 and 17.10

Main instrument cluster

1   Disconnect the cable from the negative terminal of the battery.
2   Remove the four screws from the left end and the two screws from the right end of the instrument cluster bezel and remove the bezel **(see illustrations)**.
3   Remove the two screws from the top center and the five screws from the underside of the trailing edge of the dashboard upper trim pad and remove the pad.
4   Remove the two upper left and right hand corner screws and the two lower center screws on either side of the steering column from the instrument cluster assembly.
5   Pull the left end of the instrument cluster assembly out from the dash far enough to disconnect the two electrical connectors from the right end of the cluster.

6   Remove the instrument cluster assembly.
7   Installation is the reverse of removal.

Driver information function select switch panel and screen

8   Complete Steps 1 through 4 above.
9   Remove the two screws from the left side and the two screws from the right side of the accessory trim plate and remove the plate **(see illustration)**.
10   Remove the three screws from the driver information function select switch panel and screen **(see illustration)**.
11   Pull the panel and screen far enough from the dashboard to disconnect the electrical connectors and remove the assembly.
12   Installation is the reverse of removal.

1990 and later models

Refer to illustration 17.17

13   Disconnect the cable from the negative terminal of the battery. If equipped with

airbags, disable the airbag system (see Section 30).

Main instrument cluster

14   Remove the knee bolster and trim panel.
15   Remove the steering column bolts and lower the column.
16   Remove the four cluster bezel screws and remove the bezel.
17   Remove the cluster mounting screws and bolts, then slide the cluster out far enough to disconnect the electrical connectors **(see illustration)**.
18   Remove the cluster, taking care not to scratch the lens.
19   Installation is the reverse of removal.

Driver Information Center (DIC)

20   Remove the console trim plate and the accessory trim plate (see Section 15).
21   Remove the Driver Information Center mounting screws and remove the DIC far enough to disconnect the electrical connectors.

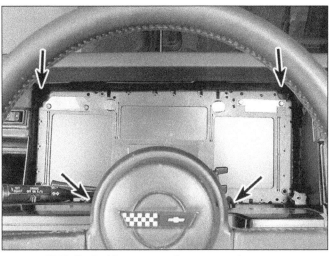

17.17 Typical instrument cluster mounting screw locations (arrows)

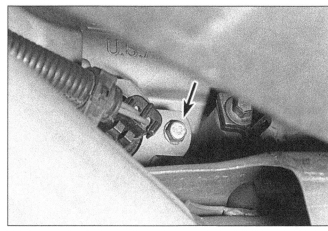

18.4 The speedometer signal generator, located on the left rear end of the transmission, is easily removed - simply disconnect the electrical connector, release the hold down bolt (arrow) and clamp and pull the generator from the transmission housing

22   Disconnect the harness from the DIC and remove the DIC.
23   Installation is the reverse of removal.

18   Speedometer signal generator - replacement

Refer to illustration 18.4
1   Disconnect the cable from the negative terminal of the battery.
2   Raise the vehicle and place it securely on jackstands.
3   The speedometer signal generator is located on the left rear end of the transmission, just in front of the catalytic converter. **Warning:** *Make sure that the catalytic converter has cooled before replacing the speedometer signal generator.*
4   Disconnect the electrical connector from the speedometer signal generator **(see illustration)**.
5   Remove the hold down bolt and clamp from the speedometer signal generator.
6   Remove the signal generator.
7   Installation is the reverse of removal.

19   Back-up/Neutral start switch (automatic transmission) - removal and installation

Refer to illustration 19.3
Warning: *Later models are equipped with airbags. Always disable the airbag system before working in the vicinity of the steering column, instrument panel or impact sensors to avoid the possibility of accidental deployment of the airbag, which could cause personal injury (see Section 30).*
1   Disconnect the cable from the negative battery terminal. If equipped with airbags, disable the airbag system (see Section 30).
2   Remove the floor console cover. Disconnect the electrical connectors from the

back-up/Neutral start switch.
3   Place the shift lever in the Neutral notch of the detent plate **(see illustration)**.
4   Remove the two screws and detach the switch.
5   Make sure that the shift lever is in the Neutral notch of the detent plate before installing the replacement switch.
6   Place the new switch in position on the shift lever. Make sure that the pin on the shift lever is in the slot of the switch and secure the switch with two screws. Tighten the screws securely.
7   Move the shift lever out of the Neutral position to shear the plastic pin.
8   Connect the electrical connectors to the switch. Apply the parking brake and start the vehicle. Check to make sure that the vehicle starts only in the Park or Neutral positions. Check that the back-up lights are on when the shift lever is in the Reverse position. Check that the seat belt warning system operates.
9   Turn off the ignition and install the floor console cover.

20   Rear window defogger - check and repair

Refer to illustration 20.11
1   The rear window is equipped with a number of horizontal elements baked into the glass surface during the glass forming operation.
2   Small breaks in the element can be successfully repaired without removing the rear window.
3   To test the grids for proper operation, start the engine and turn on the system.
4   Ground one lead of a test light and carefully touch the other lead to each element line.
5   The brilliance of the test light should increase as the lead is moved across the element. If the test light glows brightly at both

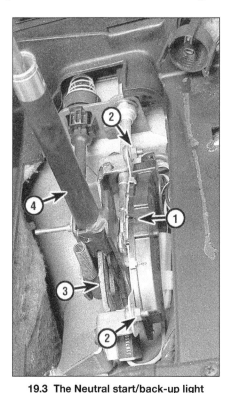

19.3 The Neutral start/back-up light switch (1) for vehicles equipped with an automatic transmission is located at the base of the shifter assembly - make sure that the shift lever (4) is in the Neutral position in the detent plate (3) before removing the old switch (2 indicates the mounting nuts)

ends of the lines, check for a loose ground wire. All of the lines should be checked in at least two places.
6   To repair a break in a line, it is recommended that a repair kit specifically for this purpose be purchased from your auto parts store. Included in the repair kit will be a decal, a container of silver plastic and hardener, a mixing stick and instructions.

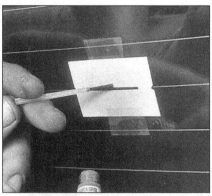

20.11 To repair a broken element, apply a strip of tape to both sides of the line, then apply the proper mixture of hardener and silver plastic with a small wooden stick or spatula

7   To repair a break, first turn off the system and allow it to de-energize for a few minutes.

8   Lightly buff the element area with fine steel wool, then clean it thoroughly with alcohol.

9   Use the decal supplied in the repair kit or apply strips of electrician's tape above and below the area to be repaired. The space between the pieces of tape should be the same width as the existing lines. This can be checked from outside the vehicle. Press the tape tightly against the glass to prevent seepage.

10   Mix the hardener and silver plastic thoroughly.

11   Using the wood spatula, apply the silver plastic mixture between the pieces of tape, overlapping the undamaged area slightly on either end **(see illustration)**.

12   Carefully remove the decal or tape and apply a constant stream of hot air directly to the repaired area. A heat gun set at 500 to 700 degrees Fahrenheit is recommended. Hold the gun one inch from the glass for two minutes.

13   If the new element appears off color, tincture of iodine can be used to clean the repair and bring it back to the proper color. This mixture should not remain on the repair for more than 30 seconds.

14   Although the defogger is now fully operational, the repaired area should not be disturbed for at least 24 hours.

21 Windshield wiper arm - removal and installation

Refer to illustrations 21.4a and 21.4b
Caution: *It's a good idea to coat the windshield with a film of water before activating the windshield wiper switch to prevent scratches on the windshield glass or abrasion of the wiper blades.*

Removal

1   Turn the ignition key to the On position and activate the windshield wiper switch. Turn off the wiper switch when the arms are about halfway through their arcs. Turn off the ignition key.

2   Disconnect the washer hose at the connector.

3   Lift up (away from the windshield) on the wiper arm until it stops.

4   To release the wiper arm assembly from the transmission spindle, pry the retainer out with a small screwdriver **(see illustrations)**.

5   Pull the wiper arm off the spindle.

Installation

6   Turn the ignition key to the On position,

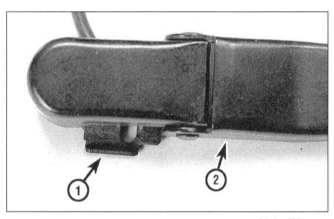

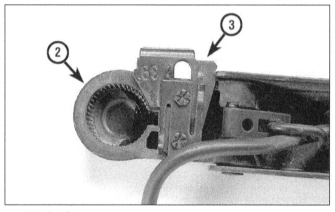

21.4a Wiper arm assembly details

1 Wiper arm retainer in released position	3 Wiper arm retainer shown from underside in
2 Wiper arm assembly	released position

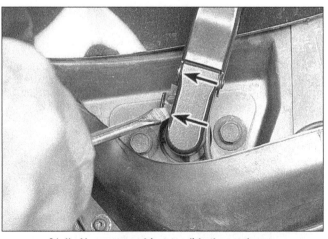

21.4b Use a screwdriver to slide the retainer to its released position

22.1 To gain access to the windshield wiper motor crank arm nut, remove these four Torx screws (arrows) and the vent grille

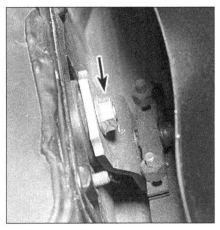

22.4 To remove the windshield wiper motor crank arm nut (arrow), activate the wiper motor and turn it off when the crank arm is pointing down like this

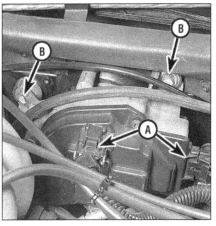

22.5 The wiper motor actually has three electrical connectors (A) and three mounting bolts (B), but only two of each are visible from the engine compartment - the third mounting bolt and electrical connector are underneath the motor

cycle the wipers once and allow them to park (come to rest in their normal position).

7   Install the wiper arm in the correct position.

8   Cycle the wipers again and check the gap between the tip of the wiper arm and the windshield post at the end of the arm's sweep. The gap between the wiper pattern and the top of the windshield molding should be between 0.0 and 40 mm on the driver's side. On the passenger's side it should be from 0.0 to 50 mm. If the gap is incorrect, remove the arm and reinstall it correctly.

9   Install the washer hose.

22 Windshield wiper motor - removal and installation

Refer to illustrations 22.1, 22.4 and 22.5

1   Remove all four Torx screws from the vent grille **(see illustration)**.

2   Remove the vent grille.

3   Turn the ignition key to the On position and activate the windshield wiper switch. When the motor crank arm points down, turn off the wiper switch.

4   Remove the crank arm nut **(see illustration)**.

5   Disconnect the two upper electrical connectors from the wiper motor **(see illustration)**.

6   Remove all three motor mounting screws (two visible in the illustration and one underneath the motor, not shown) and extract the motor far enough to disconnect the third electrical connector located on the underside of the motor. Remove the motor.

7   Installation is the reverse of the removal procedure.

23 Cruise control - general information

1984 vehicles are equipped with the Custom Cruise III system and 1985 and later vehicles have Electronic Cruise Control. Both systems will maintain a driver-selected speed under normal driving conditions.

Because of obvious safety considerations, the cruise control system should be serviced only by trained professionals. However, adjustment of the servo cable may become necessary from time to time. That procedure is detailed in the next Section.

24 Cruise control - cable adjustment

Refer to illustrations 24.3a, 24.3b and 24.5

1984

1   With the servo cable attached to the servo bracket (located in the forward void in front of the front left wheel on 1984 vehicles), place the second ball on the servo chain on the cable end.

2   With the throttle completely closed and the ignition Off, adjust the cable jam nuts until the cable sleeve at the TBI is tight (but not so tight that it holds the throttle open). Tighten the jam nuts.

1985 on

Note: *Some models are equipped with Traction Control (TCS) and Acceleration Slip Regulation (ASR). These systems interact with the anti-lock brake and the engine management systems. Therefore the throttle and cruise cables should be serviced at the dealer only.*

3   With the cable installed on the cable bracket, throttle lever, clip and servo bracket, connect the servo chain (servo assembly) to the cable assembly connector, leaving a

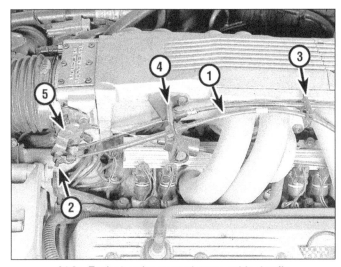

24.3a Typical cruise control servo cable details

1	Cable	4	Bracket
2	Retainer clip	5	Throttle lever
3	Cable clamp		

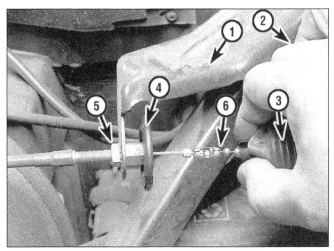

24.3b Typical cruise control details at servo end of cable

1	Servo bracket	4	Washer
2	Servo assembly (under hand)	5	Jam nut
3	Boot	6	Ball link chain

24.5 Aside from its location, the cable jam nut setup on all vehicles is basically the same - to adjust the servo cable, loosen the smaller nut, adjust the cable with the larger nut then tighten the smaller nut - the idea is to get the cable tight but not so tight that it holds the throttle plate open

27.4 To remove the power door lock switch from the backside of the remote control plate, pry it out like this with a small screwdriver - make sure that you note the orientation of the tabs on the edge of the switch in relation to the plate to ensure that the new switch is installed properly

space of four ball links **(see illustrations)**.
4   Close the throttle completely (with the ignition turned to Off).
5   Adjust the cable jam nuts **(see illustration)** until the cable sleeve at the throttle lever is tight but not holding the throttle open. Tighten the jam nuts.
6   Pull the rubber boot over the washer on the cable.

25   Theft deterrent system - general information

If activated, the theft deterrent system alarm pulsates the vehicle's horns at 50 cycles per minute. At the same time the alarm is triggered, the starter interrupt relay is energized, disabling the starter solenoid circuit. In order to conserve vehicle battery power, the alarm shuts off after three to seven minutes of operation and rearms. But once triggered, the starter interrupt relay will remain energized by the controller unit until the system has been properly disarmed.

Because of the complexity of the theft deterrent electrical circuitry, service should be performed only by trained mechanics.

26   Horn - removal and installation

1   Disconnect the cable from the negative terminal of the battery.
2   Disconnect the electrical connector from the horn assembly.
3   Remove the horn bracket bolt.
4   Remove the horn.
5   Installation is the reverse of removal.

27   Power door lock switches - replacement

Refer to illustration 27.4
1   Because of the difficulty involved in gaining access to the remote power door lock motor and because of the special tools required to remove and reinstall its mounting rivets, replacement of this component is best left to a trained professional. The door lock

switches, however, are easily replaced should they fail.
2   Disconnect the cable from the negative terminal of the battery.
3   Remove the door trim panel and remote control plate (Chapter 11).
4   Pry the switch loose from the remote control plate with a small screwdriver **(see illustration)**. Note the orientation of the tabs on the edge of the switch to insure proper installation of the new switch.
5   Carefully position the new switch and pop it into place.
6   Installation is otherwise the reverse of the removal procedure.

28   Power windows - general information

Because of the difficulty involved in gaining access to the power window motor and because of the special tools required to remove and reinstall the mounting rivets, replacement of this component should be done by a dealer service department.

29   Headlight doors - emergency (manual) operation

Refer to illustration 29.3
1   In case of headlight door motor failure, the headlights may be opened manually as follows. **Note:** *If manual operation of the headlight doors is necessary due to a system malfunction, leave the doors open. Do not reconnect the headlight motor wiring harness or attempt to force the headlights open or closed by any method other than the manual control knob.*
2   Open the hood and disconnect the headlight motor wire harness on the outer side of the headlight (the connector is colored gray).
3   Rotate the manual control knob in the appropriate direction (to open or close the door) until stiff resistance is felt **(see illustration)**.
4   Place the retainer wire over the control knob to hold the headlight door open. If the

retainer is not used, the door could be jarred out of position.
5   Close the hood and check the headlights.

30   Airbag system - general information

Warning: *Some of the models covered by this manual are equipped with Supplemental Inflatable Restraints (SIR), more commonly known as airbags. Airbag system components are located in the steering wheel, steering column, instrument panel and right and left front body side members. The airbag(s) could accidentally deploy if any of the system components or wiring harnesses are disturbed, so be extremely careful when working in these areas and don't disturb any airbag*

29.3 Each headlight door is equipped with a manual control knob that can be used to open the headlight door - be sure to fasten the retainer wire in place after the door is open

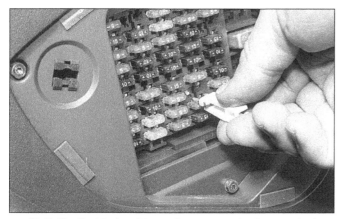

30.10 To disable the airbags, first remove the fuse marked AIRBAG from the fuse box . . .

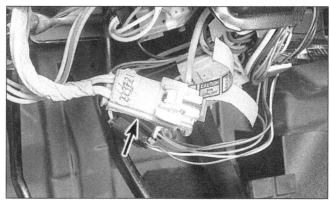

30.12 . . . then disconnect the yellow airbag module connectors located below the steering column (arrow) - if equipped with a passenger side airbag there will be two connectors

system components or wiring. You could be injured if an airbag accidentally deploys, or the airbag might not deploy correctly in a collision if any components or wiring in the system have been disturbed. The yellow wires and connectors routed through the instrument panel and steering column are for this system. Do not use electrical test equipment on these yellow wires or tamper with them in any way while working in their vicinity.

Description

1   Some models covered by this manual are equipped with a Supplemental Inflatable Restraint (SIR) system, more commonly known as an airbag system. The SIR system is designed to protect the driver and on 1994 and later models, the passenger, from serious injury in the event of a head-on or frontal collision.
2   The SIR system consists of an airbag located in the center of the steering wheel, and on 1994 and later models, another located on the right side of the instrument panel to protect the passenger. The system utilizes two impact sensors; located in the right and left front body side members; an arming sensor located in the center of the instrument panel; and a diagnostic/energy reserve module (DERM) located at the right end of the instrument panel.

Sensors

3   The system has three separate sensors; two impact sensors and an arming sensor. The sensors are basically pressure sensitive switches that complete an electrical circuit during an impact of sufficient G force. The electrical signal from the sensors is sent to the diagnostic module, which then completes circuit and inflates the airbags.
4   The arming sensor closes at a lower velocity level than the impact sensors, ensuring the system is "armed" if either impact sensor closes.

Diagnostic/energy reserve module

5   The diagnostic/energy reserve module contains an on-board microprocessor which monitors the operation of the system. It performs a diagnostic check of the system every

time the vehicle is started. If the system is operating properly, the AIRBAG warning light will blink on and off seven times. If there is a fault in the system, the light will remain on and the airbag control module will store fault codes indicating the nature of the fault. If the AIRBAG warning light remains on after staring, or comes on while driving, the vehicle should be taken to your dealer immediately for service. The diagnostic/energy reserve module also contains a back-up power supply to deploy the airbags in the event battery power is lost during a collision.

Operation

6   For the airbag(s) to deploy, an impact of sufficient G force must occur within 30-degrees of the vehicle centerline. When this condition occurs, the circuit to the airbag inflator is closed and the airbag inflates. If the battery is destroyed by the impact, or is too low to power the inflators, a back-up power supply inside the diagnostic/energy reserve module supplies current to the airbags.

Self-diagnosis system

7   A self-diagnosis circuit in the module displays a light when the ignition switch is turned to the On position. If the system is operating normally, the light should go out after seven flashes. If the light doesn't come on, or doesn't go out after seven flashes, or if it comes on while you're driving the vehicle, there's a malfunction in the SIR system. Have it inspected and repaired as soon as possible. Do not attempt to troubleshoot or service the SIR system yourself. Even a small mistake could cause the SIR system to malfunction when you need it.

Servicing components near the SIR system

8   Nevertheless, there are times when you need to remove the steering wheel, radio or service other components on or near the instrument panel. At these times, you'll be working around components and wiring harnesses for the SIR system. SIR system wiring is easy to identify; they're all covered by a bright yellow conduit. Do not unplug the connectors for the SIR system wiring, except to

disable the system. And do not use electrical test equipment on the SIR system wiring. Always disable the SIR system before working near the SIR system components or related wiring.

Disabling the SIR system

Refer to illustrations 30.10 and 26.12
9   Turn the steering wheel to the straight ahead position, place the ignition switch in Lock and remove the key.
10   Remove the airbag fuse from the fuse block **(see illustration)**.
11   Remove the knee bolster and sound insulator panel below the instrument panel.
12   Unplug the yellow Connector Position Assurance (CPA) connectors at the base of the steering column **(see illustration)**.

Enabling the SIR system

13   After you've disabled the airbag and performed the necessary service, plug in the CPA connectors. Reinstall the knee bolster and sound insulator panel.
14   Install the airbag fuse.
15   Turn the ignition Key On and watch the AIRBAG warning light, it should flash seven times then go out.

---

31   Wiring diagrams - general information

Since it isn't possible to include all wiring diagrams for every year covered by this manual, the following diagrams are those that are typical and most commonly needed.

Prior to troubleshooting any circuit, check the fuse and circuit breakers (if equipped) to make sure they're in good condition. Make sure the battery is properly charged and check the cable connections (see Chapter 1, Section 9).

When checking a circuit, make sure that all electrical connectors are clean, with no broken or loose terminals. When unplugging an electrical connector, do not pull on the wires. Pull only on the connector housings themselves.

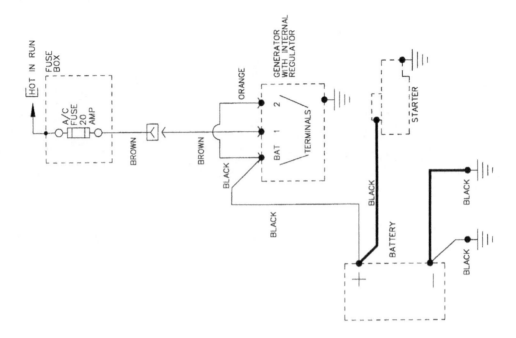

Charging system - typical

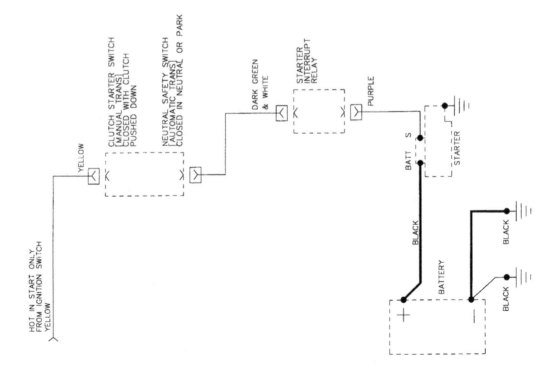

Starting system - typical

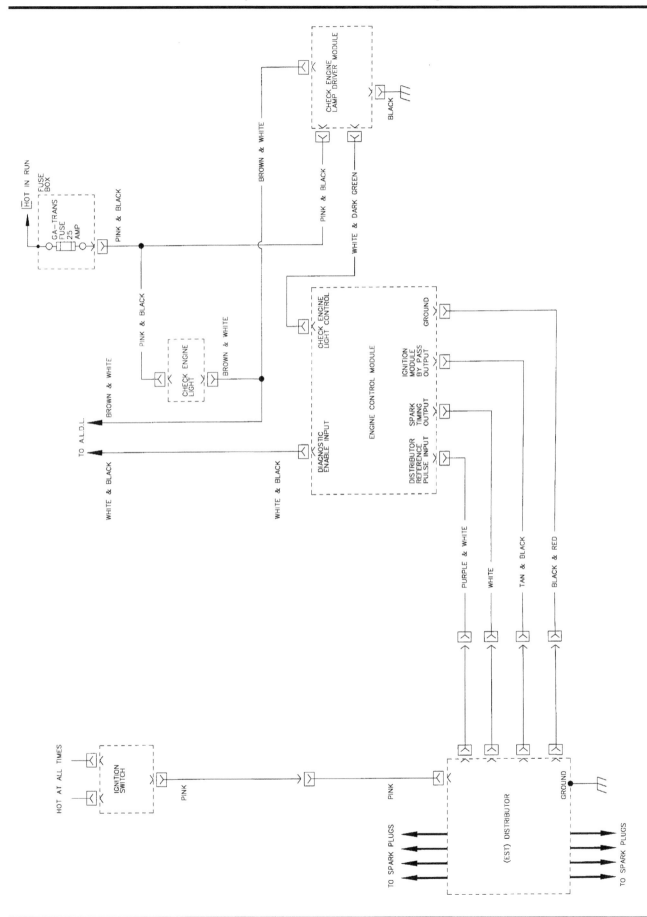

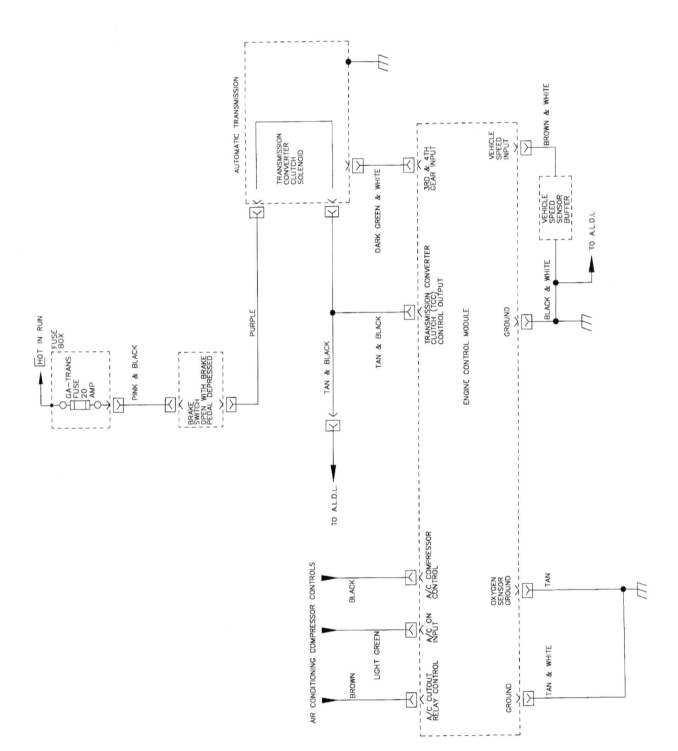

Engine control system (part 2 of 3)

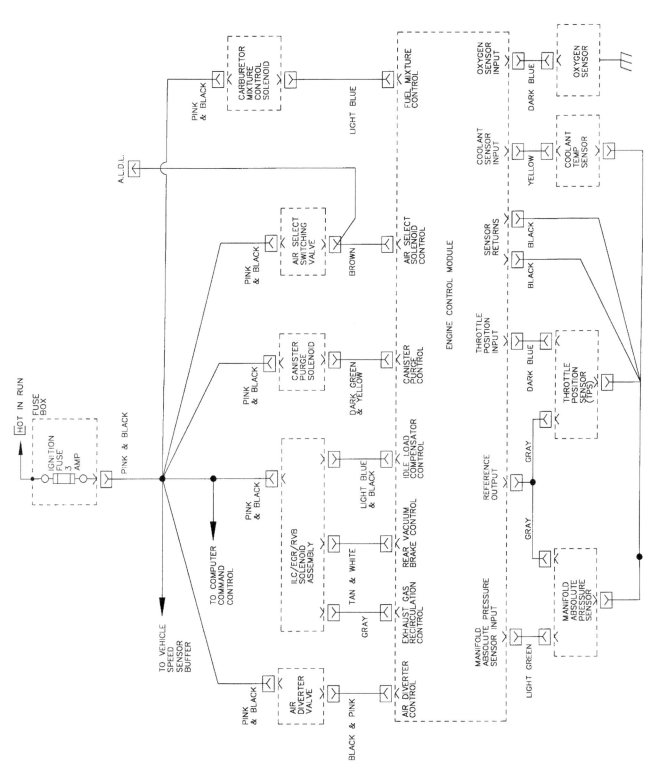

Engine control system (part 3 of 3)

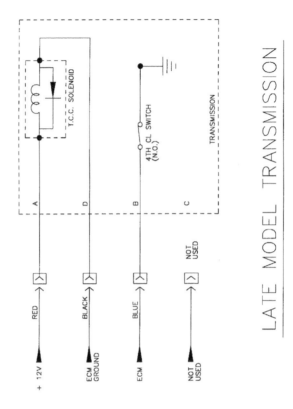

LATE MODEL TRANSMISSION

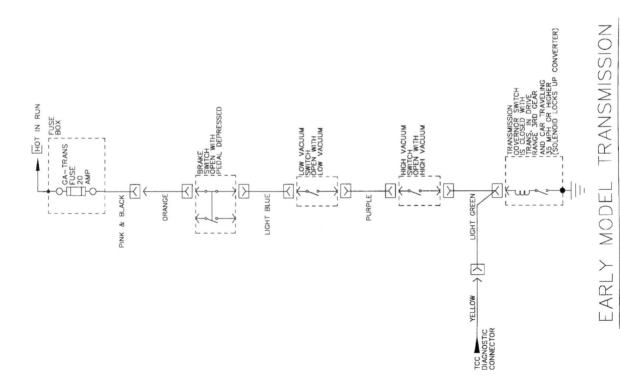

EARLY MODEL TRANSMISSION

Transmission controls

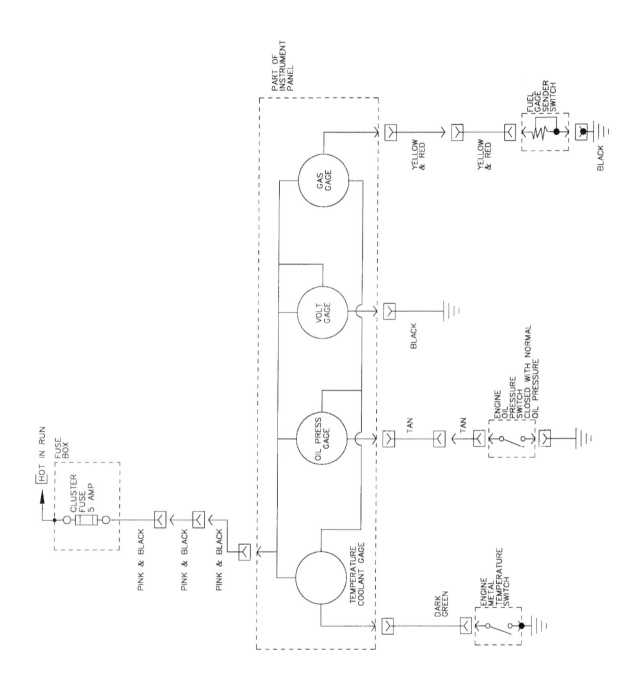

Gauges and sending units

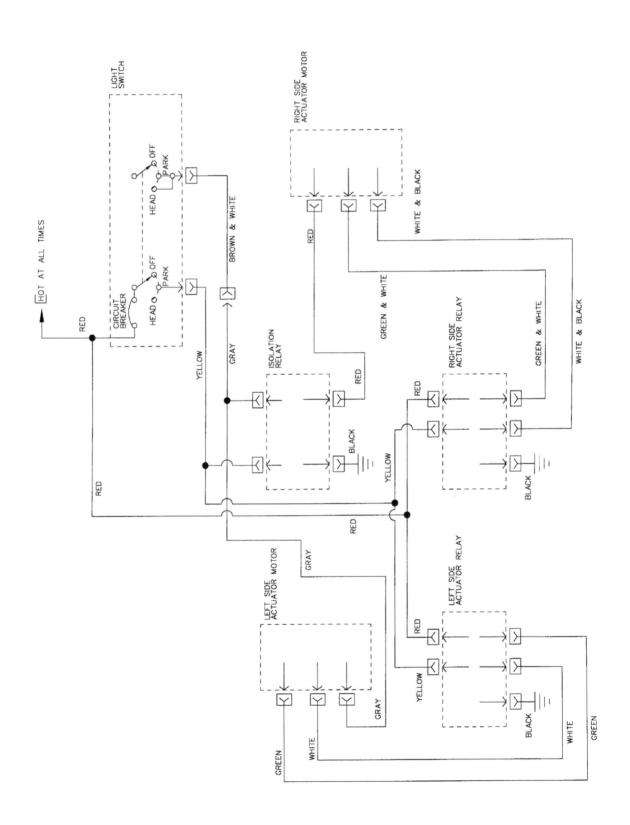

Exterior lighting - headlight door system

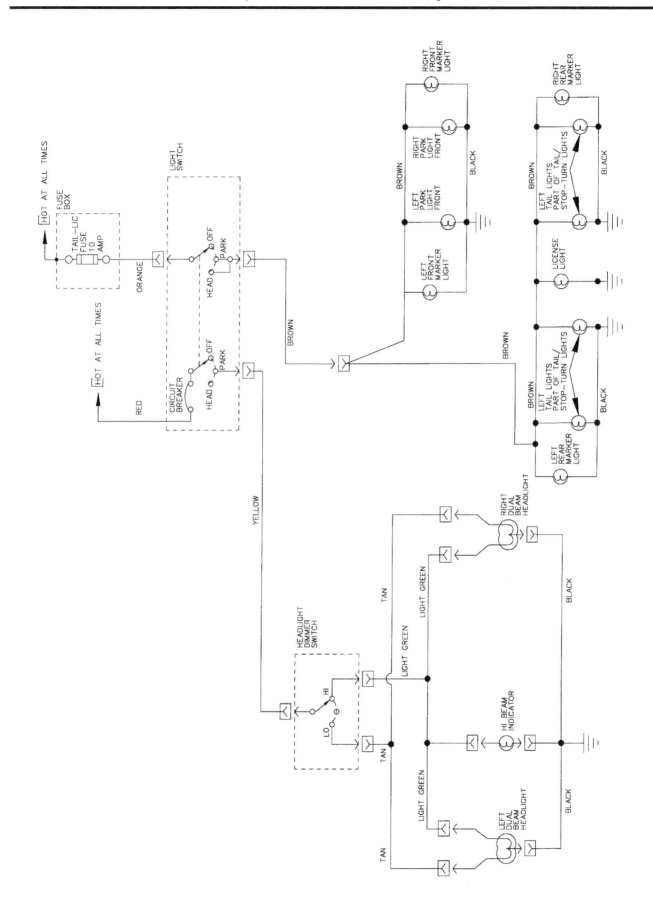

Exterior lighting system (part 1 of 2)

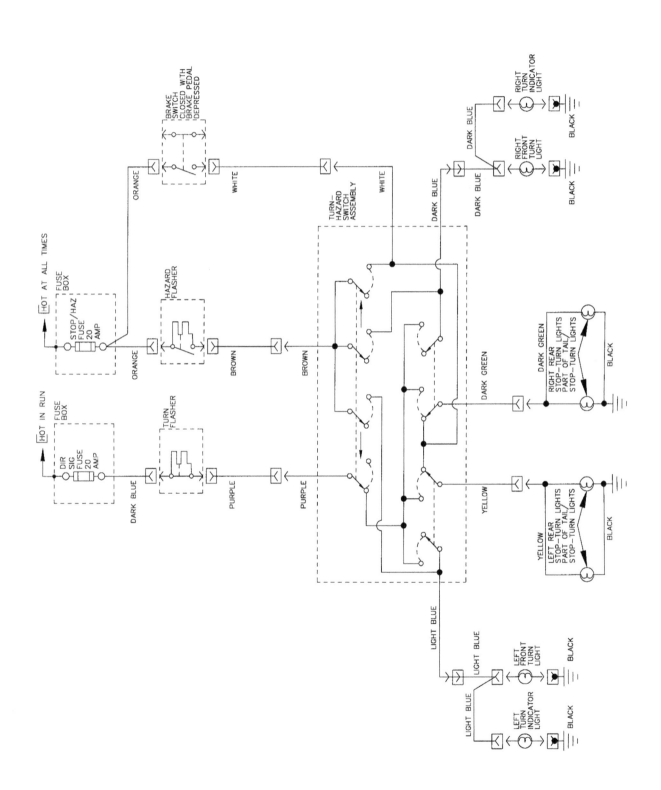

Exterior lighting system (part 2 of 2)

Fog light system

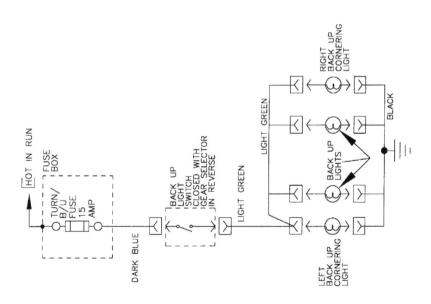

Backup light system

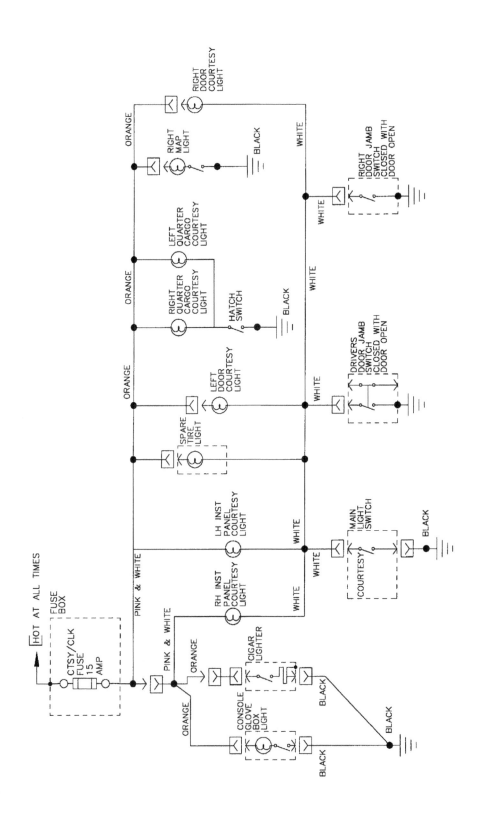

Interior lighting system

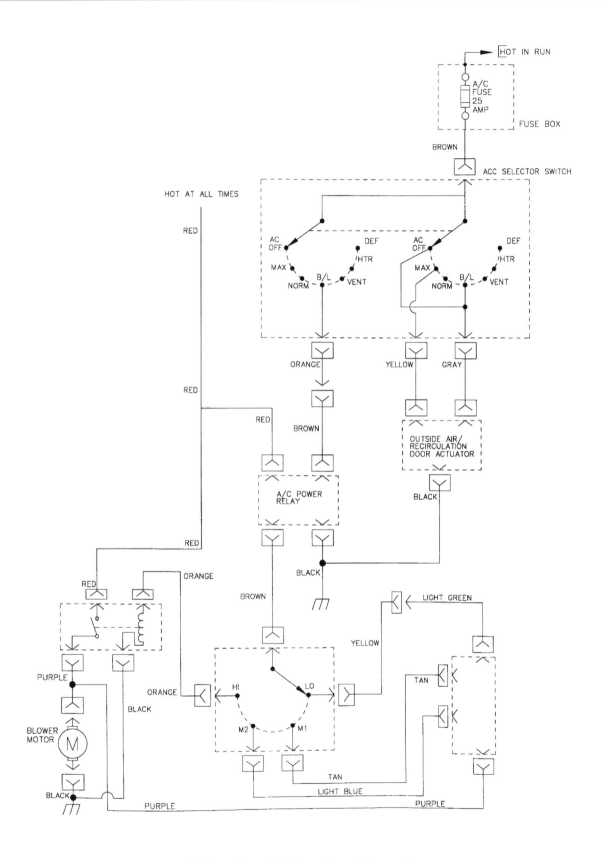

Heating, air-conditioning and ventilation system

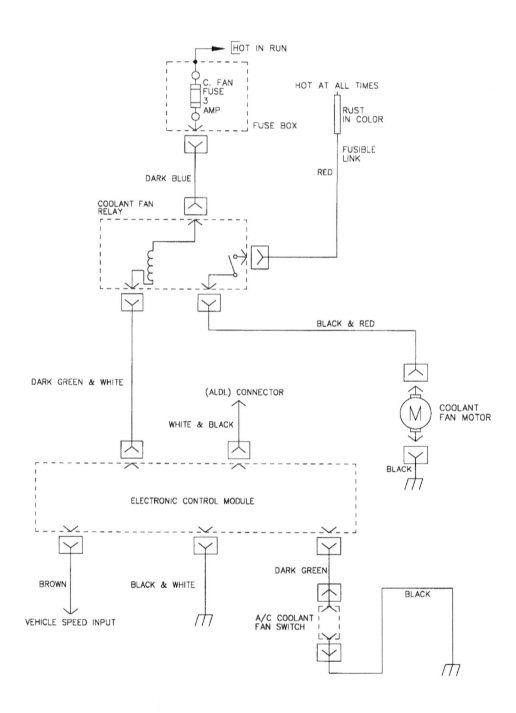

Coolant fan system - typical

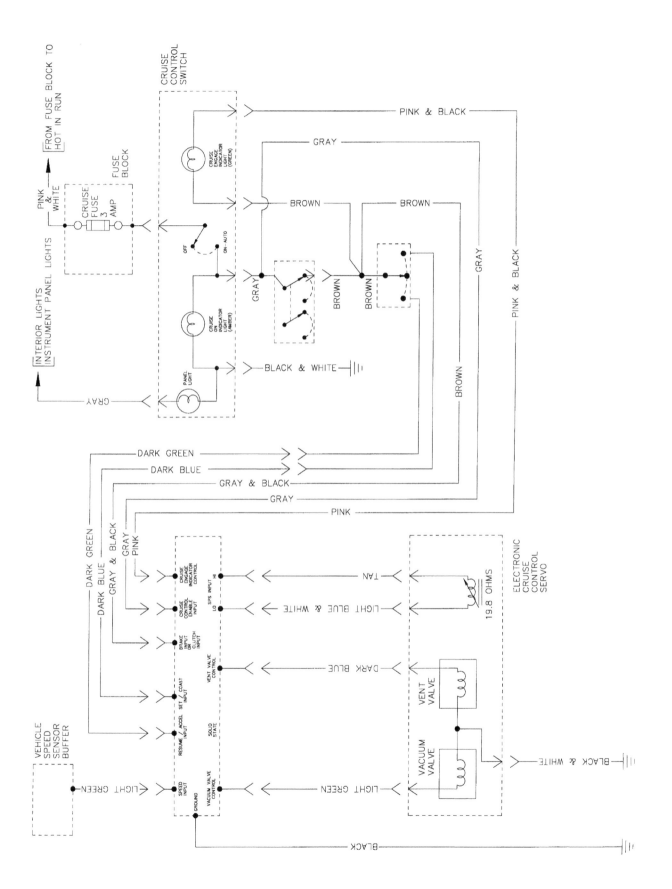

Cruise control system

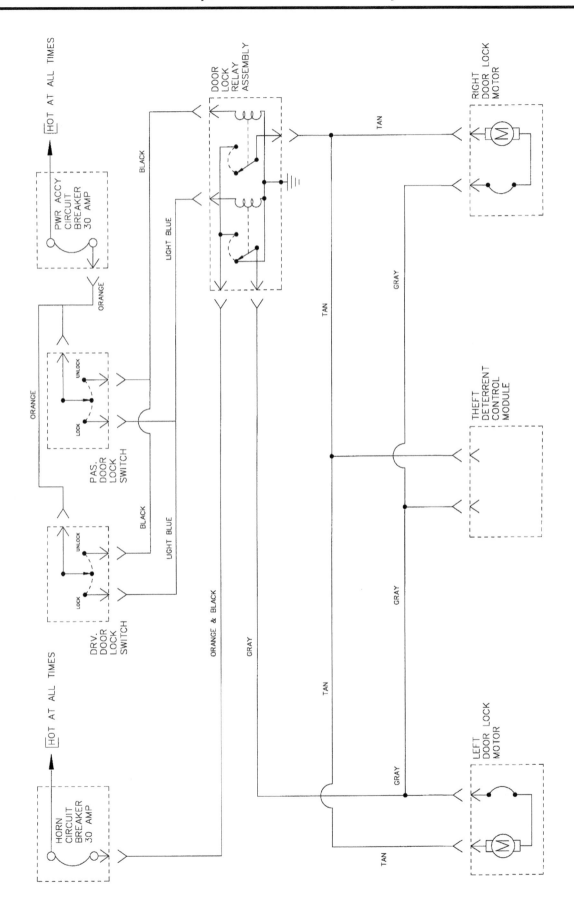

Power door lock system

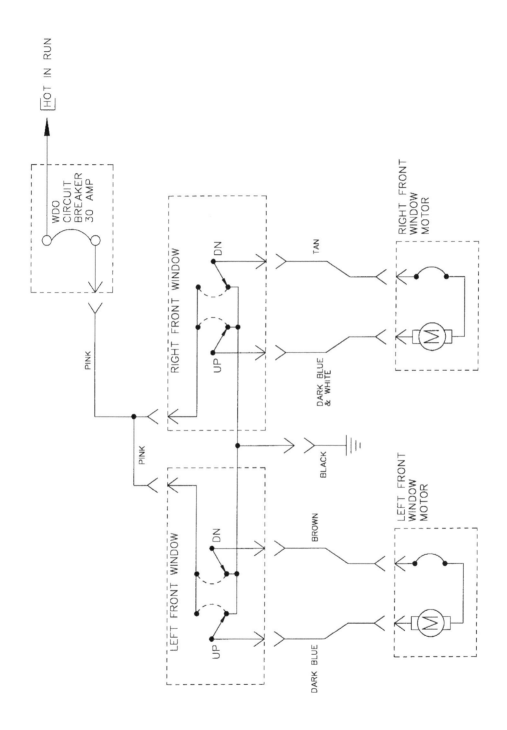

Power window system

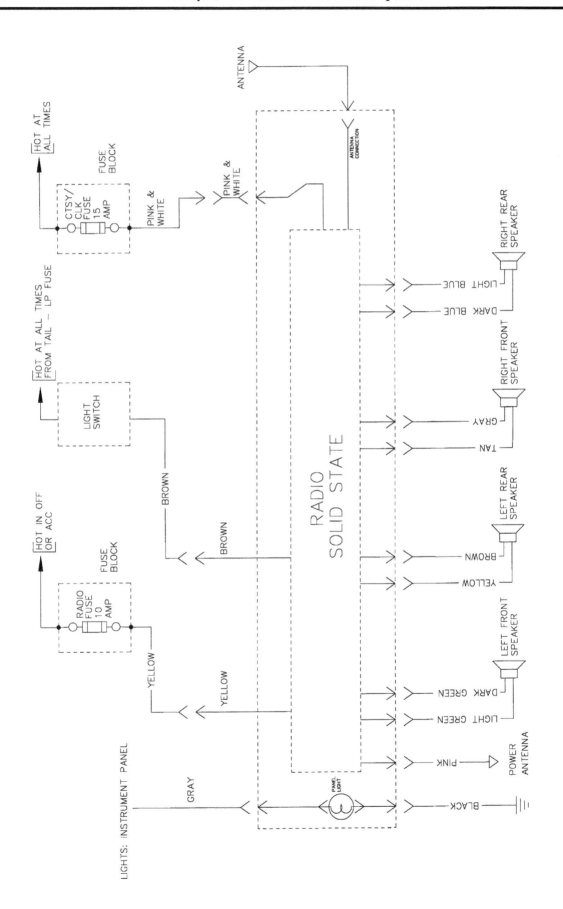

Audio system - typical base model

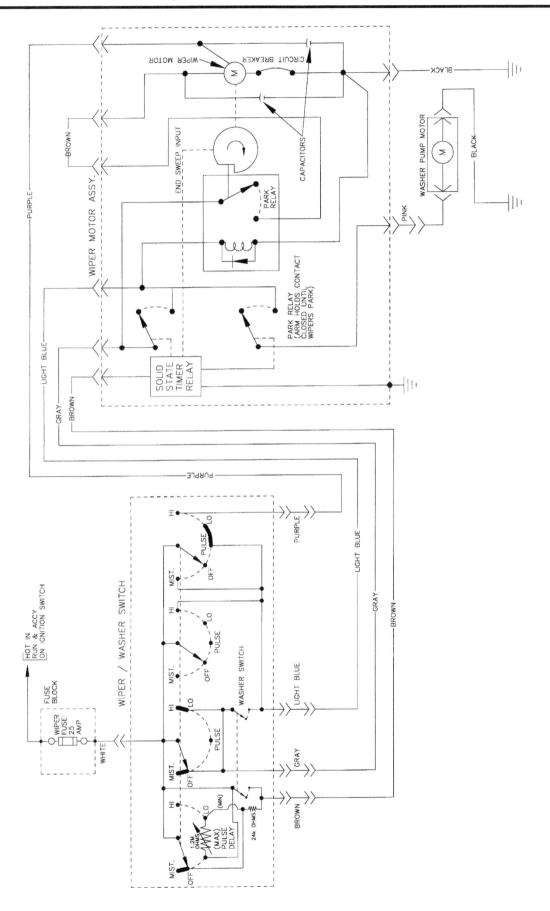

Windshield wiper and washer system

Notes

A

Active material - The material on the negative and positive battery plates that interacts with the electrolyte to produce a charge.

AC generator - An electromechanical device that generates alternating current (AC), commonly known as an alternator. Usually belt-driven off the engine. Provides maximum output at relatively low rpm. Used on all modern vehicles. Requires a rectifier to convert AC to direct current (DC), which is used by automotive electrical system.

Aftermarket parts - Components that can be added to a vehicle after its manufacture. These parts are often accessories and should not be confused with original equipment manufacturer (OEM) service or replacement parts.

Alligator clip - A long-nosed spring-loaded metal clip with meshing teeth. Used to make temporary electrical connections.

Alloy - A mixture of two or more materials.

Alternating current (AC) - An electric current, generated through magnetism, whose polarity constantly changes back and forth from positive to negative.

Alternator - A device used in automobiles to produce electric current. The alternator's AC output is rectified to direct current before it reaches the vehicle's electrical system.

Ammeter - 1. An instrument for measuring current flow. An ammeter can be designed to measure alternating or direct current. 2. An instrument panel gauge used to measure the flow rate of current into or out of the battery. The ammeter is calibrated in amperes for both charge and discharge rates, in ranges of 20, 30 or more amperes.

Amperage - The total amount of current (amperes) flowing in a circuit.

Ampere (amp) - The unit of measurement for the flow of electrons in a circuit. The amount of current produced by one volt acting through a resistance of one ohm (1 coulomb per second).

Ampere hour - A unit of measurement for battery capacity, obtained by multiplying the current (in amperes) by the time (in hours) during which the current is delivered.

Analog gauge - see "gauge."

Analog signal - A signal which varies in exact proportion to a measured quantity, such as pressure, temperature, speed, etc.

Arcing - When electricity leaps the gap between two electrodes, it is said to be "arcing."

Armature - The rotating part of a generator or motor. Actually a coil of wires wrapped in a specific pattern, which rotates on a shaft.

Atoms - The small particles which make up all matter. Atoms are made up of a positive-charged nucleus with negative-charged electrons whirling around in orbits.

B

Battery - A group of two or more cells connected together for the production of an electric current by converting chemical energy into electrical energy. A battery has two poles - positive and negative. The amount of positive and negative energy is called potential.

Battery charging - The process of energizing a battery by passing electric current through the battery in a reverse direction.

Battery ratings - Performance standards conducted under laboratory conditions to describe a battery's reserve capacity and cold-crank capacities. The amp-hour rating is no longer in widespread use. See "cold crank rating."

Battery state of charge - The available amount of energy in a battery in relation to that which would ordinarily available be if the battery was fully charged.

Battery voltage - A figure determined by the number of cells in a battery. Because each cell generates about two volts, a six cell battery has 12 volts.

Bendix inertia drive - A self-engaging and releasing starter drive mechanism. The pinion gear moves into engagement when the starter motor spins and disengages when the engine starts.

Bound electron - An electron whose orbit is near the nucleus of an atom and is strongly attracted to it.

Brush - A spring-loaded block of carbon or copper that rubs against a commutator or slip ring to conduct current. A key component in all alternators and starters.

Bulkhead connector - An OEM device used to connect wiring inside the vehicle body with wiring outside the body. Usually located at the bulkhead or firewall.

Butt connector - A solderless connector used to permanently join two wire ends together.

C

Cable - An assembly of one or more conductors, usually individually insulated and enclosed in a protective sheath.

Capacity - The current output capability of a cell or battery, usually expressed in ampere hours.

Cell - In a storage battery, one of the sets of positive and negative plates which react with electrolyte to produce an electric current.

Charge - A definite quantity of electricity.

Charge (recharge) - To restore the active materials in a battery cell by electrically reversing the chemical action.

Circuit - An electrical path - from the source (battery or generator) through the load (such as a light bulb) and back to the source - through which current flows. A typical circuit consists of a battery, wire, load (light or motor) and switch. See "simple circuit" and "single-wire circuit."

Circuit breaker - A circuit-protection device that automatically opens or breaks an overloaded circuit. The typical circuit breaker usually consists of movable points that open if the preset ampere load is exceeded. Some circuit breakers are self-resetting; others require manual resetting.

Closed circuit - A circuit which is uninterrupted from the current source, through the load and back to the current source.

Closed-end connector - Solderless connector shaped like a hat. Used to join two, three or more wires together. Similar to wire connectors used in home wiring, but installed by crimping instead of twisting.

Clutch interlock switch - A switch that prevents the vehicle from starting unless the clutch is pressed.

Coil - Any electrical device or component consisting of wire loops wrapped around a central core. Coils depend on one of two electrical properties for operation, depending on the application (electromagnetism or induction).

Cold-crank rating - The minimum number of amperes a fully charged 12-volt battery can deliver for 30 seconds at 0-degrees F without falling below 7.2 battery volts.

Commutator - A series of copper bars insulated from each other and connected to the armature windings at the end of the armature. Provides contact with fixed brushes to draw current from (generator) or bring current to (starter) the armature.

Conductance - A measure of the ease with which a conductor allows electron flow. In DC circuits, conductance is the reciprocal of resistance.

Conduction - The transmission of heat or electricity through, or by means of, a conductor.

Conductor - Any material - usually a wire or other metallic object - made up of atoms whose free electrons are easily dislodged, permitting easy flow of electrons from atom to atom. Examples are copper, aluminum and steel. Conductors are all metals. The metal part of an insulated wire is often called the conductor.

Constant voltage regulator (CVR) - A device used to maintain a constant voltage level in a circuit, despite fluctuations in system voltage. CVRs are wired into some gauge circuits so voltage fluctuations won't affect accuracy of the gauge readings.

Contact - One of the contact-carrying parts of a relay or switch that engages or disengages to open or close the associated electrical circuits.

Continuity - A continuous path for the flow of an electrical current.

Conventional theory - In this theory, the direction of current flow was arbitrarily chosen to be from the positive terminal of the voltage source, through the external circuit, then back to the negative terminal of the source.

Coulomb - The unit of quantity of electricity or charge. The electrons that pass a given point in one second when the current is maintained at one ampere. Equal to an electrical charge of 6.25×10^{18} electrons passing a point in one second. See "ampere."

Current - The movement of free electrons along a conductor. In automotive electrical work, electron flow is considered to be from positive to negative. Current flow is measured in amperes.

Cycle - A recurring series of events which take place in a definite order.

D

DC generator - An electromechanical device that generates direct current. Usually belt-driven off the engine. Because the DC generator requires high rpm for maximum output, it's no longer used in production automobiles.

Deep cycling - The process of discharging a battery almost completely before recharging.

Digital gauge - See "gauge."

Diode - A semiconductor which permits current to flow in only one direction. Diodes are used to rectify current from AC to DC.

Direct current (DC) - An electrical current which flows steadily in only one direction.

Discharge - Generally, to draw electric current from the battery. Specifically, to remove more energy from a battery than is being replaced. A discharged battery is of no use until it's recharged.

Disconnect terminal - Solderless connectors in male and female forms, intended to be easily disconnected and connected. Typically, a blade or pin (male connector) fits into a matching receptacle or socket (female connector). Many components have built-in (blade) terminals that require a specialized female connector.

Display - Any device that conveys information. In a vehicle, displays are either lights, gauges or buzzers. Gauges may be analog or digital.

DPDT - A double-pole, double-throw switch.

DPST - A double-pole, single-throw switch.

Draw - The electric current required to operate an electrical device.

Drive - A device located on the starter to allow for a method of engaging the starter to the flywheel.

E

Electric - A word used to describe anything having to do with electricity in any form. Used interchangeably with electrical.

Electrical balance - An atom or an object in which positive and negative charges are equal.

Electricity - The movement of electrons from one body of matter to another.

Electrochemical - The production of electricity from chemical reactions, as in a battery.

Electrolyte - A solution of sulfuric acid and water used in the cells of a battery to activate the chemical process which results in an electrical potential.

Electromagnet - A soft-iron core which is magnetized when an electric current is passed through a coil of wire surrounding it.

Electromagnetic - Having both electrical and magnetic properties.

Electromagnetism - The magnetic field around a conductor when a current is flowing through the conductor.

Electromechanical - Any device which uses electrical energy to produce mechanical movement.

Electrons - Those parts of an atom which are negatively charged and orbit around the nucleus of the atom.

Electron flow - The movement of electrons from a negative to a positive point on a conductor, or through a liquid, gas or vacuum.

Electron theory - States that all matter is made up of atoms which are made up of a nucleus and electrons. Free electrons moving from one atom to another, in a single direction, produce what is known as electricity.

Electronics - The science and engineering concerned with the behavior of electrons in devices and the utilization of such devices. Especially devices utilizing electron tubes or semiconductor devices.

Energy - The capacity for performing work.

EVR - Electronic Voltage Regulator; a type of regulator that uses all solid state devices to perform the regulatory functions.

F

Field - An area covered or filled with a magnetic force. Common terminology for field magnet, field winding, magnetic field, etc.

Field coil - A coil of insulated wire, wrapped around an iron or steel core, that creates a magnetic field when current is passed through the wire.

Filament - A resistance in an electric light bulb which glows and produces light when an adequate current is sent through it.

Fluorescent - Having the property of giving off light when bombarded by electrons or radiant energy.

Flux - The lines of magnetic force flowing in a magnetic field.

Flywheel - A large wheel attached to the crankshaft at the rear of the engine.

Flywheel ring gear - A large gear pressed onto the circumference of the flywheel. When the starter gear engages the ring gear, the starter cranks the engine.

Free electron - An electron in the outer orbit of an atom, not strongly attracted to the nucleus; it can be easily forced out of its orbit.

Fuse - A circuit-protection device containing a soft piece of metal which is calibrated to melt at a predetermined amp level and break the circuit.

Fuse block - An insulating base on which fuse clips or other contacts are mounted.

Fuse link - See "fusible link."

Fuse panel - A plastic or fiberboard assembly that permits mounting several fuses in one centralized location. Some fuse panels are part of, or contain, a terminal block (see "terminal block").

Fuse wire - A wire made of an alloy which melts at a low temperature.

Fusible link - A circuit protection device consisting of a conductor surrounded by heat-resistant insulation. The conductor is two gages smaller than the wire it protects, so it acts as the weakest link in the circuit. Unlike a blown fuse, a failed fusible link must be cut from the wire for replacement.

G

Gage - A standard SAE designation of wire sizes, expressed in AWG (American Wire Gage). The larger the gage number, the smaller the wire. Metric wire sizes are expressed in cross-sectional area, which is expressed in square millimeters. Sometimes the spelling "gauge" is also used to designate wire size. Using this spelling, however, avoids confusion with instrument panel displays (see "gauge").

Gassing - The breakdown of water into hydrogen and oxygen gas in a battery.

Gauge - An instrument panel display used to monitor engine conditions. A gauge with a movable pointer on a dial or a fixed scale is an analog gauge. A gauge with a numerical readout is called a digital gauge. Also refers to measuring device used to check regulator point openings.

Generator - An engine-driven device that produces an electric current through magnetism by converting rotary motion into electrical potential (see "AC generator" and "DC generator").

Grid - A lead screen that is pasted with active materials to form a negative or positive battery plate.

Grommet - A donut shaped rubber or plastic part used to protect wiring that passes through a panel, firewall or bulkhead.

Ground - The connection made in grounding a circuit. In a single-wire system, any metal part of the car's structure that's directly or indirectly attached to the battery's negative post. Used to conduct current from a load back to the battery. Self-grounded components are attached directly to a grounded metal part through their mounting screws. Components mounted to nongrounded parts of a vehicle require a ground wire attached to a known good ground.

H

Halogen light - A special bulb that produces a brilliant white light. Because of its high intensity, a halogen light is often used for fog lights and driving lights.

Harness - A bundle of electrical wires. For convenience in handling and for neatness, all wires going to a certain part of the vehicle are bundled together into a harness.

Harness ties - Self-tightening nylon straps used to bundle wires into harnesses. Available in stock lengths that can be cut to size after installation. Once tightened, they can't be removed unless they're cut from the harness.

Harness wrap - One of several materials used to bundle wires into manageable harnesses. See "loom," "split loom," "loom tape" and "harness ties").

Hot - Connected to the battery positive terminal, energized.

Hydrogen gas - The lightest and most explosive of all gases. Emitted from a battery during charging procedures. This gas is very dangerous and certain safety precautions must be observed.

Hydrometer - A syringe-like instrument used to measure the specific gravity of a battery's electrolyte.

I

IAR - Integral Alternator/Regulator; a type of regulator mounted at the rear of the alternator.

Ignition switch - A key-operated switch that opens and closes the circuit that supplies power to the ignition and electrical system.

Indicator light - An instrument-panel display used to convey information or condition of the monitored circuit or system. See "warning light."

Induced - Produced by the influence of a magnetic or electrical field.

Induced current - The current generated in a conductor as it moves through a magnetic field, or as a magnetic field is moved across a conductor.

Induced voltage - The voltage produced as a result of an induced current flow.

Inductance - That property of a coil or other electrical device which opposes any change in the existing current, present only when an alternating or pulsating current is flowing. Has no effect on the flow of direct, or static, current.

Induction - The process by which an electrical conductor becomes charged when near another charged body. Induction occurs through the influence of the magnetic fields surrounding the conductors.

Input - 1. The driving force applied to a circuit or device. 2. The terminals (or other connection) where the driving force may be applied to a circuit or device.

Instrument Voltage Regulator - See "Constant Voltage Regulator."

Insulator - A material that has few or no free electrons that readily leave their orbits. A non-conducting material used for insulating wires in electrical circuits. Cloth, glass, plastic and rubber are typical examples. Wires for modern vehicles use plastic insulation.

Integral - Formed as a unit with another part.

Intercell connector - A lead strap or connector that connects the cells in a battery.

Intermittent - Coming or going at intervals; not continuous.

Ion - An atom having an imbalanced charge due to the loss of an electron or proton. An ion may be either positively charged (have a deficiency of electrons) or negatively charged (have a surplus of electrons).

J

Jumper - A short length of wire used as a temporary connection between two points.

Junction - 1. A connection between two or more components, conductors or sections of transmission line. 2. Any point from which three or more wires branch out in a circuit.

Junction box - A box in which connections are made between different cables.

L

Lead-acid battery - A common automotive battery in which the active materials are lead, lead peroxide and a solution of sulfuric acid.

Lead dioxide - A combination of lead and oxygen, as found in the storage battery. Lead dioxide is reddish brown in color.

Lead sulfate - A combination of lead, oxygen and sulfur, as found in the storage battery.

Light - An electrical load designed to emit light when current flows through it. A light consists of a glass bulb enclosing a filament and a base containing the electrical contacts. Some lights, such as sealed beam headlights, also contain a built-in reflector.

Load - Any device that uses electrical current to perform work in a vehicle electrical system. Lights and motors are typical examples.

Loom - A harness covering. Older vehicles used woven-cloth loom; most modern vehicles use a corrugated-plastic loom or split loom.

Loom tape - A nonadhesive tape used as a harness wrap. Adhesive-type tapes, including electrical tapes, are not recommended for wrapping harnesses. Often, a piece of shrink wrap is used at tape ends to keep the tape from unraveling.

M

Magnet - A material that attracts iron and steel. Temporary magnets are made by surrounding a soft-iron core with a strong electromagnetic field. Permanent magnets are made with steel.

Magnetic field - The field produced by a magnet or a magnetic influence. A magnetic field has force and direction.

Magnetic poles - The ends of a bar or horseshoe magnet.

Magnetism - A property of the molecules of certain materials, such as iron, that allows the substance to be magnetized.

Meter - An electrical or electronic measuring device.

Module - A combination of components packaged in a single unit with a common mounting and providing some complete function.

Motor - An electromagnetic apparatus used to convert electrical energy into mechanical energy.

Multimeter - A test instrument with the capability to measure voltage, current and resistance.

N

Negative charge - The condition when an element has more than a normal quantity of electrons.

Negative ion - An atom with more electrons than normal. A negative ion has a negative charge.

Negative terminal - The terminal on a battery which has an excess of electrons. A point from which electrons flow to the positive terminal.

Neutral - Neither positive nor negative, or in a natural condition. Having the normal number of electrons, i.e. the same number of electrons as protons.

Neutral start switch - On vehicles with an automatic transmission, a switch that prevents starting if the vehicle is not in Neutral or Park.

Neutron - A particle within the nucleus of an atom. A neutron is electrically neutral.

Nichrome - A metallic compound containing nickel and chromium, used in making high resistances.

North pole - The pole of a magnet from which the lines of force are emitted. The lines of force travel from the north to the south pole.

Nucleus - The core of an atom. The nucleus contains protons and neutrons.

Nylon ties - See "harness ties."

O

OEM (Original Equipment Manufacturer) - A designation used to describe the equipment and parts installed on a vehicle by the manufacturer, or those available from the vehicle manufacturer as replacement parts. See "aftermarket parts."

Ohm - The practical unit for measuring electrical resistance.

Ohmmeter - An instrument for measuring resistance. In automotive electrical work, it's often used to determine the resistance that various loads contribute to a circuit or system.

Ohm's Law - The electrical formula that describes how voltage, current and resistance are related. The basic formula is E (electrical pressure in volts) = I (current flow in amperes) X R (resistance in ohms). What does it mean? Simply put, amperage varies in direct ratio to voltage and in inverse ratio to resistance.

Open circuit - An electrical circuit that isn't complete because of a broken or disconnected wire.

Open circuit voltage - The battery voltage when the battery has no closed circuit across the posts and is not delivering or receiving voltage.

Orbit - The path followed by an electron around the nucleus.

Output - The current, voltage, power or driving force delivered by a device or circuit. The terminals or connections where the current can be measured.

Overrunning clutch - A device located on the starter to allow for a method of engaging the starter with the flywheel. The overrunning clutch uses a shift lever to actuate the drive pinion to provide for a positive meshing and de-meshing of the pinion with the flywheel ring gear.

P

Parallel circuit - A method or pattern of connecting units in an electrical circuit so that they're connected negative-to-negative and positive-to-positive. In a parallel circuit, current can flow independently through several components at the same time. See "series circuit."

Permanent magnet - A magnet made of tempered steel which holds its magnetism for a long period of time.

Pinion - A small gear that either drives or is driven by a larger gear.

Plate - A battery grid that's pasted with active materials and given a forming charge which results in a negative or positive charge. Plates are submerged, as elements, in the electrolyte and electricity is produced from the chemical reactions between the plates and the electrolyte.

Polarity - The quality or condition of a body which has two opposite properties or directions; having poles, as in an electric cell, a magnet or a magnetic field.

Polarity-protected connector - A multiple-cavity connector that can be connected in only one way, either to a mating connector or to a component.

Pole - A positive (or negative) terminal in a cell or battery; the ends of a magnet (north or south).

Pole shoe - The part of a starter that's used to hold the field coils in place in their proper positions; consists of a soft-iron core which is wound with heavy copper ribbons.

Positive - Designating or pertaining to a kind of electrical potential.

Positive terminal - The battery terminal to which current flows.

Post - A round, tapered battery terminal that serves as a connection for battery cables.

Potential - Latent, or unreleased, electrical energy.

Power supply - A unit that supplies electrical power to a unit. For example, a battery.

Printed circuit - An electrical conductor consisting of thin metal foil paths attached to a flexible plastic backing. Also called a PC board. PC boards are used primarily in OEM instrument clusters and other electronic devices.

Proton - A positively-charged particle in the nucleus of an atom.

Q

Quick charger - A battery charger designed to allow the charging of a battery in a short period of time.

R

Rectification - The process of changing alternating current to direct current.

Rectifier - A device in the electrical system used to convert alternating current to direct current.

Regulator - A device used to regulate the output of a generator or alternator by controlling the current and voltage.

Relay - An electromagnetic device that opens or closes the flow of current in a circuit.

Resistance - The resistance to electron flow present in an electrical circuit, expressed in ohms.

Resistor - Any conductor that permits electron movement but retards it. Tungsten and nickel are typical resistors.

Rheostat - A variable resistor, operated by a knob or handle, used to vary the resistance in a circuit. A rheostat consists of a coil of resistance wire and a movable contact or wiper that creates more or less resistance in the circuit, depending on how many coil windings it allows the current to pass through. The dimmer control for instrument panel illumination is an example.

Ring terminal - A conductor used to attach a wire to a screw or stud terminal. The ring is sized to the mating screw. Ring terminals are the connectors least likely to vibrate loose in rugged applications. Comes in soldered and unsoldered versions.

Rotor - That part of an alternator which rotates inside the stator.

S

Schematic - A drawing system for portraying the components and wires in a vehicle's electrical system, using standardized symbols.

Secondary - The output winding of a transformer, i.e. the winding in which current flow is due to inductive coupling with another coil called the primary.

Sending unit - Used to operate a gauge or indicator light. In indicator light circuits, contains contact points, like a switch. In gauge circuits, contains a variable resistance that modifies current flow in accordance with the condition or system being monitored.

Separator - A thin sheet of non-conducting material that is placed between the negative and positive plates in an element to prevent the plates from touching.

Series circuit - A circuit in which the units are consecutively connected or wired positive to negative and in which current has to pass through all components, one at a time.

Series/parallel circuit - A circuit in which some components are wired in series, others in parallel. An example: Two loads wired in parallel with each other, but in series with the switch that controls them.

Short circuit - An unintentional routing of a current, bypassing part of the original circuit.

Shorted winding - The winding of a field or armature that's grounded because of accidental or deliberate reasons.

Shrink wrap - An insulating material used to protect wire splices and junctions at terminals. Upon application of open flame or heat, the wrap shrinks to fit tightly on the wire or terminal.

Shunt - 1. Connected in parallel with some other part. 2. A precision low value resistor placed across the terminals of an ammeter to increase its range.

Simple circuit - The simplest circuit includes an electrical power source, a load and some wire to connect them together.

Single-wire circuit - Generally used in production vehicles, in which one wire brings current to the load and the vehicle's frame acts as the return path (ground).

Slip ring - A device for making electrical connections between stationary and rotating contacts.

Snap-splice connector - Solderless connector used to tap an additional wire into an existing wire without cutting the original. Often used in installing trailer wiring to a tow vehicle.

Solder - An alloy of lead and tin which melts at a low temperature and is used for making electrical connections.

Solderless connector - Any connector or terminal that can be installed to a wire without the use of solder. They're usually crimped in place using a special crimping tool. Ring terminals, spade terminals, disconnect terminals, butt connectors, closed-end connectors and snap-splice connectors are typical examples. Ring and spade terminals also come in soldered versions.

Solenoid - An electromechanical device consisting of a tubular coil surrounding a movable metal core, or plunger, which moves when the coil is energized. The movable core is connected to various mechanisms to accomplish work.

Spade terminal - A terminal used to connect a wire to a screw or stud terminal. The spade terminal has two forked ends, either straight or with upturned tips. They're more convenient to install than ring terminals, but slightly less secure for rugged applications. Comes in soldered and unsoldered versions.

SPDT - A single-pole, double-throw switch.

Specific gravity - The measure of a battery's charge, made by comparing the relative weight of a volume of its electrolyte to the weight of an equal volume of pure water, which has an assigned value of 1.0. A fully charged battery will have a specific gravity reading of 1.260. See "hydrometer."

Split loom - Flexible, corrugated conduit used to bundle wires into harnesses.

SPST - A single-pole, single-throw switch.

Starter - A device used to supply the required mechanical force to turn over the engine for starting.

Stator - In an alternator, the part which contains the conductors within which the field rotates.

Sulfuric acid - A heavy, corrosive, high-boiling liquid acid that is colorless when pure. Sulfuric acid is mixed with distilled water to form the electrolyte used in storage batteries.

Supply voltage - The voltage obtained from the power supply to operate a circuit.

Switch - An electrical control device used to turn a circuit on and off, or change the connections in the circuit. Switches are described by the number of poles and throws they have. See "SPST, SPDT, DPST and DPDT."

T

Tachometer - A device that measures the speed of an engine in rpm.

Terminal - A device attached to the end of a wire or to an apparatus for convenience in making electrical connections.

Terminal block - A plastic or resin assembly containing two rows of terminals screws. Used to join the circuits in several wiring harnesses.

Test light - A test instrument consisting of an indicator light wired into the handle of a metal probe. When the probe contacts a live circuit, current flows through the light, lighting it, and to ground through an attached lead and alligator clip. Used to test for voltage in live circuits only.

Test light (self-powered) - A test device containing an indicator light and a built-in battery. Used to test continuity of circuits not containing voltage at the time of the test. Used to test continuity in a harness before it's installed in the vehicle. Also called a continuity tester.

Thermal relay - A relay actuated by the heating effect of the current flowing through it.

Thermistor - The electrical element in a temperature sending unit that varies its resistance in proportion to temperature. Unlike most electrical conductors, in which resistance increases as temperature rises, resistance in a thermistor decreases. Thermistors are made from the oxides of cobalt, copper, iron or nickel.

Tracer - A stripe of a second color applied to a wire insulator to distinguish that wire from another one with the same color insulator.

Transformer - An apparatus for transforming an electric current to a higher or lower voltage without changing the total energy.

V

Variable resistor - A wire-wound or composition resistor with a sliding contact for changing the resistance. See "rheostat."

Variable transformer - An iron-core transformer with a sliding contact which moves along the exposed turns of the secondary winding to vary the output voltage.

Volt - A practical unit for measuring current pressure in a circuit; the force that will move a current of one ampere through a resistance of one ohm.

Voltage drop - The difference in voltage between two points, caused by the loss of electrical pressure as a current flows through an impedance or resistance. All wire, no matter how low the resistance, shows at least a trace of voltage drop.

Voltage regulator - An electromechanical or electronic device that maintains the output voltage of a device at a predetermined value.

Voltmeter - 1. A test instrument that measures voltage in an electrical circuit. Used to check continuity and determine voltage drop in specific circuits of vehicle electrical systems. 2. An instrument panel gauge that measures system voltage. When the engine's not running, the voltmeter indicates battery voltage, which should be 12 to 13 volts in a 12-volt system. When the engine's running, the voltmeter indicates total system voltage, or the combined voltage output of the alternator and the battery.

VOM (Volt-Ohmmeter) - A two-in-one test instrument. For convenience, a voltmeter and an ohmmeter are mounted in the same case and share a common readout and test leads.

W

Warning light - An instrument panel display used to inform the driver when something undesirable has happened in the monitored circuit or system, such as an overheated engine or a sudden loss of oil pressure.

Watt - The unit for measuring electrical energy or "work." Wattage is the product of amperage multiplied by voltage.

Winding - One or more turns of a wire, forming a coil. Also, the individual coils of a transformer.

Wire - A solid or stranded group of cylindrical conductors together with any associated insulation.

Index

H

Headlight
assembly, removal and installation, 11-4
doors, emergency (manual) operation, 12-14
switch, replacement, 12-5

Headlights
removal and installation, 12-6
adjustment, 12-6

Heater and air conditioning control assembly, removal and installation, 3-11

Heater blower motor, removal and installation, 3-11

Heater core, removal and installation, 3-16

Heater water control valve replacement, 3-16

Hinges and locks, maintenance, 11-3

Hood, removal, installation and adjustment, 11-3

Horn, removal and installation, 12-14

Hub and bearing, removal and installation
front, 10-6
rear, 10-9

I

Idle speed check and adjustment, 1-23

Ignition
coil, test, removal and installation, 5-9
key lock cylinder, replacement, 12-4
module, replacement, 5-7
pick-up coil, test and replacement, 5-8
switch, replacement, 12-4
timing check and adjustment (1991 and earlier models), 1-29

Ignition system
check, 5-3
general information and precautions, 5-1

Information sensors, 6-6
air conditioning compressor clutch engagement signal (1984 models only), 6-8
Air conditioning On signal (1985 and later models), 6-8
cold start module (1984 models only), 6-8
crank signal (1984 models only), 6-8
engine coolant temperature sensor, 6-6
Intake Air Temperature (IAT) sensor (1985 and later models), 6-7
Mass Air Flow (MAF) sensor (1985 through 1990, 1995 and 1996 models), 6-7
oxygen sensor, 6-7
Park/Neutral switch (automatic transmission-equipped models only), 6-8
Throttle Position Sensor (TPS), 6-7
vehicle speed sensor (1985 and later models), 6-8

Initial start-up and break-in after overhaul, 2B-24

Instrument panel, removal and installation, 12-9

Intake Air Temperature (IAT) sensor (1985 and later models), 6-7

Intake manifold, removal and installation, 2A-5

Intermediate steering shaft, removal and installation, 10-15

Introduction to the Chevrolet Corvette, 0-5

J

Jacking and towing, 0-16

K

Knuckle
rear (suspension) removal and installation, 10-7
steering, removal and installation, 10-10

M

Main and connecting rod bearings, inspection, 2B-18

Maintenance schedule, 1-8

Maintenance techniques, tools and working facilities, 0-7

Manual transmission and overdrive unit, 7A-1 through 7A-6
oil level check, 1-19
output seal, replacement, 7A-5
overdrive switch, removal and installation, 7A-4
overdrive unit, inspection and overhaul, 7A-5
shifter, removal and installation, 7A-3
steering column lock cable, adjustment, 7A-3
transmission and overdrive unit, removal and installation, 7A-4
transmission shift linkage, adjustment, 7A-2

Mass Air Flow (MAF) sensor (1985 through 1990, 1995 and 1996 models), 6-7

Master cylinder, removal, overhaul and installation, 9-9

Mounts, engine, check and replacement, 2A-20

N

Neutral start switch (manual transmission), removal and installation, 8-6

O

Oil and oil filter change, 1-16

Oil pan, removal and installation, 2A-17

Oil pump, removal and installation, 2A-17

On Board Diagnostic (OBD) system and trouble codes, 6-2

Outside mirror, removal and installation, 11-6

Overdrive switch, removal and installation, 7A-4

Overdrive unit
fluid change, 1-24
fluid level check (manual transmission only), 1-20
inspection and overhaul, 7A-5

Oxygen sensor, 6-8

Notes

Haynes Automotive Manuals

NOTE: If you do not see a listing for your vehicle, please visit **haynes.com** *for the latest product information and check out our* ***Online Manuals!***

ACURA

12020 Integra '86 thru '89 & Legend '86 thru '90
12021 Integra '90 thru '93 & Legend '91 thru '95
Integra '94 thru '00 - *see HONDA Civic (42025)*
MDX '01 thru '07 - *see HONDA Pilot (42037)*
12050 Acura TL all models '99 thru '08

AMC

14020 Mid-size models '70 thru '83
14025 (Renault) Alliance & Encore '83 thru '87

AUDI

15020 4000 all models '80 thru '87
15025 5000 all models '77 thru '83
15026 5000 all models '84 thru '88
Audi A4 '96 thru '01 - *see VW Passat (96023)*
15030 Audi A4 '02 thru '08

AUSTIN-HEALEY

Sprite - *see MG Midget (66015)*

BMW

18020 3/5 Series '82 thru '92
18021 3-Series incl. Z3 models '92 thru '98
18022 3-Series incl. Z4 models '99 thru '05
18023 3-Series '06 thru '14
18025 320i all 4-cylinder models '75 thru '83
18050 1500 thru 2002 except Turbo '59 thru '77

BUICK

19010 Buick Century '97 thru '05
Century (front-wheel drive) - *see GM (38005)*
19020 Buick, Oldsmobile & Pontiac Full-size (Front-wheel drive) '85 thru '05 Buick Electra, LeSabre and Park Avenue; Oldsmobile Delta 88 Royale, Ninety Eight and Regency; Pontiac Bonneville
19025 Buick, Oldsmobile & Pontiac Full-size (Rear wheel drive) '70 thru '90 Buick Estate, Electra, LeSabre, Limited, Oldsmobile Custom Cruiser, Delta 88, Ninety-eight, Pontiac Bonneville, Catalina, Grandville, Parisienne
19027 Buick LaCrosse '05 thru '13
Enclave - *see GENERAL MOTORS (38001)*
Rainier - *see CHEVROLET (24072)*
Regal - *see GENERAL MOTORS (38010)*
Riviera - *see GENERAL MOTORS (38030, 38031)*
Roadmaster - *see CHEVROLET (24046)*
Skyhawk - *see GENERAL MOTORS (38015)*
Skylark - *see GENERAL MOTORS (38020, 38025)*
Somerset - *see GENERAL MOTORS (38025)*

CADILLAC

21015 CTS & CTS-V '03 thru '14
21030 Cadillac Rear Wheel Drive '70 thru '93
Cimarron - *see GENERAL MOTORS (38015)*
DeVille - *see GENERAL MOTORS (38031 & 38032)*
Eldorado - *see GENERAL MOTORS (38030)*
Fleetwood - *see GENERAL MOTORS (38031)*
Seville - *see GM (38030, 38031 & 38032)*

CHEVROLET

10305 Chevrolet Engine Overhaul Manual
24010 Astro & GMC Safari Mini-vans '85 thru '05
24013 Aveo '04 thru '11
24015 Camaro V8 all models '70 thru '81
24016 Camaro all models '82 thru '92
24017 Camaro & Firebird '93 thru '02
Cavalier - *see GENERAL MOTORS (38016)*
Celebrity - *see GENERAL MOTORS (38005)*
24018 Camaro '10 thru '15
24020 Chevelle, Malibu & El Camino '69 thru '87
Cobalt - *see GENERAL MOTORS (38017)*
24024 Chevette & Pontiac T1000 '76 thru '87
Citation - *see GENERAL MOTORS (38020)*
24027 Colorado & GMC Canyon '04 thru '12
24032 Corsica & Beretta all models '87 thru '96
24040 Corvette all V8 models '68 thru '82
24041 Corvette all models '84 thru '96
24042 Corvette all models '97 thru '13
24044 Cruze '11 thru '19
24045 Full-size Sedans Caprice, Impala, Biscayne, Bel Air & Wagons '69 thru '90
24046 Impala SS & Caprice and Buick Roadmaster '91 thru '96
Impala '00 thru '05 - *see LUMINA (24048)*
24047 Impala & Monte Carlo all models '06 thru '11
Lumina '90 thru '94 - *see GM (38010)*
24048 Lumina & Monte Carlo '95 thru '05
Lumina APV - *see GM (38035)*
24050 Luv Pick-up all 2WD & 4WD '72 thru '82
24051 Malibu '13 thru '19
24055 Monte Carlo all models '70 thru '88
Monte Carlo '95 thru '01 - *see LUMINA (24048)*
24059 Nova all V8 models '69 thru '79
24060 Nova and Geo Prizm '85 thru '92
24064 Pick-ups '67 thru '87 - Chevrolet & GMC
24065 Pick-ups '88 thru '98 - Chevrolet & GMC
24066 Pick-ups '99 thru '06 - Chevrolet & GMC
24067 Chevrolet Silverado & GMC Sierra '07 thru '14
24068 Chevrolet Silverado & GMC Sierra '14 thru '19
24070 S-10 & S-15 Pick-ups '82 thru '93, Blazer & Jimmy '83 thru '94,
24071 S-10 & Sonoma Pick-ups '94 thru '04, including Blazer, Jimmy & Hombre
24072 Chevrolet TrailBlazer, GMC Envoy & Oldsmobile Bravada '02 thru '09
24075 Sprint '85 thru '88 & Geo Metro '89 thru '01
24080 Vans - Chevrolet & GMC '68 thru '96
24081 Chevrolet Express & GMC Savana Full-size Vans '96 thru '19

CHRYSLER

10310 Chrysler Engine Overhaul Manual
25015 Chrysler Cirrus, Dodge Stratus, Plymouth Breeze '95 thru '00
25020 Full-size Front-Wheel Drive '88 thru '93
K-Cars - *see DODGE Aries (30008)*
Laser - *see DODGE Daytona (30030)*
25025 Chrysler LHS, Concorde, New Yorker, Dodge Intrepid, Eagle Vision, '93 thru '97
25026 Chrysler LHS, Concorde, 300M, Dodge Intrepid, '98 thru '04
25027 Chrysler 300 '05 thru '18, Dodge Charger '06 thru '18, Magnum '05 thru '08 & Challenger '08 thru '18
25030 Chrysler & Plymouth Mid-size front wheel drive '82 thru '95
Rear-wheel Drive - *see Dodge (30050)*
25035 PT Cruiser all models '01 thru '10
25040 Chrysler Sebring '95 thru '06, Dodge Stratus '01 thru '06 & Dodge Avenger '95 thru '00
25041 Chrysler Sebring '07 thru '10, 200 '11 thru '17 Dodge Avenger '08 thru '14

DATSUN

28005 200SX all models '80 thru '83
28012 240Z, 260Z & 280Z Coupe '70 thru '78
28014 280ZX Coupe & 2+2 '79 thru '83
300ZX - *see NISSAN (72010)*
28018 510 & PL521 Pick-up '68 thru '73
28020 510 all models '78 thru '81
28022 620 Series Pick-up all models '73 thru '79
720 Series Pick-up - *see NISSAN (72030)*

DODGE

400 & 600 - *see CHRYSLER (25030)*
30008 Aries & Plymouth Reliant '81 thru '89
30010 Caravan & Plymouth Voyager '84 thru '95
30011 Caravan & Plymouth Voyager '96 thru '02
30012 Challenger & Plymouth Sapporro '78 thru '83
30013 Caravan, Chrysler Voyager & Town & Country '03 thru '07
30014 Grand Caravan & Chrysler Town & Country '08 thru '18
30016 Colt & Plymouth Champ '78 thru '87
30020 Dakota Pick-ups all models '87 thru '96
30021 Durango '98 & '99 & Dakota '97 thru '99
30022 Durango '00 thru '03 & Dakota '00 thru '04
30023 Durango '04 thru '09 & Dakota '05 thru '11
30025 Dart, Demon, Plymouth Barracuda, Duster & Valiant 6-cylinder models '67 thru '76
30030 Daytona & Chrysler Laser '84 thru '89
Intrepid - *see CHRYSLER (25025, 25026)*
30034 Neon all models '95 thru '99
30035 Omni & Plymouth Horizon '78 thru '90
30036 Dodge & Plymouth Neon '00 thru '05
30040 Pick-ups full-size models '74 thru '93
30042 Pick-ups full-size models '94 thru '08
30043 Pick-ups full-size models '09 thru '18
30045 Ram 50/D50 Pick-ups & Raider and Plymouth Arrow Pick-ups '79 thru '93
30050 Dodge/Plymouth/Chrysler RWD '71 thru '89
30055 Shadow & Plymouth Sundance '87 thru '94
30060 Spirit & Plymouth Acclaim '89 thru '95
30065 Vans - Dodge & Plymouth '71 thru '03

EAGLE

Talon - *see MITSUBISHI (68030, 68031)*
Vision - *see CHRYSLER (25025)*

FIAT

34010 124 Sport Coupe & Spider '68 thru '78
34025 X1/9 all models '74 thru '80

FORD

10320 Ford Engine Overhaul Manual
10355 Ford Automatic Transmission Overhaul
11500 Mustang '64-1/2 thru '70 Restoration Guide
36004 Aerostar Mini-vans all models '86 thru '97
36006 Contour & Mercury Mystique '95 thru '00
36008 Courier Pick-up all models '72 thru '82
36012 Crown Victoria & Mercury Grand Marquis '88 thru '11
36014 Edge '07 thru '19 & Lincoln MKX '07 thru '18
36016 Escort & Mercury Lynx all models '81 thru '90
36020 Escort & Mercury Tracer '91 thru '02
36022 Escape '01 thru '17, Mazda Tribute '01 thru '11, & Mercury Mariner '05 thru '11
36024 Explorer & Mazda Navajo '91 thru '01
36025 Explorer & Mercury Mountaineer '02 thru '10
36026 Explorer '11 thru '17
36028 Fairmont & Mercury Zephyr '78 thru '83
36030 Festiva & Aspire '88 thru '97
36032 Fiesta all models '77 thru '80
36034 Focus all models '00 thru '11
36035 Focus '12 thru '14
36045 Fusion '06 thru '14 & Mercury Milan '06 thru '11
36048 Mustang V8 all models '64-1/2 thru '73
36049 Mustang II 4-cylinder, V6 & V8 models '74 thru '78
36050 Mustang & Mercury Capri '79 thru '93
36051 Mustang all models '94 thru '04
36052 Mustang '05 thru '14
36054 Pick-ups & Bronco '73 thru '79
36058 Pick-ups & Bronco '80 thru '96
36059 F-150 '97 thru '03, Expedition '97 thru '17, F-250 '97 thru '99, F-150 Heritage '04 & Lincoln Navigator '98 thru '17
36060 Super Duty Pick-ups & Excursion '99 thru '10
36061 F-150 full-size '04 thru '14
36062 Pinto & Mercury Bobcat '75 thru '80
36063 F-150 full-size '15 thru '17
36064 Super Duty Pick-ups '11 thru '16
36066 Probe all models '89 thru '92
Probe '93 thru '97 - *see MAZDA 626 (61042)*
36070 Ranger & Bronco II gas models '83 thru '92
36071 Ranger '93 thru '11 & Mazda Pick-ups '94 thru '09
36074 Taurus & Mercury Sable '86 thru '95
36075 Taurus & Mercury Sable '96 thru '07
36076 Taurus '08 thru '14, Five Hundred '05 thru '07, Mercury Montego '05 thru '07 & Sable '08 thru '09
36078 Tempo & Mercury Topaz '84 thru '94
36082 Thunderbird & Mercury Cougar '83 thru '88
36086 Thunderbird & Mercury Cougar '89 thru '97
36090 Vans all V8 Econoline models '69 thru '91
36094 Vans full size '92 thru '14
36097 Windstar '95 thru '03, Freestar & Mercury Monterey Mini-van '04 thru '07

GENERAL MOTORS

10360 GM Automatic Transmission Overhaul
38001 GMC Acadia '07 thru '16, Buick Enclave '08 thru '17, Saturn Outlook '07 thru '10 & Chevrolet Traverse '09 thru '17
38005 Buick Century, Chevrolet Celebrity, Oldsmobile Cutlass Ciera & Pontiac 6000 all models '82 thru '96
38010 Buick Regal '88 thru '04, Chevrolet Lumina '88 thru '04, Oldsmobile Cutlass Supreme '88 thru '97 & Pontiac Grand Prix '88 thru '07
38015 Buick Skyhawk, Cadillac Cimarron, Chevrolet Cavalier, Oldsmobile Firenza, Pontiac J-2000 & Sunbird '82 thru '94
38016 Chevrolet Cavalier & Pontiac Sunfire '95 thru '05
38017 Chevrolet Cobalt '05 thru '10, HHR '06 thru '11, Pontiac G5 '07 thru '09, Pursuit '05 thru '06 & Saturn ION '03 thru '07
38020 Buick Skylark, Chevrolet Citation, Oldsmobile Omega, Pontiac Phoenix '80 thru '85
38025 Buick Skylark '86 thru '98, Somerset '85 thru '87, Oldsmobile Achieva '92 thru '98, Calais '85 thru '91, & Pontiac Grand Am all models '85 thru '98
38026 Chevrolet Malibu '97 thru '03, Classic '04 thru '05, Oldsmobile Alero '99 thru '03, Cutlass '97 thru '00, & Pontiac Grand Am '99 thru '03
38027 Chevrolet Malibu '04 thru '12, Pontiac G6 '05 thru '10 & Saturn Aura '07 thru '10
38030 Cadillac Eldorado, Seville, Oldsmobile Toronado & Buick Riviera '71 thru '85
38031 Cadillac Eldorado, Seville, DeVille, Fleetwood, Oldsmobile Toronado & Buick Riviera '86 thru '93
38032 Cadillac DeVille '94 thru '05, Seville '92 thru '04 & Cadillac DTS '06 thru '10
38035 Chevrolet Lumina APV, Oldsmobile Silhouette & Pontiac Trans Sport all models '90 thru '96
38036 Chevrolet Venture '97 thru '05, Oldsmobile Silhouette '97 thru '04, Pontiac Trans Sport '97 thru '98 & Montana '99 thru '05
38040 Chevrolet Equinox '05 thru '17, GMC Terrain '10 thru '17 & Pontiac Torrent '06 thru '09

GEO

Metro - *see CHEVROLET Sprint (24075)*
Prizm - *'85 thru '92 see CHEVY (24060), '93 thru '02 see TOYOTA Corolla (92036)*
40030 Storm all models '90 thru '93
Tracker - *see SUZUKI Samurai (90010)*

(Continued on other side)

Haynes North America, Inc. • (805) 498-6703 • www.haynes.com

Haynes Automotive Manuals (continued)

NOTE: If you do not see a listing for your vehicle, please visit **haynes.com** for the latest product information and check out our **Online Manuals!**

GMC
- **Acadia** - see GENERAL MOTORS (38001)
- **Pick-ups** - see CHEVROLET (24027, 24068)
- **Vans** - see CHEVROLET (24081)

HONDA
- 42010 **Accord CVCC** all models '76 thru '83
- 42011 **Accord** all models '84 thru '89
- 42012 **Accord** all models '90 thru '93
- 42013 **Accord** all models '94 thru '97
- 42014 **Accord** all models '98 thru '02
- 42015 **Accord** '03 thru '12 & **Crosstour** '10 thru '14
- 42016 **Accord** '13 thru '17
- 42020 **Civic 1200** all models '73 thru '79
- 42021 **Civic 1300 & 1500 CVCC** '80 thru '83
- 42022 **Civic 1500 CVCC** all models '75 thru '79
- 42023 **Civic** all models '84 thru '91
- 42024 **Civic & del Sol** '92 thru '95
- 42025 **Civic** '96 thru '00, **CR-V** '97 thru '01 & **Acura Integra** '94 thru '00
- 42026 **Civic** '01 thru '11 & **CR-V** '02 thru '11
- 42027 **Civic** '12 thru '15 & **CR-V** '12 thru '16
- 42030 **Fit** '07 thru '13
- 42035 **Odyssey** all models '99 thru '10
  - **Passport** - see ISUZU Rodeo (47017)
- 42037 **Honda Pilot** '03 thru '08, **Ridgeline** '06 thru '14 & **Acura MDX** '01 thru '07
- 42040 **Prelude CVCC** all models '79 thru '89

HYUNDAI
- 43010 **Elantra** all models '96 thru '19
- 43015 **Excel & Accent** all models '86 thru '13
- 43050 **Santa Fe** all models '01 thru '12
- 43055 **Sonata** all models '99 thru '14

INFINITI
- **G35** '03 thru '08 - see NISSAN 350Z (72011)

ISUZU
- **Hombre** - see CHEVROLET S-10 (24071)
- 47017 **Rodeo** '91 thru '02, **Amigo** '89 thru '94 & '98 thru '02 & **Honda Passport** '95 thru '02
- 47020 **Trooper** '84 thru '91 & **Pick-up** '81 thru '93

JAGUAR
- 49010 **XJ6** all 6-cylinder models '68 thru '86
- 49011 **XJ6** all models '88 thru '94
- 49015 **XJ12 & XJS** all 12-cylinder models '72 thru '85

JEEP
- 50010 **Cherokee, Comanche & Wagoneer Limited** all models '84 thru '01
- 50011 **Cherokee** '14 thru '19
- 50020 **CJ** all models '49 thru '86
- 50025 **Grand Cherokee** all models '93 thru '04
- 50026 **Grand Cherokee** '05 thru '19 & **Dodge Durango** '11 thru '19
- 50029 **Grand Wagoneer & Pick-up** '72 thru '91 Grand Wagoneer '84 thru '91, Cherokee & Wagoneer '72 thru '83, Pick-up '72 thru '88
- 50030 **Wrangler** all models '87 thru '17
- 50035 **Liberty** '02 thru '12 & **Dodge Nitro** '07 thru '11
- 50050 **Patriot & Compass** '07 thru '17

KIA
- 54050 **Optima** '01 thru '10
- 54060 **Sedona** '02 thru '14
- 54070 **Sephia** '94 thru '01, **Spectra** '00 thru '09, **Sportage** '05 thru '20
- 54077 **Sorento** '03 thru '13

LEXUS
- **ES 300/330** - see TOYOTA Camry (92007, 92008)
- **ES 350** - see TOYOTA Camry (92009)
- **RX 300/330/350** - see TOYOTA Highlander (92095)

LINCOLN
- **MKX** - see FORD (36014)
- **Navigator** - see FORD Pick-up (36059)
- 59010 **Rear-Wheel Drive Continental** '70 thru '87, **Mark Series** '70 thru '92 & **Town Car** '81 thru '10

MAZDA
- 61010 **GLC** (rear-wheel drive) '77 thru '83
- 61011 **GLC** (front-wheel drive) '81 thru '85
- 61012 **Mazda3** '04 thru '11
- 61015 **323 & Protegé** '90 thru '03
- 61016 **MX-5 Miata** '90 thru '14
- 61020 **MPV** all models '89 thru '98
  - **Navajo** - see Ford Explorer (36024)
- 61030 **Pick-ups** '72 thru '93
  - **Pick-ups** '94 thru '09 - see Ford Ranger (36071)
- 61035 **RX-7** all models '79 thru '85
- 61036 **RX-7** all models '86 thru '91
- 61040 **626** (rear-wheel drive) all models '79 thru '82
- 61041 **626 & MX-6** (front-wheel drive) '83 thru '92
- 61042 **626** '93 thru '01 & **MX-6/Ford Probe** '93 thru '02
- 61043 **Mazda6** '03 thru '13

MERCEDES-BENZ
- 63012 **123 Series Diesel** '76 thru '85
- 63015 **190 Series** 4-cylinder gas models '84 thru '88
- 63020 **230/250/280** 6-cylinder SOHC models '68 thru '72
- 63025 **280 123 Series** gas models '77 thru '81
- 63030 **350 & 450** all models '71 thru '80
- 63040 **C-Class:** C230/C240/C280/C320/C350 '01 thru '07

MERCURY
- 64200 **Villager & Nissan Quest** '93 thru '01
  - *All other titles, see FORD Listing.*

MG
- 66010 **MGB** Roadster & GT Coupe '62 thru '80
- 66015 **MG Midget, Austin Healey Sprite** '58 thru '80

MINI
- 67020 **Mini** '02 thru '13

MITSUBISHI
- 68020 **Cordia, Tredia, Galant, Precis & Mirage** '83 thru '93
- 68030 **Eclipse, Eagle Talon & Plymouth Laser** '90 thru '94
- 68031 **Eclipse** '95 thru '05 & **Eagle Talon** '95 thru '98
- 68035 **Galant** '94 thru '12
- 68040 **Pick-up** '83 thru '96 & **Montero** '83 thru '93

NISSAN
- 72010 **300ZX** all models including Turbo '84 thru '89
- 72011 **350Z & Infiniti G35** all models '03 thru '08
- 72015 **Altima** all models '93 thru '06
- 72016 **Altima** '07 thru '12
- 72020 **Maxima** all models '85 thru '92
- 72021 **Maxima** all models '93 thru '08
- 72025 **Murano** '03 thru '14
- 72030 **Pick-ups** '80 thru '97 & **Pathfinder** '87 thru '95
- 72031 **Frontier** '98 thru '04, **Xterra** '00 thru '04, & **Pathfinder** '96 thru '04
- 72032 **Frontier & Xterra** '05 thru '14
- 72037 **Pathfinder** '05 thru '14
- 72040 **Pulsar** all models '83 thru '86
- 72042 **Roque** all models '08 thru '20
- 72050 **Sentra** all models '82 thru '94
- 72051 **Sentra & 200SX** all models '95 thru '06
- 72060 **Stanza** all models '82 thru '90
- 72070 **Titan pick-ups** '04 thru '10, **Armada** '05 thru '10 & **Pathfinder Armada** '04
- 72080 **Versa** all models '07 thru '19

OLDSMOBILE
- 73015 **Cutlass** V6 & V8 gas models '74 thru '88
  - *For other OLDSMOBILE titles, see BUICK, CHEVROLET or GENERAL MOTORS listings.*

PLYMOUTH
- *For PLYMOUTH titles, see DODGE listing.*

PONTIAC
- 79008 **Fiero** all models '84 thru '88
- 79018 **Firebird** V8 models except Turbo '70 thru '81
- 79019 **Firebird** all models '82 thru '92
- 79025 **G6** all models '05 thru '09
- 79040 **Mid-size Rear-wheel Drive** '70 thru '87
  - **Vibe** '03 thru '10 - see TOYOTA Corolla (92037)
  - *For other PONTIAC titles, see BUICK, CHEVROLET or GENERAL MOTORS listings.*

PORSCHE
- 80020 **911** Coupe & Targa models '65 thru '89
- 80025 **914** all 4-cylinder models '69 thru '76
- 80030 **924** all models including Turbo '76 thru '82
- 80035 **944** all models including Turbo '83 thru '89

RENAULT
- **Alliance & Encore** - see AMC (14025)

SAAB
- 84010 **900** all models including Turbo '79 thru '88

SATURN
- 87010 **Saturn** all S-series models '91 thru '02
  - **Saturn Ion** '03 thru '07 - see GM (38017)
  - **Saturn Outlook** - see GM (38001)
- 87020 **Saturn L-series** all models '00 thru '04
- 87040 **Saturn VUE** '02 thru '09

SUBARU
- 89002 **1100, 1300, 1400 & 1600** '71 thru '79
- 89003 **1600 & 1800** 2WD & 4WD '80 thru '94
- 89080 **Impreza** '02 thru '11, **WRX** '02 thru '14, & **WRX STI** '04 thru '14
- 89100 **Legacy** all models '90 thru '99
- 89101 **Legacy & Forester** '00 thru '09
- 89102 **Legacy** '10 thru '16 & **Forester** '12 thru '16

SUZUKI
- 90010 **Samurai/Sidekick & Geo Tracker** '86 thru '01

TOYOTA
- 92005 **Camry** all models '83 thru '91
- 92006 **Camry** '92 thru '96 & **Avalon** '95 thru '96
- 92007 **Camry, Avalon, Solara, Lexus ES 300** '97 thru '01

(TOYOTA continued)
- 92008 **Camry, Avalon, Lexus ES 300/330** '02 thru '06 & **Solara** '02 thru '08
- 92009 **Camry, Avalon & Lexus ES 350** '07 thru '17
- 92015 **Celica Rear-wheel Drive** '71 thru '85
- 92020 **Celica Front-wheel Drive** '86 thru '99
- 92025 **Celica Supra** all models '79 thru '92
- 92030 **Corolla** all models '75 thru '79
- 92032 **Corolla** all rear-wheel drive models '80 thru '87
- 92035 **Corolla** all front-wheel drive models '84 thru '92
- 92036 **Corolla & Geo/Chevrolet Prizm** '93 thru '02
- 92037 **Corolla** '03 thru '19, **Matrix** '03 thru '14, & **Pontiac Vibe** '03 thru '10
- 92040 **Corolla Tercel** all models '80 thru '82
- 92045 **Corona** all models '74 thru '82
- 92050 **Cressida** all models '78 thru '82
- 92055 **Land Cruiser** FJ40, 43, 45, 55 '68 thru '82
- 92056 **Land Cruiser** FJ60, 62, 80, FZJ80 '80 thru '96
- 92060 **Matrix** '03 thru '11 & **Pontiac Vibe** '03 thru '10
- 92065 **MR2** all models '85 thru '87
- 92070 **Pick-up** all models '69 thru '78
- 92075 **Pick-up** all models '79 thru '95
- 92076 **Tacoma** '95 thru '04, **4Runner** '96 thru '02 & **T100** '93 thru '08
- 92077 **Tacoma** all models '05 thru '18
- 92078 **Tundra** '00 thru '06 & **Sequoia** '01 thru '07
- 92079 **4Runner** all models '03 thru '09
- 92080 **Previa** all models '91 thru '95
- 92081 **Prius** all models '01 thru '12
- 92082 **RAV4** all models '96 thru '12
- 92085 **Tercel** all models '87 thru '94
- 92090 **Sienna** all models '98 thru '10
- 92095 **Highlander** '01 thru '19 & **Lexus RX330/330/350** '99 thru '19
- 92179 **Tundra** '07 thru '19 & **Sequoia** '08 thru '19

TRIUMPH
- 94007 **Spitfire** all models '62 thru '81
- 94010 **TR7** all models '75 thru '81

VW
- 96008 **Beetle & Karmann Ghia** '54 thru '79
- 96009 **New Beetle** '98 thru '10
- 96016 **Rabbit, Jetta, Scirocco & Pick-up** gas models '75 thru '92 & Convertible '80 thru '92
- 96017 **Golf & Jetta** '93 thru '98, **Cabrio** '95 thru '02
- 96018 **Golf, GTI, Jetta** '99 thru '05
- 96019 **Jetta, Rabbit, GLI, GTI & Golf** '05 thru '11
- 96020 **Rabbit, Jetta & Pick-up** diesel '77 thru '84
- 96021 **Jetta** '11 thru '18 & **Golf** '15 thru '19
- 96023 **Passat** '98 thru '05 & **Audi A4** '96 thru '01
- 96030 **Transporter 1600** all models '68 thru '79
- 96035 **Transporter 1700, 1800 & 2000** '72 thru '79
- 96040 **Type 3 1500 & 1600** all models '63 thru '73
- 96045 **Vanagon Air-Cooled** all models '80 thru '83

VOLVO
- 97010 **120, 130 Series & 1800 Sports** '61 thru '73
- 97015 **140 Series** all models '66 thru '74
- 97020 **240 Series** all models '76 thru '93
- 97040 **740 & 760 Series** all models '82 thru '88
- 97050 **850 Series** all models '93 thru '97

TECHBOOK MANUALS
- 10205 **Automotive Computer Codes**
- 10206 **OBD-II & Electronic Engine Management**
- 10210 **Automotive Emissions Control Manual**
- 10215 **Fuel Injection Manual** '78 thru '85
- 10225 **Holley Carburetor Manual**
- 10230 **Rochester Carburetor Manual**
- 10305 **Chevrolet Engine Overhaul Manual**
- 10320 **Ford Engine Overhaul Manual**
- 10330 **GM and Ford Diesel Engine Repair Manual**
- 10331 **Duramax Diesel Engines** '01 thru '19
- 10332 **Cummins Diesel Engine Performance Manual**
- 10333 **GM, Ford & Chrysler Engine Performance Manual**
- 10334 **GM Engine Performance Manual**
- 10340 **Small Engine Repair Manual, 5 HP & Less**
- 10341 **Small Engine Repair Manual, 5.5 thru 20 HP**
- 10345 **Suspension, Steering & Driveline Manual**
- 10355 **Ford Automatic Transmission Overhaul**
- 10360 **GM Automatic Transmission Overhaul**
- 10405 **Automotive Body Repair & Painting**
- 10410 **Automotive Brake Manual**
- 10411 **Automotive Anti-lock Brake (ABS) Systems**
- 10420 **Automotive Electrical Manual**
- 10425 **Automotive Heating & Air Conditioning**
- 10435 **Automotive Tools Manual**
- 10445 **Welding Manual**
- 10450 **ATV Basics**

Over a 100 Haynes motorcycle manuals also available

10/22